Seed Inoculation, Coating and Precision Pelleting

Science, Technology and Practical Applications

Seed Inoculation, Coating and Precision Pelleting

Science, Technology and Practical Applications

Gerald M. Bennett
Agricultural Consultant
Christchurch, New Zealand

Contributor/Critic
John M. Lloyd
CEO, Axis Associates
Nelson, New Zealand

 CRC Press
Taylor & Francis Group
Boca Raton   London   New York

CRC Press is an imprint of the
Taylor & Francis Group, an **informa** business

A SCIENCE PUBLISHERS BOOK

CRC Press
Taylor & Francis Group
6000 Broken Sound Parkway NW, Suite 300
Boca Raton, FL 33487-2742

© 2016 by Taylor & Francis Group, LLC
CRC Press is an imprint of Taylor & Francis Group, an Informa business

No claim to original U.S. Government works

Printed on acid-free paper
Version Date: 20150824

International Standard Book Number-13: 978-1-4987-1643-7 (Hardback)

Visit the Taylor & Francis Web site at
http://www.taylorandfrancis.com

and the CRC Press Web site at
http://www.crcpress.com

Preface

This publication will serve not only advanced agricultural systems by creating a better understanding of the technology involved in past and present seed processing methods, it will also "open some doors" for underdeveloped agriculture around the world, where opportunities remain largely emergent.

Mankind in general has spiraled toward personal independence hardly dreamed of a short time ago—yet, sadly, many millions of people remain socially and economically excluded by lack of education—even basic literacy, by unstable government, endemic illness, starvation and religious prejudice, so debilitated, many are simply unaware of the natural resources and land use opportunity which is their birth right. In many forms, seed offers hope and remediation.

This book identifies some of the opportunities—also identifies investigative error and misunderstanding, important because that, while expensive of time and money, also contributes to knowledge accumulation—the building blocks of success and the escape route from poverty.

Technological development in seed, one of the world's largest and most important industries has been a significant contributor toward better living standards and personal independence making the world a better place. While some of the knowledge related here is already superceded by leading edge development, much of it has nevertheless remained largely unpublished until now and offers new participants a secure starting point and importance for all of a full understanding of the science, technology and practice of seed management—an industry vital to all mankind. This brief account of development in pre-sowing seed treatment will assist not only students, researchers, industry workers and astute Farmers and Growers, but may also assist those determined to break free of the shackles of economic paralysis and actually grow something to sell.

While international seed sales exceed NZ$38 billion annually [excluding farmer to farmer sales and internal use], the facts are that of seed bought and surface sown for forage grazing, much of it does not grow. Such seed has been tossed out over hills, plains and fields for hundreds of years and while some of it has grown—today with better seed at higher cost, establishment qualities need serious enhancement, and that is being achieved using modern seed treatments including those described in this book. Proof of treatment value is beyond dispute—much of it documented here. It is also just the beginning of major progress with top-quality plant breeding, transfer of DNA qualities and advanced microbiology being just some of the exciting prospects ahead.

In an industry laced with huge sums of money plus rapid technical development, it has been a challenging task to be able to illustrate what has taken place in this sector and right up to the present, by legal removal of former extreme claims of confidentiality.

Importantly, it will serve as a useful reference for a wide range of interested people.

Note: It has been decided to show $ values in this book in New Zealand dollars only. Reason is, there are Tables here which [unusually] quote all ACTUAL costs of products used in research and other treatments back as far as the 1960's/70's allowing substantially accurate cost/benefit comparisons between various methods and treatments right up to today. Currency values fluctuate daily but for conversion purposes here we have adopted NZ$.8 [80 cents] to the US dollar, and NZ $.9 [90 cents] to the Australian dollar at May 2015.

Gerald M. Bennett

Acknowledgments

The old cliché that *"The author could not have written this book without the assistance of...."*, has never been more true than here where associate John Lloyd, contributor [in Chapter 2] and valuable critique has made its publication possible. He has outstanding ability in both the science of formulation and in skilled application of it. A great spirit of co-operation and a very good memory are additional qualities of this exceptional man. Thank you John.

Also essential in assembling a coherent and full account of seed processing development as experienced in New Zealand has been the generosity and encouragement of many copyright holders for permission to quote their text and show illustrations—all free of charge. Appreciation and thanks are also due to Dr. A.H.C. Roberts who is Chief Scientific Officer of "Ravensdown", the premium New Zealand superphosphate and fertilizer supplier, who has expertly explained that very important issue in farming of covalent bonding described in Chapter 3 herein. I had hoped this book would be distributed by UN's FAO {Food and Agriculture Organisation} in key languages as a charitable contribution to world poverty which however cannot happen unless the book is first commissioned by that organization which in turn cannot happen until it is published. This book is an advancement on previous FAO publications on this subject and contains much information not available from formal teaching institutions and technical writers. It is still my wish to have it circulated importantly into underdeveloped agricultural and horticultural economies.

Those scientists who permitted their photographs to be included after considerable badgering by me are specially thanked as are the many farmers, growers, staff of research organisations in New Zealand, Australia and the US, seed firm sales people and technicians plus those special friends in Celpril [California] and Celanese Corpn. [Summit, New Jersey] USA—all of whom allowed me to visit and/or work with them toward better results from processing various new formulations and/or explain the ones then available.

Left: T.C. Brash CBE, first Secretary of the New Zealand Dairy Board.{Author's grandfather]. New Zealand is now the largest Dairy Exporting Nation in the world.
[Bennett library]

Photo from *"History of the New Zealand Dairy Industry 1840–1935. Philpott, Dairy Div. New Zealand Department of Agriculture"*

My appreciation for selection into a 5 year Government sponsored Rural Field Cadet [RFC] scholarship at age 18 yrs triggered by working school holidays on various farms of relatives and others–all being experiences which contributed to this book. So did my time as Farm Advisory Officer, Department of Agriculture including secondment under the Colombo Plan to the Kingdom of Nepal, for a 6 months investigation and report into sheep farming presented to the New Zealand Govt. and King of Nepal. According to Ministry of Foreign Affairs, the author was the first New Zealand'er to work officially in Nepal. From attending Diplomatic functions in Katmandu to tramping the remoteness of the Himalayas at age 28, apparently the youngest New Zealand'er sent alone overseas under the Colombo Plan—all contributed to this book by providing a better understanding of agricultural systems both in New Zealand and around the world.

T.C. Brash, Secretary New Zealand Dairy Board from ITS Inception in 1923.

Though brought up in the City [with 4 years boarding at prep. school from age 9] the author left home permanently at age 17, first to work a year at remote Mt. Cook (sheep and cattle) Station—not too different from the mountainous native grasslands of temperate Nepal—followed by 3½ months of compulsory military training.

Why City upbringing then to farming? Possibly a mistake – but maybe there has been a little bit of ancestral agricultural blood passed on by my grandfather T.C. Brash CBE who helped lay the foundations for New Zealand's most important single development in agriculture—i.e., now the largest Dairy exporting nation in the world. I remember him as a most friendly man with a very warm smile, a lay Moderator of the Presbyterian Church and President of the New Zealand Fruitgrowers Federation, a most honourable man and a pillar of New Zealand society. He was never late!—punctuality was his hallmark and he rose from a humble Dairy factory worker to the peak of his career path.

While I cannot aspire to those heights, his inspiration has also been a driver where hopefully both John Lloyd's and my own best efforts in seed processing have and will continue to do some good, not only for New Zealand (where tussock grasslands nationwide have now largely been transformed by hugely successful oversowing) but for farming everywhere a need exists.

My thanks also to a very patient wife and adult family while distracted preparing this manuscript.

Registration acknowledgments:

Of the following Trademark names Registered in New Zealand:

Prillcote™ which belongs to PGG Wrightson Seeds Ltd.
Rhizocote™ which belonged to I.C.I. New Zealand, but is now expired.
Dolomite [Golden Bay Dolomite™] belongs to Solly's Freight (1978) Ltd.
Procoat™ is a Registered Trade Name of Food Investments Ltd.
Propell™ is a Registered Trade Name of Food Investments Ltd.
Polycote™ is a Registered Trade Name of Nuplex Industries (Aust) Pty Limited
Spectrum™ is a Regd. Trade Name.
Counter™ 20 g is a Regd. Trade Name of Amvac Corpn.

My apology is unreserved should there be any omissions to this list, or errors in the book. The author has no vested interest in any Registered Company or Trade Name in New Zealand, or elsewhere.

Gerald Bennett: [formerly]
Regd. Agricultural Consultant NZIASc [1990–96]
Farm Advisory Officer, New Zealand Department of Agric.
Regd. Public Valuer NZIV
Real Estate Principal AREINZ
Resource Planning Consultant
[retired}

Contents

Synopsis

Seed is one of the world's great multi-billion dollar industries. Mankind and most other living organisms are totally dependent on availability of seed.

It is not surprising therefore that VALUABLE INFORMATION in this book—recently developed, about the science of seed processing, *machinery, methods, formulae* and financial rewards has been held confidential for several years. That technology restraint is now terminated—to this author [by New Zealand High Court Order] who has added to its importance with cost benefit analysis never previously available.

This narrative investigation reveals the challenging development of seed processing pioneered in New Zealand—now used worldwide. Promotion and distribution of the book internationally will be rewarding to publisher and a wide range of readers.

The amazing science involved, its associated highly specialized technology and its practical application have potential to make a large impact on availability of food supply—specially in developing agricultural economies, but also importantly in all primary production systems. Quite astounding research results and advances in understanding are illustrated in this important book.

Throughout the world, excluding the two arctics, there are huge areas of undeveloped native grassland either unoccupied or unused, which could contribute significantly to more food production, much of it sparsely grazed [if at all] by nomadic sheep, goats and cattle whose herders are constantly on the move because the poor quality of grazing does not sustain foraging for long. Many advanced agricultural nations also have vast areas of wasteland where this technology can be applied.

The book focusses on one key limitation to better grasslands and that is the availability of nitrogen. A universally deficient major nutrient without which plant growth is severely restricted. It can be expensively manufactured—or it can be inexpensively transformed from the billions of tonne of atmospheric N^2 [which is unavailable to plants in that form]—into plant food available nitrogenous compounds via the activity of *rhizobia* nitrogen fixing bacteria—if those bacteria already exist in soil—or are effectively introduced in combination with their symbiotic host legume seed—chiefly clovers. The air we breathe is nearly 80% nitrogen. It is this atmospheric goldmine around the planet that is targeted here.

Intensive research in many countries has studied *rhizobia* to the point we are able to culture effective strains of the bacteria, apply this to symbiotic legume seed when sown on *Rhizobium* barren soils which then infect the root hairs of developing clovers and begin transforming N^2 in the soil to plant food within those clovers—and other legumes. In New Zealand the standard of living is now dependent on it.

There are many physical and technical problems associated with successful delivery of effectively inoculated legume seed which, together with remedies, are all revealed in this book. Coating of the seed was introduced both for protection of sensitive *Rhizobium* nitrogen fixing bacteria—because most seed carries natural toxins against soil decay which destroy *rhizobia* as well, and to provide precision placed finely milled nutrients right where seed is deposited in or on infertile soil. Thus, nutrients are fed to seedlings, not to acres or hectare.

Coating was then expanded to grass seed [ryegrass, cocksfoot, etc.] which does not involve inoculation but can carry important nutrients providing ballistic properties when aerial oversown. The coating can contain or seal in chemicals for protection of seedling development—not only clovers, lucerne and grasses, but a wide range of other crop seed as well.

The book relates further development into—not just coating of seed, but precision pelleting of any crop or horticultural seed which is of difficult shape or size. Recently developed precision farm drills select single seeds for accurate placement in the seed row but precision pelleting converts discoid [parsnip], elongate [lettuce] or pitted [onion] seed into specific and larger sizes mostly shaped like ball bearings which machines can then select individually. Resultant crops are free of inter-plant competition [as in scatter sowings], more uniformly shaped and sized product with a higher percentage of top priced yield in the marketplace.

How these developments were discovered and approved by scientist investigation then farmer adoption is all revealed including the establishment in New Zealand of a jointly owned new specialist company for commercial manufacture of these specialized and exciting products. This company, named Coated Seed Ltd. [CSL] established jointly by two parent contributors—Fruit growers Chemical Co. [FCC Ltd.—the processor] and Wrightson [WS—the seed merchant] became the first manufacturer worldwide to make a separate industry out of this technology naming its products "Prillcote™".

The expertise developed by CSL propelled "Prillcote" into world leadership in most aspects of seed processing during the 1970's and 1980's including sale of its original technology in the 1970's to a major development company in the USA.

Author Bennett, Farm Advisory Officer of the New Zealand Department of Agriculture risked his career at appointment to development manager of CSL amid controversy as to authentication of the techniques at commencement before Prillcote had been proven and accepted, only resigning after 14 years when the industry had become firmly established.

The author then left to begin his own company—not in seed or processing, but intending some part time assistance to seed companies who wished to produce their own processed products.

Author Bennett naturally requested termination of confidentiality restraint contained in his employment contract, which was granted but [quote] "in the opinion of the [FCC Ltd.] board actual termination would not take effect for a further 5 years". Although 3 years is generally considered a reasonable restraint period for employers to recoup costs of development, Bennett worked on for 5 more years respecting that restraint, then before departing, as a further precaution obtained a written legal opinion confirming that he was now free to consult. Then was stopped by ex partè injunction from disclosing any technology to anyone else anywhere in the world and for all time.

Subsequent action in the New Zealand High court removed that injunction limiting further restraint to a short time (1983 in Australasia and 1990 world wide) also excluding restraint on any technology published (being much of the original information), and excluding all grass seed coating (which was ruled not confidential). The author did not contest [as invited by the court] these further excessive restraint periods which were requested by the companies, because opportunity for consultancy had gone. The book barely touches on this legal aspect as it is not part of the biological success story of this newly developed New Zealand technology, however the publisher needs to know, not only why and how the author has long since been free of all restraint, but also that this freedom is exclusive to this author who spent some years and thousands of dollars on legal expense to win that freedom.

Since the author's departure, all three—FCC, CSL and WS no longer exist, but fortunately, there are successors.

The progress above suggests a covert but technically well managed development, however it involved 20 years of pioneering growth, from early controversy, through intensive research and development, much proving by trial and error, new manufacturing processes and coating formulae—equipment, adhesives, mineral coatings, field machinery and aerial technology for delivery of the products, plus a manufacturing process for new high quality inoculants. Of critical importance was the independent official proving and commercial acceptance of all this technology. It is fully related here, not revealed previously in one complete narrative publication which only the author and his contributor could do because no one else alive worked through this experience.

While other Nations have picked up and indeed advanced on some of this knowledge today, nevertheless the facts, discoveries, successes and failures recorded here remain an important historical basis—not only for a sound grasp of the science but as an advanced starting point and reference for those intending to engage in the industry. It will be valuable to the International seed industry, to agronomists, farmers and growers, libraries and student study, for engineering interests, polymer manufacturers, chemical and finely milled mineral manufacturers—and not least for better agricultural and horticultural research worker understanding. It is a disclosure of technology and an historical record of the important part played by those who's expertise helped develop seed processing into an International industry in an astonishingly short period of time from its early controversial days of development in New Zealand.

Gerald M. Bennett

GLOSSARY

Note: This is an explanation of words and phrases commonly used in New Zealand agricultural and horticultural industries. Technical or unusual names and phrases are selected to the level thought necessary—but proprietary items such as chemicals, adhesives, fungicides and insecticide formulations are separately described by their manufacturers with the product and are usually also described by searching the Internet. Technical names of seed and plant diseases and botanical names are also widely published elsewhere.

Words and Phrases in this book	What do they mean?
12 replications of each treatment	repeated 12 times for accuracy
Acidic soil	soils of low pH (say pH > 5.5)
Agitated by a wheel	seed kept "fluid" by agitation
Agronomist in seed	agricultural expert with specialist seed skills
Aircraft hoppers	holding bin (seed or fertiliser) in aircraft
Anion storage capacity (ASC)	soil binding of phosphate, sulphur, etc.
Anomalies	mistakes or errors
Antithesis	the opposite of—or sharply opposed to
Anti biotic	subdues or kills bacteria
Anti-biotic tannins	tannins in seed which destroy bacteria
Arthropod	organisms with exterior skeletons
Arthropod larvae	grub or maggot stage of arthropods
ASL	above sea level

Biological nitrogen fixation	bacterial transfer of atmospheric [N_2] [nitrogen] to roots of legume plants—mainly by rhizobia
Bradyrhizobia	slow growing acid tolerant nitrogen fixing bacteria which commonly nodulate tropical plants
Carcinogenic	disorderly cell growth. May spread cancer
Challenging tussock grasslands sites	localities hostile to pasture establishment
Chemical desiccation	chemical destruction of existing growth
Chromosome arrangement such as in tetraploid enhanced varieties	There are four sets of chromosomes in tetraploids which include many of the cropped brassicas. They have special qualities (e.g., are unable to breed with more common diploids with two sets of chromosomes)—the ancestors of tetraploids
Classic field research	mathematically sound key field studies by highly skilled independent scientists
Contiguous	adjoining along boundaries
Covalent bonding	ability of pairs of electrons with comparable electronegativity to bond from separate atoms—not necessarily of the same elements as in metal to metal bonding, but where iron and aluminium etc bond with phosphate or sulphur, which can block plant food availability
Cps	centipoises. A measure of dynamic viscosity
Cultivars	plants of the same strain (i.e., rye grasses) but with differing genetic structures
Dissident	dissenting, non conforming, rebellious
Disease vectors	any agent which transmits disease
DNA	Deoxyribonucleic acid. A molecule that encodes genetic instructions for all cell growth and functioning
Dolomite	magnesium carbonate
Double or even triple tool bar drill	two or even three axles in tandem to which drill units are attached—each offset [to desired row width]
"Doubles"	multiple seeds attached in one coating

Dressing plant offal	waste material, foreign seed, straw, etc. discarded in the process of cleaning seed
Empirical	realistic, practical or observed
Empirical results	based on experience, not on formal theory
Endophyte	bacterium or fungus living healthily in plants
Endophyte protection	a plant living inside another and not necessarily parasitic
Engendered	stimulated, produced or created
Exponentially	increasing at a fast rate—as in mathematics
Extrapolated	calculated by deduction—from the facts
FCC/CSL	Fruitgrowers Chemical Co. Ltd. and Coated Seed Ltd. [both long since terminated]
Finely milled nutrient	plant food element ground to 300# or smaller
Frontier town	service centre for rural development
Fundamental research	investigation of key elements in nature
Gafsa/dolomite	African gafsa phosphate and dolomite 50/50
Gagging	prevent discussion or publication
Glaciated debris	rock and shale broken up by pressure of ice
GMO grain	genetically modified organism—grain
Grasslands Huia	Plant breeders name for New Zealand white clover
Header tanks	Primary tank for collection and dissemination
Hostile hill country conditions	Harder climate due to height exposure
Host specific to the seed	optimum rhizobia strain/seed variety match
Hybrid seed	seed from parents of different genus
Hydroxy benzoic acid	a phenolic acid. Also a plant hormone
Hygroscopic	tendency to absorb moisture, gets damp
Hypotheses	theories or suggestions
Infect root hairs	rhizobia enter legumes via root hairs
Imperial tons	long ton in the avoirdupois system
Imperative	vital, necessary and important
Independent auditing	external examination of financial records

Inherently hygroscopic	naturally prone to attract moisture
Interplant competition	plants competing for sun, moisture, nutrients
Invermay	Govrenment Research Station at Taieri, Otago, New Zealand
Ionic bonding	electrons able to attach to soil elements
Irradiated	subjected to radiation for sterilisation
IT	Internet technology
Jettison system	procedure for rapid drop from aircraft
Jurisprudence	study of system of laws in any place
Largesse	Liberality and generosity specially of money
Lincoln University	Agricultural University at Lincoln, New Zealand
Liquid inoculants	rhizobia applied to soil as water based culture
Mathematical design	Field experiments mathematically designed
Manual bowl	seed coating open bowl for personal use
Mesh size	number of holes per [inch/cm] in a sieve
Moratorium	suspension, pause or cessation
NIT	"New inoculation technique" This was the big technical leap forward—the "gamechanger"
Nodulation failure	legumes which fail infection by rhizobia
NZIAE	New Zealand Institute of Agricultural Engineering
Obese	above average weight related to height
Organisms	living animals or plants
Oversown grassland	grassland from surface sown seed
Pastoral revolution	development of quality pasture without Cultivation by oversowing processed seed
Pasture pellets	several grass seeds in a nutrient pellet
Phenolic compounds	hydroxyl compounds with ether, alcohol and ester properties all toxic to bacteria
Phosphate "fixation"	rendering phosphate unavailable to plants
Phytotoxic	toxic to the host (plant)
Ploidy science	science of the number of chromosome sets in the nucleus of a biological cell

Polymers, co-polymers and homopolymers	natural and synthetic compounds with variable chemical structures (incl. adhesives)
Polyphenols	multiple phenolic anti-biotic compounds
Polysaccharide	a class of carbohydrate, sweet and insoluble
Proven in essence	proven at fundamental level. Indisputable
Pulsator	facility for regularly emitting enhanced pulsing electric charges
R&D	Research and Development
R. japonicum	rhizobia specific to the host Soyabean
R. meliloti	rhizobia specific to the host Lucerne [alfalfa]
Raymond roller mill	Steel crusher. Fine grinding of particles
Reseeding	clover and grass maturing to produce seed
Reticulation	a connecting network structure
Rhizomatous	plants whose rhizomes spread below ground
Rhizospheric or systemic action	action around roots or within the plant
Sanguine	colour of blood, ruddy and florid
Serial dilution agar tubechecks	seed inoculated in agar gel with increasing dilutions of rhizobia as a check on culture quality
Small scale trials	full scale procedures practised in miniature
Static electricity ignition of unstable micro-dust particles	explosion from atmospheric electricity igniting combustible dust particles
Stoloniferous ground cover	plants which spread above the ground surface
Strains selected legume inoculant cultures	Rhizobium of selected strain/s cultured into one inoculant for optimum performance
Surfactants	compounds that lower the surface tension of a liquid, the interfacial tension between two liquids, or that between a liquid and a solid
Symbiotic	mutually beneficial bacteria/plant support
Synthetic latex	an artificially made white flexible adhesive
Systemic	carried within the plant system
Technician	a person skilled in complex technology

Terminal velocity	maximum speed of drop related to gravity, weight and volume
Tertiary	education level 3—University or Polytech
Toxicologically effective	effective as a poison
Transconjugants	An organism (specially a bacterium) that has incorporated DNA from another via conjugation
Trocar	sharply pointed hollow tube for draining fluid (inserted into cattle abdomen to release dangerous frothing gasses causing bloat]
μm	micrometer
Uniform distribution of seed by	seed sown evenly—by species and by coverage
Vacuum refrigeration driers	drying by condensing moisture in a vacuum
Vector borne disease	disease borne by host and introduced to it by insect or parasite
Volatilisation	loss of gas through vapourisation to air
Water sensitive polyelectrolyte	polyanions and polycations which disassociate in water thus becoming "charged". Used to treat drinking water
Weakly plasticized polymers	polymers which are almost like plastic when dry

Gerald M. Bennett

After a year working at Mt. Cook high country sheep and cattle station, Gerald Bennett was selected a Govt. sponsored Rural Field Cadet at 18 yrs of age (one of about a dozen young men selected from the whole of New Zealand each year for a specialist training in agriculture) with 5 years spent at both Massey and Lincoln Agricultural Universities plus practical work on selected farms.

Then, bonded to the New Zealand Department of Agriculture for a further 5 years as Farm Advisory Officer, first in Hawkes Bay—then sole officer for North/West Southland he initiated field research and investigation which identified among other facts, the overwhelming need for clover seed inoculation for thousands of acres of new grassland planned to be sown on a major Govt. land development project in the Lakes Te Anau & Manapouri districts of Southland.

In 1959 at age 27 yrs he was selected from 13 Farm Advisory Officer applicants to represent New Zealand in Nepal where he tramped some 400 miles from Katmandu and Pokhara through Himalayan high country investigating nomadic sheep and goat herds and the native grasses they fed on preparing a Report for both the King of Nepal and the New Zealand Government as to recommended assistance from New Zealand in developing sheep and cattle farming in temperate Nepalese mountainous grasslands.

From that visit, New Zealand grass and clover seed was gifted to Nepal. Nepalese Department of Agriculture Officer Keshav Raj Keshary (extreme right in the picture below) was sent to New Zealand where he completed a Masters Degree in Animal Husbandry. Other Nepalese staff were trained at New Zealand's "Flock House" in general farming and particularly animal management. The huge opportunities available from better animal feed in the hill country of Nepal helped drive the need for this book.

Preceding a period of vocational consolidation during which time he became Regd. Public Valuer and Resource Planning Consultant, Gerald Bennett took the professional risk of appointment as project officer, later promoted to development manager at Coated Seed Ltd. [CSL] from technically fragile early beginnings where he spent 14 years toward developing it into a robust and successful Company particularly in the processing of legume seed to a high standard of inoculation. Coating and pelleting for the majority of the New Zealand seed industry CSL was strongly supported by FCC Ltd. [Fruitgrowers Chemical Co.] and staff.

At Katmandu airport on departure after 6 months in Nepal, Gerald Bennett (wearing the garland) is farewelled by Senior Nepalese Department of Agriculture Officers {at left} Janicki Pradahn (Livestock) and next to author Bennett - Netra Bahadur Basnyat (Agronomy), a Departmental Officer alongside Mrs. Pradahn (from Australia where Janicki Pradahn completed his Degree), then extreme right Keshav Raj Keshary.

At departure from CSL he set up his own two Companies and qualified as Registered Consultant of the *New Zealand Institute of Agricultural and Horticultural Science*, as well as a Real Estate Principal and AREINZ, by exam. He continued to work in close association with agriculture and is author [& a co-author] of research publications and articles in the *New Zealand Journal of Agriculture, New Zealand Journal of Agricultural Research, New Zealand Farming and Forestry Review, The Christchurch Press*—and produced the first substantial *Technical Manual* on inoculated, coated and precision pelleted seed published in New Zealand which received high commendation from FCC Ltd. managing director T.J. McKee.

A recent study found there have been almost 50,000 research articles published to date around the world on *Rhizobium* alone, the vast majority associated with genetics—which, while extremely important in mankind's search for methods of releasing the undervalued vast potential in mighty rhizobia, do not offer a comprehensive narrative explanation of the basics of seed inoculation, coating and precision pelleting. While this book cannot of course record all the findings of those thousands of research papers, it does embrace many key elements of success to date, explains difficulties encountered, unforeseen failures, positives, triumphs and future possibilities as well as providing those practical steps required in producing effective products and economically successful results.

In New Zealand, farmers and run holders were kept well informed about these new inoculation and coated seed developments by a series of very well attended local meetings which co-author Bennett addressed using both his own professional collection of colour transparencies from the 1950's plus others taken to assist CSL promotion.

A typical meeting of Otago farmers and run holders organized by Donald Reid & Co., Dunedin [Grain & Seed Department Manager Graeme Dixon] expressly to hear a 1 ½ hour address by CSL's Gerald Bennett on new developments with inoculated and coated seed. Interest was high. Aerial application of Prillcote™ seed and fertilizer was widely adopted here. Similar meetings were held throughout New Zealand.

John M. Lloyd

It is fair to record that while [retired] author Bennett has written the majority of this book—contributor Lloyd has been a busy active Director of his Partnership Company, expert in the manufacture and testing of inoculants, also in formulations, and hugely experienced in processing of research samples of seed for independent scientist use. He has made an invaluable contribution in the capacity of critic of this manuscript. John Lloyd's contribution has been of huge importance to the integrity of the publication.

His background experiences are:

A career devoted almost entirely to the development of new and novel formulations for use in agriculture and horticulture. During the past 50 years he has been responsible for developing more than 80 major commercial products including a significant contribution to the establishment and growth of the New Zealand seed coating industry as well as agricultural chemicals industries of Australia, New Zealand and elsewhere. His company continues to produce chemicals and formulations including for protection of inoculants and seed.

He is the inventor of numerous patents, a recipient of the ICI Australia 1992 and 1993 *Advanced Sciences Group National Finalist Technical Innovation Awards* and in 1998 received the *Grassland Memorial Trust Service Industries Award* 'In Recognition of Exceptional Contribution to the Advancement of New Zealand Grassland Farming Through Developing New Technology'. A [Rotary] Paul Harris Fellow and recipient of the personal certificate of former New Zealand Prime Minister, the late Rt. Hon. Sir Wallace Rowling, John maintains an active interest in seed coating, microbiological products and pesticide formulation development through the Richmond [Nelson, New Zealand] based Research & Development company Axis Associates Limited of which he is a part owner.

John does not seem to need the Ph.D. which authenticates other scientists. He has been able to contribute so ably to agricultural science by use of an astounding natural and instinctive ability for problem solving and solution finding plus remarkable precision in all his work which have shaped his notable career.

He has developed many practical "tricks" [skilled handling methods] in chemical, mineral and microbiological management plus seed handling being a "hands on" technologist in addition to a senior Manager, a Director and an experienced high profile negotiator with widespread overseas experience.

He would be an asset to any board of directors which might be lucky enough to secure him. We are most fortunate to have some of his expertise recorded here while in his very active 70's.

1

Introduction

There are more than seven billion people in the world[1] which, due to improving health and increasing birth rate, is expected by some analysts to double in less than 30 years in developing nations and in less than 100 years in the developed world. Apparently, as people become more financially secure and educated, birth rates do not increase at such a steep rate and while the world rate of increase has slowed since the 1960's, analysts still estimate there will be an 8 to 10.5 billion world wide population by the year 2050. Sadly, it is the developing nations that need substantially reduced birth rates, at least until food production safely matches populations.

Production of food based on current technology could not be expected to keep pace with today's explosive human situation; due to inequitable distribution, food supply is already failing to meet the needs of some 20% of the current world population including an estimated 6 million children[2] under 5 years of age dying from starvation (or food related disease) annually.

Urgent measures are required to deal with this disgraceful state of affairs.

While simply abandoning health support is not an option, reducing birth rates is; however, as we know, that has religious, cultural and biological complications.

The most readily attainable option in overcoming starvation would appear to be a more equitable distribution of world food, the production of which is now in fact sufficient, at present, to feed everyone adequately (with even greater potential) if it could be distributed both equally and economically. However, this too has limitations because many developing nations cannot afford to buy, import and distribute this food. Also, growers in exporting nations cannot produce food without recompense as they have bills to pay and costs to meet.

There is another way.

Some developing nations could do much more toward self sufficiency if they adopted methods and technology already discovered and refined (often at high cost) in more developed countries. The transfer of new strains of plants, animals, micro-flora and associated technology to maximize their potential in developing nations is a prime objective of 'The Green Revolution in Agriculture' as promoted by the American scientist, the late Norman E. Borlaug[3], whose programme for knowledge transfer around the world is being supported by many nations, as well as the UN.

Seed inoculation, coating and precision pelleting are now well established and important technologies in more developed countries and are contributing significantly to increased production in agriculture and horticulture. Unfortunately the adoption of these practices has not been universal and *huge scope exists to use them for improving food crop production in many parts of the world.*

This book is intended to assist everyone interested in these technologies, particularly those developing nations that may be able to use the procedures recorded here to improve production at home. The technologies can be adapted to enable improved forage and crop yields on difficult terrains and under adverse climatic conditions.

The publication of this book volunteers a contribution primarily from research and practice in New Zealand, Australia and USA toward alleviating problems of food scarcity because this subject promotes more and better food production. Universal spread of technological understanding will, in many situations, also help increase food production at lower cost, result in lower food prices which enables more equitable distribution of food to those in need.

This book describes methods of increasing food production through various recent advances in technology and, in particular, by better harnessing of the world wide, freely available, abundant supply of a natural component of the air we breathe that can be converted to essential plant food: nitrogen gas.

Importance of World Wide Nitrogen Fixation

Nitrogen is essential to the survival of plants, animals and mankind.

It is created in the universe by the fusion processes of stars, then becomes part of satellite atmospheres (including that of planet Earth). It is estimated to be the seventh largest mass constituent of the universe as discovered in interstellar space by astronomers using special detection equipment. It comprises a major part of the stellar atmosphere which surrounds some of the greater planets.

Back here on Earth, while the air we breathe contains oxygen, it is nevertheless predominantly comprised of elemental nitrogen (almost 80% by volume of dry air), a colourless, odourless and tasteless gas in the chemical form of N_2.

Yes, this is the same basic nitrogen that makes our crops and vegetables nice and green with vigorous healthy growth. But in that form, N_2 is unavailable to most living organisms (the molecule is structurally triple bonded almost to inertia) and must be 'fixed'—or combined with other elements creating (for example) ammonia [NH_4] and nitrate [NO_3]—before it can change to a form of nitrogen available for plant uptake and growth.

There are essentially two ways nitrogen can be 'fixed' to become available for growth: biological fixation and non-biological fixation.

Non-biologically available nitrogen comes from sources such as manufacture of nitrogenous compounds and fertilizers, from combustion and from electrical generation; for example, a small volume through lightning generated by stormy weather.

By far the greatest production of available nitrogen, however, comes from biological processes. Sources including the sea, forests and native flora contribute a significant volume, but easily the most important producer of all is that achieved by microorganisms, bacteria which convert N_2 to plant food available nitrogenous compounds. Of these, *Rhizobium* is the most common and the most important to mankind. Rhizobia are microscopic bacteria more sensitive to desiccation than most other bacteria but can live in suitably moist, mildly acid and reasonably organic soils until they are able naturally to infect the root hairs of legume plants (clover, peas, beans, soy, lucerne, etc.) and begin the process of converting nitrogen gas (N_2) contained in soil air, to plant food available nitrogenous compounds.

As long ago as 1965, eminent Professor of Soil Science, T.W. Walker[4] at New Zealand's Lincoln University wrote:

> "*Few people realize how dependent they are on the activities of microbes, and not many New Zealanders know that they owe their standard of living to the bacteria in the nodules on the roots of white clover plants*" [Foreword 'Legume Inoculation in New Zealand' Department of Scientific and Industrial Research Information Series No. 58].

That is even more true today than it was in 1965. We also know a great deal more about these nitrogen fixing microbes now, than we did then. He added:

> '*If New Zealand has one lesson to teach the rest of the world, it is that legumes are of great significance in agriculture.*'

Nitrogen fixation by legumes is the foundation of New Zealand's highly successful agricultural economy.

It is estimated by scientists that the weight of biological nitrogen fixed around the world is more than double that of non-biological nitrogen supply annually and that perhaps up to 150 million metric tonne of plant usable nitrogen is produced annually by all contributors. While the price of ammonia fluctuates a lot, largely due to prices of natural gas (methane) used in the fusion process during manufacture under intense heat, when converted to the value of nitrogen fertilizer at (say) $875 per ton of (anhydrous) ammonia, this would be worth about NZ$130 billion annually—and $10.4 Trillion over the lifespan of a man. That is a major pillar of world agriculture, forestry and plant life.

Let's look at one small farming nation. Whereas New Zealand comprises approximately 27 million hectares, much of it is mountainous leaving less than half that area in pasture and crop. Of the roughly 12 million hectare in grass and crop, the majority of which is growing clover with nodule bacteria of variable efficiency, over one million tonne of biologically fixed nitrogen is produced at the rate of as much as 400 kg per hectare of sulphate of ammonia equivalent per year, saving many millions of dollars annually in equivalent nitrogen fertilizer cost.

New Zealand does not use artificial nitrogen to the extent it is used in the USA, Europe or Asia. For one thing, apart from cost, artificial nitrogen applied to pasture actually suppresses clover growth; it is almost as if the rhizobia are saying to the farmer. "OK, if you don't need me, I will go into hibernation." But there is, of course, still a place for artificially manufactured nitrogen fertilizer—in non-leguminous crop production and, for instance, where dairy farmers traditionally use it on the Canterbury Plains in New Zealand with an application of around 55 kg per hectare of urea for a late autumn boost of grass to carry into winter.

New Zealand farmers, however, predominantly use rhizobia bacteria for their nitrogen supplies.

Of the six major plant food elements—*phosphate, sulphur, calcium, magnesium, potassium* and *nitrogen*, the absence or restriction of any one of these essential ingredients in soil, if not supplied by mankind as plant food fertilizer, will lead to reduced growth and if any are severely deficient, that will give us stunted production. In the case of nitrogen, this will result not only in yellow foliage but even in death of plants if they are totally deprived

Fig. 1 This Lucerne (alfalfa) is well nodulated, green and healthy on the left of the picture, but lacking rhizobia nodules on the right; in fact, almost a nodulation failure with stunted yellow growth.

of it. Grazing animals provide usable nitrogen in the form of urea, but if the pasture or grazing lucerne is poor and yellow, there is little edible forage available for livestock to graze; thus minimal urea is applied and the cycle of poor discoloured plant growth continues.

This necessity to supplement the most deficient plant food major element (or other essentials like water and temperature) before plants can grow any better (no matter how much more of any other element is applied) is nicely demonstrated by the analogy of a barrel of productivity [see Fig. 2].

SYMBOLIC BARREL OF PRODUCTIVITY

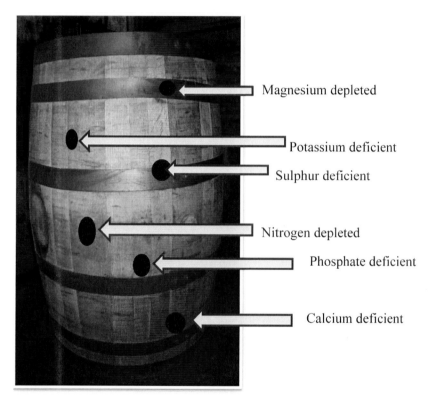

Fig. 2 We have to plug each hole (i.e., an elemental deficiency) in the barrel before productivity can rise to the next most limiting factor. Other limitations also apply—such as water.

Maximum plant growth cannot be achieved until the barrel of nutrients and essentials is full. If there are holes in the barrel (representing the limited availability of one elemental plant food or other necessity) then, until that hole is plugged (i.e., that item is provided), plants cannot grow any better than they presently are no matter how much extra (other) plant food nutrients or essential sunlight, water or temperature is provided. To fill 'the barrel of productivity' in order to achieve maximum plant growth, we must progressively plug each hole—starting, of course, from the bottom one (the most serious deficiency) and so on, up to the next hole in the barrel, the next most limiting factor, before we can expect plants to show improved growth.

Other issues can apply to this simplified but authentic analogy, that is, even if all nutrients are in adequate supply along with essential water, temperature and sunlight, they can all be subjugated by an over-riding plant disease or insect problem. However, assuming these are all under control, as they should be in a well-managed environment specially with an expensive crop to establish or an extensive grassland pasture to manage, we can expect maximum production from every hectare of land when soil or plant analysis determines deficiencies and these are corrected.

While there are areas of deficiency of sulphur and calcium, throughout the world, so are there concentrated deposits of these nutrients which are mined or quarried, crushed and modified for spreading on those deficient soils—but there are no 'deposits' of nitrogen in the ground.

There is, however, a huge 'deposit' of raw nitrogen in the atmosphere we breathe, and that source—a colourless gas, is available to plants predominantly by conversion and almost exclusively in economic terms via the mighty rhizobia nitrogen fixing bacteria.

There is also, of course, animal manure nitrogen which is valuable once pastures or forage crops have become established—but we cannot get access to that source of nitrogen until we first grow something for the animals to eat. If the foraging is sparse and spindly (not nutritious), then the quantity and value of animal manure nitrogen will be limited. If the pasture is vigorous, green and nutritious, with clean drinking water freely available, the animal manure value will be high.

Various strains of microorganisms are capable of converting elemental nitrogen to plant food; however, rhizobia are the most predominant in nature. They too, like plants and animals, contain some strains within the group which are more productive than others. If the most productive nitrogen fixing rhizobia were able to be successfully introduced to all leguminous (nitrogen fixing) plants throughout the world, food production could be lifted to more than double the present level; but at current microbiological understanding, that is an unattainable goal. Even greater production than that would be possible given adequacy of other major or minor elements, rainfall or irrigation, disease and pest control and superior cultivars. In other words, current world food production is only running at a fraction of its maximum potential.

Most bacterial infection of plants causes disease, but not in the case of rhizobia.

All legume plants (family Leguminosae) which have a symbiotic association with root nodule bacteria need to become beneficially 'infected' (mainly) in the roots to form nodules. All the clovers (white, red, subterranean, lotus, etc.), in addition to lucerne, vetch, cowpea, beans, peas, lupins, hardy

Fig. 3 Bright pink legume root nodules indicate optimum nitrogen fixation.

forageable shrubs like tree lucerne and large trees like the wattles, Tasmanian Blackwood, Coral and Strawberry trees, Pink Siris and the Japanese Pagoda tree, along with many others, when sown or planted, soon associate with rhizobia when they are present in the soil.

These bacteria are often present in previously farmed soils, even including uncultivated land where domestic animals and birds have carried them, but some of these strains of rhizobia possess a limited ability to fix nitrogen, that is, to transfer air in the soil (containing >80% nitrogen) to plant food nitrogen.

This limiting factor in the production of biological nitrogen is a disappointing feature of nature because these ineffective strains appear more able to infect the root hairs of legumes than efficient strains, so we have 'effective infectors' which are 'ineffective nitrogen fixers'.

That is a major world problem of far more widespread importance than much of the more visually spectacular phenomena of nature such as floods and droughts which are usually temporary, localised and at least partially manageable. As yet, microbiological scientists have not found a method for maximizing the infectiveness of effective strains so that they might supplant ineffective strains.

That goal, if and when achieved, and providing it is at modest cost to the farmer, will be of world-wide importance and the discoverer, at the very least, would be worthy of a Nobel Prize in microbiology for a huge contribution to world food production and the saving of millions of young lives. As things stand, they are likely to die horribly of starvation in the years to come. Introduction of rhizobia is inexpensive. If the strains used could also be effective in the face of pathogens (by cell protection, which does exist), and of competitors (which does not) also introduced in association with productive legumes in undeveloped areas of the world, then a giant step will have been achieved for mankind.

More widespread introduction of élite strains of rhizobia is just one example of the potential for increased food production on earth, yet mankind has only relatively recently begun to move in that direction, even in advanced agricultural systems.

With a greater farmer appreciation of the importance of biological nitrogen fixation and particularly of rhizobia which effect the conversion, even at present levels of understanding, mankind can substantially lift production of food for animals and for people in many parts of the world. This book is a contribution to that goal. Development has been slow due to the 'mystery of bacteria'.

Many years ago, it was not unusual for farmers about to sow a paddock (or field) into lucerne (*medicago*) species or soybean (*glycine*) species where rhizobia do not persist for long in the absence of the host plant, to cart and spread soil from an existing paddock of that legume onto the area about to be sown. They did not fully understand why they did it, but knew it produced a better crop. Unfortunately, it also spread any plant diseases which tend to become increasingly established with any crop as it matures over time (one of the reasons you should rotate your crops: to avoid a build up of host-specific disease).

Spreading soil was a cumbersome method of infection compared to the much easier, more hygienic and more positive application via inoculation of seed to be sown in the new field, despite the fact that this method too has significant problems, the rectification of which forms a large part of the technology described in this book.

Today, it is standard practice in many agriculturally advanced countries to inoculate the seed of crops such as lucerne (alfalfa), clovers, soybean, peanuts, cowpea, lentil, vetch, peas and beans, including certain host specific clovers (i.e., subterranean clover in Australia and New Zealand). This is because host plants do not grow as vigorously in the absence of those specifically host-élite strains of rhizobia which more efficiently fix nitrogen in the roots of those plants. Some plants, and even whole crops, may die without any effective rhizobia unless relatively expensive nitrogen fertilizer is applied or an 'injection' of urea via grazing animals can be sustained, but in difficult circumstances (i.e., during the sensitive establishment phase of grass or crop), when there is little foraging value at that early stage anyway—there is significant potential for plant damage.

Ineffective strains cause pale to white nodules which indicate a poor level of nitrogen fixation.

When they are also highly 'infective' and more competitive than efficient strains in occupying infection sites, we have a problem.

Seed inoculation with high quality, approved legume inoculants is relatively inexpensive so legume seed should always be inoculated, especially where that crop has not been grown for some time in the proposed soil or when it is known that effective rhizobia for that legume species are absent in that soil. Even if such crop has been grown in that soil within three or four years and residual rhizobia may be available to infect the new plants, inoculation of the seed is recommended to ensure maximum yields and as insurance against expensive nodulation failure.

Furthermore, let's hope there are no rhizobia resident in your soil because you want only the élite strain to be present in order to get the optimum chance of infection by this best performing strain. Soil conditions which favour temperate rhizobia survival usually include being moist, cool, only slightly acid (pH range about 5.2 to 6.5) and not subject to extremes of climatic change, particularly temperature and moisture.

If the rhizobia in your soils are from a previously inoculated crop (i.e., are an élite strain), maybe you do not need to inoculate. But how can the average farmer be sure?

There is a simple test. Take a hand trowel, sterilize it with a wash in cold water, then promptly immerse it in boiling water and don't allow it to touch anything at all (except perhaps a sterile plastic bag to carry it in) until you go to the field which is to be sown with a legume crop. Dig small samples of soil at about 25 mm deep from four or five locations in the field, placing each sample in a clean (preferably sterilized) plastic bag, or directly into a sterilized (boiling water treated) pot plant container. Sow a few grains of the seed you are about to grow (with a sterile spoon and without inoculation, of course) and, kept moist and warm (but not hot, and if cold, kept indoors), in just a few weeks the plants will have grown sufficiently for you to carefully remove one at a time until you can see the first few small nodules on their roots. Hopefully, they will be pink in colour. This needs to be done a couple of months ahead of sowing the crop of course, but with cold nights or if too hot by day, select a protected place to hurry growth along, which we cannot do with the main crop of course. If no nodules develop by the time the lucerne, soybean, clover or other legume reaches the first two leaf stage, then inoculation of the crop seed would seem to be essential to obtaining a good crop, unless you resort to more expensive artificial nitrogen fertilizer.

Other Processes and Formulations

In addition to processing inoculated legume seed, there are details in subsequent chapters about the coating of grass seed which are not legumes and do not associate with rhizobia or convert atmospheric N_2 to plant food nitrogen. They are coated for ballistics and precision nutrient supply which goes right where the seed drops on or in the soil. Better ballistic properties apply when surface sown (as from aircraft) and in providing protection from premature germination, bird predation (which can be severe in some locations, birds love nutritious seed which is highly proteinous) and against dampness in storage. Included with grasses, are the fine turf species which are not only coated but also coloured for identification, both of the variety and on the ground to highlight even sowing distribution. Fungicide and bird repellant can be included in the coating. Lawns, croquet greens, cricket pitches, tennis courts and bowling greens are sown with colour coated, chemically protected, fine turf seed.

There is a chapter on precision pelleting of horticulture crop seed, forestry seed and any other seed type which is of variable shape and size, unable to be sown individually because of that variability, but which pelleting builds up in size while also rounding the shape to a uniform product which is further refined as to shape and size by passing through a series of sieves. Any oversize lumps of seed and coating are eliminated over a top screen, the uniform marketable size passes through the upper mesh but is retained on the next mesh size while the dust and any broken pellets pass through the main product mesh size and are discarded off the dust screen.

Large quantities of valuable hybrid onion seed are processed this way as are carrot, parsnip, lettuce, etc. while NZ forestry nurseries have, for many years, drilled their nursery crops, mainly of pinus radiata, with precision pelleted seed, drilled in precise rows by precision drills creating uniformity of spacing and of inter-plant competition, with resultant uniformity of seedling tree size, better control of weeds and of ultimate tree lifting numbers for sale after tap root cutting by subterranean blade.

Fig. 4 A modern precision drill. This is a top class model Becker Kongskilde Aeromat Profi-Line machine. It has a classic eight row electronically controlled (from the tractor cab) hydraulic fan (compressed air) metering system. Twin side units are retractable (as shown above on the left side) for passing through gates, etc. This is an ideal machine for large farms or contractor use.

All these processes are described together with equipment, materials, formulae and factory or farm requirements for successful production. Details of special handling of product, of storage, distribution plus costing and invoicing are set out, explaining the need for awareness of individual requirements. Tips on sale and commercial promotion are stated for those selling finished products who may be seed industry associated or manufacturers who buy in seed for processing or those who process on contract to seed merchants or farmers directly.

Another chapter is devoted to various commercial considerations, modern methods and machinery. The limitations of various processes are discussed, as is the future of the industry.

Finally, while there is a reasonable need for pioneering private firms to protect confidentiality of their research and developments, there is also a need for employees to beware of excessive professional restraint (where their qualifications and experience 'hired' by an employer can become entangled in that employer's claimed confidentiality). It has happened in this industry due to employer misunderstanding and has the capacity to seriously damage career prospects and opportunities, even restricting alternative professional opportunities. Worse, if an employee cannot progress to an alternative opportunity because it is a competitive or similar activity, she/he also faces remuneration stagnation by remaining where they are because the employer is aware the employee cannot easily leave. Not many people are multiply qualified, allowing them to shift from one professional position to another, completely different activity, mid career, with impunity.

Seed treatment is a developing industry where new advancements and products are highly likely in the future, as they have arisen in the past, where employees need to be aware that they should ensure their own qualifications, professionalism and integrity are carefully protected as well as employers needing to protect legitimate proprietorship of genuinely acquired new technologies. A few simple clauses in employment contracts can protect both parties but where rejected by a potential employer, these are signals that it might be better for the employee to look elsewhere.

In contrast to that, this book is written in a spirit of fair, even generous, co-operation with everyone interested in and wishing to become involved in this fascinating industry. The aim is toward a better world, hopefully toward elimination of poverty and starvation, because it is abundantly clear from the history of mankind that the reasons for deprivation, rebellion and war derive not only from lack of education and opportunity, but also from lack of food and resources.

Our central objective is aimed at modest development of the greatest resource of all - mankind, through education, creating opportunity, leading hopefully toward more food, followed by better health, stable government, tolerance and industry.

Production: Simple or Large Scale

The author's aim is to bring together the essence of a huge volume of knowledge in scientific agricultural and horticultural research concerning seed inoculation, legume/*Rhizobium* symbiosis, legume inoculants production and use, coating non-leguminous seeds plus the pelleting of forestry and horticultural seed for precision pelleting, and their associations, in this narrative presentation. We hope it will be read and understood by those who can make it all work, the farmers, growers and processing contractors of the world.

Service providers also play an important role: the seed industry, microbiologists, machinery engineers, manufacturers of adhesives, finely milled minerals, nutrient and inorganic coatings and chemical additives plus the financiers with sufficient understanding to allow progress.

Most of these individual processes can be carried out quite simply on a farm or combined into a seed/manufacturing operation which can be carried out on a large scale in a factory (with laboratory support) where transport and distribution to farming districts allow. They can also be carried out by a contractor on successive farms (provided inoculants and test facilities are available). It can be a very inexpensive operation where supplies of inoculants, nutrient and other coatings are readily available (indeed, some coating materials like sand or limestone may be available naturally in certain localities, and others as waste from manufacturing activities).

The Essence (from a huge volume of research publications and commercial experience)

Both author and contributor expertise is in seed, inoculants, coating, granulation and pelleting—industries which, in combination, comprise an increasingly important contribution to world food supply. While this book is intended as a practical guide to those who are able to use the technology for their own and mankind's betterment, it is purposely not a formal research publication which would not attract the readers we hope to reach. The New Zealand authors are, however, aware of much of the research to date, some of the most important of which is embraced in these chapters.

Most of the world's seed, when surface sown, *does not survive*

Yet survival and establishment of seed are critical to the wellbeing of mankind.

The genetic capability of diverse plant seed is one of the great wonders of the world.

Added to these marvels of seed is the mystical phenomenon of atmospheric nitrogen conversion within legume plants whose seed was inoculated with rhizobia bacteria, or where those bacteria were resident in soil.

The science of incredible seed remains a major study by scientists worldwide, so it is not surprising mankind is learning better methods of handling and of growing plants from seed, than the time honoured method of simply random sprinkling of seed on the ground or in furrows, then optimistically awaiting the resultant growth. While this ancient method is often reasonably successful, specially for experienced growers, it is nevertheless often wasteful of seed which today, in many countries, is more likely to be genetically enhanced by natural selection, protected by plant breeders' rights, of higher germination quality, some of it disease resistant and therefore more expensive to use but with potential for a superior pasture or crop.

The old methods leave seed to fend for itself against a wide variety of pathogens—fungal, bacterial and viral, insects, also birds, hostile soil conditions and weather, particularly so when seed is sown on the surface of the ground, a hostile environment for any seed.

Fig. 5 Tussock hill country in New Zealand, previously in native grasses devoid of clover, has been oversown here with inoculated and coated clover and coated grass seed. White clover is now dominant. The livestock feed value is increased enormously, largely due to a big 'injection' of atmospheric N_2 from rhizobia root nodules.

Increasingly, land unable to be cultivated economically because it is steep, broken, wet, inaccessible or boulder strewn, can now be converted from growing poor quality low fertility native grasses for sparse intermittent animal foraging, into improved grass and legume oversown permanent pasture. This has been widely achieved, for instance, in New Zealand hill and mountainous country. Hardy native tussocks provide important shelter to more vigorous, but sensitive, introduced grass and clover species in these locations, specially during the susceptible establishment phase and up to 3,500 ft altitude.

Introduced seed is now simply oversown onto existing native grassland, either by aircraft fitted with suitable spreading equipment or by an air blower on a vehicle where accessibility is available (i.e., from farm tracks, sometimes along a ridge on steeper hill country).

Or, as practiced 100 years ago by pioneer farmers in New Zealand to get bare seed onto overburnt scrub and fern, it can still, as effectively as ever, be sown from horseback with handfuls of seed grabbed from a pikau (Maori name for a simple two pocket saddle bag made of sacks stitched together and tossed astride the horses lower neck). The seed is thrown around at random as the horse picks its way along the hillside leaving the rider to concentrate on achieving an even (uniform) spread of seed.

Today, we have access to improved plant species using higher quality seed, producing higher quality grassland. We can also improve the chances of survival of more valuable higher quality seed sown (often) into more difficult environments by careful planning which requires critical time of sowing, pre-sowing ground preparation and pre-treatment of seed. This allows us to reduce the quantity of seed required with less to purchase and sow because a higher percentage now grows, achieving good establishment results, much improved vigour and palatability, all at a reduced overall cost.

Plant breeding to raise the quality of various cultivars is now a major worldwide industry and can improve the quality and yield of forage grassland as well as crops. The cost of such hybrid or enhanced seed does and will continue to encourage procedures which reduce the quantity of seed we need to achieve a good result. In other words, better breeding of seed, as well as its pre-treatment, are paid for by avoiding seed waste and gaining much more profitable pastures and crops.

There are many books about seed breeding and some about seed dressing (removing weed seed and foreign matter) but this book is about (pre-sowing) seed treatment of which much less has been published. In a technical article titled SEED TREATMENT, Associate Professor Dr. Gregory Welham of Virginia Tech. VA, USA, wrote, *'the exact composition of [seed] coating material [and treatments generally] is a carefully guarded secret by the companies who develop them'*, which has been very true. However, the publication of this book you are reading substantially changes that situation.

A Brief History of the New Zealand Experience
(Seed inoculation, then coating, then pelleting from the 1950's to beyond 2,000)

The authors spent most of their working careers in New Zealand where the history of seed processing developed within the following background during the course of their careers.

The first major commercial development, following encouragement by DSIR (New Zealand Government's Department of Scientific and Industrial Research) and others to inoculate and protectively coat legume seed, was the formation of Coated Seed Ltd., an equal partnership of a major seed firm and a chemical and mineral processor.

Ownership of the manufacturing technology was retained by joint venture partner, Fruitgrowers Chemical Company Ltd. [FCC Ltd.] of Port Mapua Nelson, which was responsible for carrying out essential quality control plus research and formulation development for Coated Seed Ltd. (CSL). FCC Ltd. undertook responsibility for developing and manufacturing legume inoculants for use by Coated Seed Ltd. In support of this investment, it also operated a dedicated laboratory, glass house and plant growth chamber, all in support of the commercial production of Coated Seed Ltd.

Wright Stephenson and Co. Ltd. was the seed merchant partner, each company owning 50% of Coated Seed Ltd. Technical development activities of CSL were originally carried out under the direction of Mr. G.G. Taylor M. Agr. Sc. (Technical Manager for Fruitgrowers Chemical Co. Ltd.)] an experienced plant pathologist. He had earlier been Scientific Liaison Officer for New Zealand at its Embassy in the USA.

Mr. Taylor's technician, John Lloyd (a contributor to this book), managed the seed coating pilot plant at Mapua, overseeing inoculant production and coating laboratories there. He worked closely with Geoff Taylor and the author, Gerald Bennett (the latter employed by FCC Ltd. and seconded to CSL), who together developed essential expertise in seed inoculation, coating, precision pelleting and field use techniques which pioneered the commercial production of coated seed varieties collectively called Prillcote™ (a registered trade name) which became both technically and commercially successful.

On the retirement of Mr. Taylor, John Lloyd became Technical Manager of inoculants and seed processing for FCC which continued to carry out formulation and process development activities for Coated Seed Ltd. under contract and for many years.

In addition to the formulation and processing expertise offered by FCC Ltd., its associated company Lime and Marble Ltd. (L&M), a leading New Zealand minerals processing company, contributed further expertise in the development and manufacture of specialised seed coating materials. L&M provided finely milled lime coatings as well as other nutrients processed through its large Raymond Roller and hammer mills, both at Mapua and Christchurch factories.

Wrightson contributed its wealth of experience in the seed trade and all this, together with co-author Bennett's previous farm advisory experience, including seed inoculation and coating, plus his familiarity with research centres and scientists around New Zealand, placed CSL in a technically strong position from the outset. This commercial development was, of course, built on earlier research of others into seed inoculation and coating, particularly that of the Plant Diseases Division of the DSIR in Auckland which also supplied

Fig. 6 The inexpensive commercial research plant propagation house and climate control facility of FCC Ltd.

the vital mother culture of carefully selected strains of rhizobia to FCC Ltd. for the manufacture of 'Rhizocote' commercial inoculants.

That ground work, in both legume inoculant production as well as in coating seed, by the Department of Scientific and Industrial Research was driven by far sighted expert staff members including Technical Officer Mr. Athol Hastings and scientists Dr. D.W. Dye and Dr. C.N. Hale. They selected and classified the elite strains of *Rhizobium* which is today part of the extremely valuable New Zealand ICMP (International collection of microorganisms from plants). They also significantly advanced the practical and technical requirements of the industry with their 1966 Information series Bulletin No. 58 titled *'Legume Inoculation in New Zealand'* with a foreword by Prof. T.W. Walker. This government booklet became 'state of the art' in seed processing in New Zealand until further research by Agriculture Department scientist, Dr. Lowther, surpassed it.

In Australia, undoubtedly among the world's leading nations in *Rhizobium* research expertise, scientists such as J.M. Vincent, J.F. Loneragan, F. Hely, D.O. Norris, J.A. Thompson, F.J. Bergeson, R.A. Date, J. Brockwell, Dave Herridge, Rodney Roughley and many other highly skilled experts have revealed former mysteries associated with vitally important *Rhizobium* and their association with legumes which have contributed substantially to world knowledge. Much of that important research has been directed into highly specific and complex genetic, chemical and biological study at the cutting edge of intensive bacteriology and soil microbiology, in particular in relation to *Rhizobium*. Due to its huge importance, this scientific research has been extensive and intensive, much of it highly complex and well beyond the understanding of most readers.

Mankind has benefited substantially from such research, but there remains a vast potential to obtain further large gains from improved selection, cultivation, infection and survival of rhizobia, as has already been achieved for instance in plant and livestock research.

Of huge importance is the fact that rhizobia can not only be so productive but can also be quite harmful.

Harmful? Yes, where poor strains of *Rhizobium* prevent top strains from working.

Studies[5] have shown quite clearly that poor quality indigenous strains of rhizobia which, in many situations, fix little nitrogen thus restricting growth, can, in high numbers, inhibit infection by, or replace, high quality rhizobia introduced into the rhizosphere (by whatever method) limiting or preventing nodulation by those valuable strains thus impeding production.

There remains a huge potential for increased agricultural and horticultural production from research into how we may be able to reverse that situation.

Confidentiality—Who Owns the Technology?

Government agencies in technically advanced countries have spent multi-millions (of dollar equivalents) on research by highly skilled scientists into the marvels and mysteries of *Rhizobium* and their hugely beneficial relationship with plants in converting atmospheric nitrogen (N_2) to plant food available nitrogen.

These scientists are encouraged to publish results of their studies with no more thought about recovery of cost by their administrators than charging a small fee for print copies and an awareness that this work, in accumulation, will chiefly benefit the farmer/taxpayers of their nation. This research will also be added to the pool of world knowledge and be of benefit to all mankind. It is a satisfying and warming thought that by and large, while that cost is largely borne by the taxpayer/user, in fact the whole world will benefit in the long run from this excellent research. Once published, the technology belongs to mankind.

Everything in this book, now published, belongs to everyone, and anyone may use its contents to their very best advantage.

However, in private commercial enterprises, particularly in this industry, development can be cloaked in secrecy which this publication will substantially redress as cost recovery time has long since expired. The successful technology is now more readily available and is progressing in a more open, healthy and competitive way; many firms and manufacturers display their well illustrated seed processing equipment with offers to explain materials and methods in practice and on the Internet in addition to commercial brochures published around the world.

There are now methods for seed processing which use expensive machinery such as computer controlled spinning disc technology, rotary coaters, vacuum refrigeration driers and highly specified coating pans which process large volumes of seed rapidly, or precisely pelleted batches expertly for highly industrialized agriculture. Much is simply step by step improved design of equipment formerly used in application of chemical protection to seed.

This book, however, explains the equipment for and the development of inoculation and coating procedures from early development because they are now proven to be effective in the field, can be much less expensive to set up than highly sophisticated procedures and for some products or applications, have not been superseded.

There remains one major problem: those who can use this technology profitably, the farmers of the world, are largely unaware of it because they do not generally read the published results of scientists. There is, after all, not much use reading it if one does not understand it; very few people would understand much of it, so the author has endeavoured to sift out the most agronomically and economically important findings from the seed treatment industry and record them here in an easily understood narrative form which is also suitable for translation. We sincerely hope it will be suitable for reading by farmers, pastoralists and servicing firms in developing countries and for the benefit of agriculturists everywhere. New processors need to start conservatively with the basics, then modernize and expand.

Those who do not progress will stagnate, which carries its own cost.

Having spent some 60 years associated with this work during which time a great deal of progress has been made by many people, the author expects this information will not only contribute to essential advancements in agriculture but hopefully also toward alleviation of world hunger and poverty.

References

(1) [USCB] United States Census Bureau. As of today [about 2014] world population is estimated to number 7.146 billion having exceeded 7 billion on March 12, 2012 {From Wikipedia, free Encyclopedia: http://en.wikipedia.org/wiki/Population].

(2) [UNICEF] 2013. Child mortality, also known as under-5 mortality, refers to the death of infants and children under the age of five. In 2012, there were 6.6 million such deaths and more in previous years [From Wikipedia, free Encyclopedia: http://en.wikipedia.org/wiki/Child_mortality].

(3) Borlaug, N.E. 1914–2009. Born at Cresco, Iowa, died September 12, 2009 (aged 95) at Dallas, Texas.
    During the mid-20th century, Dr. Borlaug led the introduction of high-yielding wheat varieties combined with modern agricultural production techniques to Mexico, Pakistan, and India. As a result, Mexico became a net exporter of wheat by 1963. Between 1965 and 1970, wheat yields nearly doubled in Pakistan and India, greatly improving the food security in those nations. These collective increases in yield have been labelled the *Green Revolution*, and Borlaug is often credited with saving over a billion people worldwide from starvation. He was awarded the Nobel Peace Prize in 1970 in recognition of his contributions to world

peace through increasing food supply. Later in his life, he helped apply these methods of increasing food production to Asia and Africa.

In addition to the Nobel Peace Prize, he was also awarded:

Presidential Medal of Freedom – 1977

Public Welfare Medal – 2002

National Medal of Science – 2004

Congressional Gold Medal – 2008

Padma Vibhushan – 2008 (India)

Fellow of the Royal Society

[From Wikipedia, free Encyclopedia: http://en.wikipedia.org/wiki/Norman_Borlaug].

(4) Walker, T.W. 1916–2010. Former Emeritus Professor, Lincoln University, New Zealand, was the nation's first professor of soil science in 1952, retiring in 1978. Born in Shepshed, England, in 1916, died in Christchurch New Zealand aged 94. The recipient of many awards and honours during an illustrious lifetime in science—including (an Officer of) the New Zealand Order of Merit.

(5) Dudman, W.F. and J. Brockwell. 1968. Australian Journal of Agricultural Research 19: 739.

2

Legume Seed Inoculants

Outline

The enormous potential for nitrogen fixing rhizobia bacteria has been largely obscured in mysterious science. That potential is gradually becoming more widely understood and more accessible through excellent and intensive research carried out by a relatively small number of highly skilled scientists in many parts of the world.

Bacteria exist in almost all environments and play a vital role in the welfare of mankind. Able to divide every twenty to thirty minutes or so, some strains can produce a billion cells from an original one in just 15 hours. Some are pathogenic (harmful), such as those which cause tuberculosis, leprosy, typhoid fever, diphtheria, undulant fever, dysentery, syphilis, plague, cholera and tetanus. On the other hand, many others are hugely beneficial in breaking down animal and plant tissue into more readily available and recyclable nutrients. A further group of immense importance comprises those bacteria which are vital to the manufacture of food products such as cheese, pickles, yoghurt, etc., while another group plays an essential role in the manufacture of textiles, leather, detergents, farm ensilage and pharmaceuticals. Then there are bacteria called *Rhizobium* which beneficially infect leguminous plants and convert ('fix') nitrogen gas from the atmosphere to nitrogenous plant food compounds.

Rhizobium (Nitrogen Fixing Bacteria)

Microorganisms have a huge effect on the nature, structure and composition of plants, animals and many other objects, especially soil. Not only does one group convert the organic remains of dead plants and animals plus inorganic materials, even rock particles, to growth promoting nutrients, the enormously valuable group of bacteria called rhizobia have the amazing capacity to form a symbiotic association with leguminous plants, and convert huge volumes of atmospheric nitrogen to plant food, feeding the leguminous plant which in

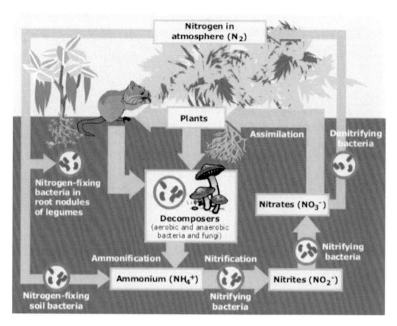

Fig. 7 Diagram and the following quote are from Wikipedia Free Encyclopedia.

'Air is about 78% nitrogen. Nitrogen is needed for life. It is an important part of proteins, DNA and RNA. In plants, nitrogen is needed for photosynthesis and growth. Nitrogen fixation is needed to change the nitrogen in air (N_2) into forms that can be used by life. Most nitrogen fixation is done by microorganisms called bacteria. These bacteria have an enzyme that combines N_2 with hydrogen gas (H_2) to make ammonia (NH_3). Some of these bacteria live in the roots of plants (mostly legumes). In these roots, they make ammonia for the plant and the plant gives them carbohydrates.'

turn feeds the rhizobia. Without rhizobia, the world would be a very different place and the effect on people and animals would be disastrous.

The diagram above explains the basic processes of nitrogen fixation which, in finer detail, are much more technically complex and well beyond the understanding of most readers—including the author who would have preferred to see a cow, sheep or goat rather than a rodent as the animal waste provider in the diagram; however, any animal represents the role animals play.

While mankind has historically striven to improve its health, wealth and happiness, it has done so at a fast pace in some countries and at a much slower pace in others. Education and understanding propagate rapid advances, but we can't all be scientists and fortunately, we do not all have to be to benefit from this new technology. Natural processes often work autonomously, but must be introduced where absent and then allowed to do their work.

Rhizobium research is largely quite specific, involving intensive studies of small aspects of the biology and microbiology which, in combination, create a more complete picture of why, how and where this science works. Most of us, including farmers and growers, do not understand these intensive studies but nevertheless, it is the farmer who turns the science into practice out in the field. This book is written for those people; its chief aim is to modify technical complexity into language any capable farmer or grower will understand.

Fortunately today, scientists in many countries have discovered and de-mystified a huge volume of the intricacies of symbiotic nitrogen fixation and while there remains much to explore, we now have at our disposal a vast, largely untapped potential available to agriculture and horticulture world wide. An indication of that untapped potential can be gained from estimates by people who are expert in this work. However, even they can only approximate from the evidence available to them thus the mystery and marvels of *Rhizobium* may never be fully quantified, but we can at least get some idea of the importance of this group of bacteria.

Value of Rhizobia

Of the world's land area of over 209,000,000 square kilometres [x 100 to make it hectare] about 260 million hectare currently grow legumes which various experts have calculated fix from 20 to 70 million tonnes of atmospheric nitrogen [N^2] to plant available nitrogen annually—which, even at the credible median of about 45 million tonnes is worth around \$50,000,000,000 of ammonium sulphate equivalent per year. This legume-rich farmland however, is less than the total land area of Argentina and only slightly more than the land area of Western Australia. It is but a fraction of the area it could be—so, sadly, even allowing a wide margin for error, it follows that the world is converting only a tiny proportion of the N^2 available to it, consequently missing out on an enormous source of wealth which impoverished nations absolutely cannot afford to ignore.

Take the small nation of New Zealand as an example of what is being achieved. It comprises about 26.8 million hectare, more than half of which is Alpine or in mountainous National Parks, lakes, forests and native bush. Yet it is essentially a farming nation, commonly using legume seed inoculants, supplying élite strains of *Rhizobium* chiefly to lucerne (alfalfa) and clover seed when developing new grasslands. In many soil situations there are high numbers of naturally occurring rhizobia ranging from efficient to inefficient nitrogen converters. The higher quality pasture soils of New Zealand may carry more than 1 million rhizobia per gram (of soil) and overall combined quality pastures benefit from receiving about 1 million tonnes of that worldwide (estimated 45 million tonnes of atmospheric nitrogen fixed annually). This represents about 2.2% of the estimated world current biological nitrogen supply obtained from less than 0.12% of the earth's land area.

Less than half of that 0.12% New Zealand land area is farmed so it is more accurately 2.2% of world biological nitrogen 'fixed' on less than .06% of the earth's land surface area; or let us say about 1/40th of the world's biological nitrogen fixed on about 1/20th of the world's legume-rich farmland. These statistics are estimates only but suggest that New Zealand farmers are benefitting at above average of world nitrogen fixation levels, of the land areas where legumes are grown. If and when science allows farmers to re-inoculate their grasslands (i.e., inoculating the soil, not just the seed) with newly developed super-elite strains of *Rhizobium* which are able to infect legume roots to the exclusion (or domination) of lower grade strains (that day will come as long as we keep funding good research)—then we will see a major increase in pastoral and crop production.

At about the land area of Nevada (USA) and only slightly larger than Oregon, this New Zealand (over half of which is not farmed), was, in 2010, not only the world's fifth largest dairy products producer, but also the largest dairy exporter in the world. These rising dairy exports were worth $12.75 billion in 2009, elevating the New Zealand dairy industry into a leading role, even ahead of mighty '*Dairy Farmers of America*' in production.

In addition to that, it exports large volumes of meat, wool, grain and seed, all containing and supported by the fabulous and amazing biological nitrogen supply received free from the atmosphere. It confirms that New Zealand farmers, in general, are quite knowledgeable with many holding agricultural, horticultural or science diplomas and degrees. Others, most of whom regularly use computers and modern technology to aid their work, have developed considerable expertise from practical life experience. They are also aware of the need to provide a balanced diet of the 5 major plant food elements, including biological nitrogen or face up to buying sulphate of ammonia at $895/tonne ex depot (2015). They are also aware, not only of the hefty cost of manufactured nitrogen but also that it is mostly manufactured from natural gas, using an environmentally unfriendly process that requires a great deal of energy.

Biological Nitrogen May Be Capped—Without More Research

In a recent paper delivered by W.L. Lowther and G.A. Kerr to the New Zealand Grasslands Association (Proceedings Volume 73, 93–102), titled '*White Clover Seed Inoculation and Coating in New Zealand*', the authors have recorded a considerable volume of *Rhizobium* research directed at identifying the effectiveness of strains and their ability to infect clover varieties in the presence or absence of resident 'native' rhizobia in New Zealand, where large volumes of clover seed have been commercially inoculated over many years.

The essence of these studies and the consequent view of those authors is that much of New Zealand pastoral grassland now has not just 'adequate' levels of rhizobia in soils, but resident strains and their levels of fixation which are incontestable, except in undeveloped grassland where clover is absent, in virgin land cleared directly from scrub and on arable land where crops (specially maize) have been continuously grown for 10 years or more. The claim that introduced strains of *Rhizobium* on inoculated seed may improve levels of nitrogen fixation where resident strains exist is not supported by their investigations. However, authors Lowther and Kerr issued a warning in this publication that some results obtained defied logic—thus it is clear that this is a complex microbiological research target which will demand intensive bacterial study to progress.

If as suggested by Lowther and Kerr, New Zealand soils are now largely 'occupied' by *Rhizobium trifolii* (are those strains of adequate quality to support optimum growth where all else is satisfactory?). What percentage of resident rhizobia are 'efficient', what percentage are of 'mediocre' quality and what percentage are of poor nitrogen fixing capability in any grassland situation? We doubt anyone knows.

If the vast majority of resident rhizobia in New Zealand's soils are of 'mediocre' nitrogen fixing ability, which is suspected from observation of largely pale grey to white nodules dominant over a wide variation of New Zealand soils and over many years by this timeworn author, and if those dominant strains have the competitive ability to exclude (or more dominantly infect) better strains which are more efficient nitrogen fixers then a whole new challenge to mankind's technological ability lies ahead in either suppression, domination or elimination of inferior strains, alternatively increasing the infective ability of superior strains, dominant in both infectiveness and effectiveness, or a combination of all.

The development of 'superbug' strains is not beyond the realms of possibility by isolation, selection and genetic modification but if achievable, one can only suspect that mankind is currently still in the 'Stone Age' of enhanced biological nitrogen fixation potential.

What is clear, is that parts of New Zealand have deep green, clover rich grasslands with abundant nitrogen supply, some of it from animal effluent but also some from very efficient nitrogen fixation via bright pink nodules on the roots of those clovers.

This is not confined to naturally super-fertile pastures such as are found in New Zealand's lush Waikato dairy lands, but also in areas developed in the 1960's out of stunted fern and tussock totally devoid of legumes, on raw glaciated virgin soils with high iron and aluminium ionizing capability, barren as to major element fertility (particularly phosphate) but which were oversown with clover seed inoculated with introduced èlite strains of *Rhizobium* supplied to the inoculant manufacturer by the (formerly named) New Zealand Department of Scientific and Industrial Research.

The soils of these pastures now largely harbour introduced superior strains, being devoid of resident *Rhizobium* altogether at the time of development. Today, such areas, particularly in the Te Anau basin of Southland, though recently developed, with limited organic residue support and neither matching the dry matter productivity of Waikato pastures nor benefitting from the same level of livestock grazing intensity, are nevertheless among the most brilliantly lush green pastures in New Zealand. This phenomenon is a 'message' from nature which does not appear to have been recognised by some agriculturists; there seems little doubt that it remains unrealised largely because it is a consequence of rare opportunity, most areas are 'contaminated' by ordinary strains of *Rhizobium* but these newly developed areas were not.

The following extract from this Lowther and Kerr research paper provides a summary of their findings:

> "It has been hypothesized that inoculating clover seed may improve clover growth through introducing more effective rhizobia, however trial work suggests that the likelihood of any significant response from doing this, with our current technology, is very unlikely. Resident populations of rhizobia in New Zealand pastoral soils can be up to 1,000,000 per gram of soil (or 3,000,000 in a teaspoon of soil), within each soil there are a range of rhizobial strains which vary widely in their nitrogen fixing ability, and are well adapted to local conditions. Results show that the rhizobia from inoculated seed are likely to form such a small portion of nodules they will have little or no effect on clover growth. An additional factor is that clover cultivars have all been selected with existing soil populations of rhizobia, and there is no guarantee that the individual strain(s) used in a clover inoculant will be the most effective for any cultivar of white clover.
>
> In addition to improving survival of rhizobia, there are some situations where seed coating can provide other advantages to clover establishment. Lime coating can enhance nodulation through a localized increase in pH on acid soils where it is uneconomic to broadcast lime. There is also considerable evidence to support the use of seed coating through both targeted application of fertilizer to enhance growth of the germinating seedling, and insecticide to help protect seedlings from pest damage."

There is, of course, a danger that where chemicals are being included in or on the coat of inoculated and coated seed, that some of these may adversely affect the viability or the longevity of rhizobia to the extent that the number of cells per seed are simply insufficient to compete with 'residents' at point of infection. Further complications arise if and where stresses are applied under commercial conditions including delays in sowing, dampness in storage, post sowing field conditions such as partial dampening, then drought, and other phenomena.

With respect, the author of and the main contributor to this book do not agree that the only useful potential remaining for clover seed inoculation in New Zealand are these three situations suggested by Lowther and Kerr: *undeveloped grassland where clover is absent; virgin pastoral land cleared directly from scrub and arable land where crops (specially maize) have been continuously grown for 10 years or more.* There are other situations where *Rhizobium trifolii* may be severely depleted if not entirely absent [such as oversowing slip and slump damaged pastures] and where an adequate loading of viable rhizobia per seed may productively infect newly established clovers. It is also possible that in those situations, and even where low grade resident rhizobia exist, using for instance newly developed 'pasture pellets', where clover seed can be surrounded not only by a high loading of élite, highly host specific *Rhizobium*, but also wholly enclosed in nutrient pellet material. This comprised milled phosphate, lime, peat, etc. separating introduced seed and inoculant from its soil environment at least until the seed has germinated and the seedling rhizosphere [within the pellet] has been infected by the introduced strain (having been shielded from "resident" rhizobia so that introduced *Rhizobium trifolii* may succeed. Hopefully dominating the subsequent level of N² fixation at least for a period of time. These hypotheses remain to be tested.

Those depleted resident rhizobia situations are likely to be areas of hill country which have been subject to prolonged severe drought—as we see periodically in New Zealand (i.e., Hawkes Bay, Marlborough, McKenzie Basin, Central and North Otago) and quite commonly in Australia, where grass is sometimes seriously depleted, even perished white in colour and clover is sparse or gone. They could also be areas of controlled burning of scrub in New Zealand and rogue bush fires as are frequently experienced in Australia or those subjected to controlled rejuvenation burning which can reach temperatures of over 500°C above ground and over 35°C down to 25 mm deep in some soils for 10 minutes and more as found in studies of biological fire damage.[1] *Rhizobium trifolii* are unlikely to survive those temperatures for that period of time.

In addition, after a Lucerne stand has been sole occupant of soils for some years, annually cut for hay or silage but not grazed—and in other prolonged cash cropping/fallow situations in the farm management regime, may also be responsive to EFFECTIVE seed inoculation—but exactly where, farmers do not know. There is a soundly based axiom in New Zealand:

If in doubt, because it is low-cost, the seed should be inoculated.

International Opinion

Larger nations such as Australia and the USA have scientists working in the field of biological nitrogen fixation through the legume/*Rhizobium* symbiosis, chiefly, of course, for the benefit of their own farmers but, generously, for the benefit of developing nations (as well as a form of overseas aid, like this book). The following important information from USA explains the history of development of an expert facility called 'NifTAL' [Nitrogen Fixation by

Tropical Agricultural Legumes] which has assisted and promoted legume seed inoculation worldwide.

[Quoted with kind permission of Dr. Paul Singleton, Director of NifTAL (1992 to 2005 when its USAID funding ended) currently at CTAHR (College of Tropical Agriculture and Human Resources), University of Hawaii, located at Maui Agricultural Research Centre, Department of Tropical Plant and Soil Science, at Kula, on Maui, Hawaii, USA.]

In 1975 at the peak of the oil crisis, national economies worldwide contended with high costs of fossil fuels. The nitrogen-fixing symbiosis between legumes and rhizobia was recognized as an attractive, low-cost alternative to petroleum-based nitrogen fertilizers for increased crop production in the tropics. Legumes are one of the three most important plant-based food sources globally, and many families rely upon legumes as a primary source dietary protein. Since nitrogen is a key element of all protein, BNF [biological nitrogen fixation] could have immediate benefits to the welfare of families in developing countries.

The United States Agency for International Development (USAID) drew on expertise in the University of Hawaii's College of Tropical Agriculture and Human Resources to establish an interdisciplinary unit called NifTAL (Nitrogen Fixation by Tropical Agricultural Legumes). NifTAL was dedicated to the application of technologies based on BNF to international development goals. NifTAL's ultimate purpose was to help farmers maximize BNF inputs to their cropping systems, and thereby increase the production and quality of high protein foods while reducing their dependence on expensive nitrogen fertilizers.

NifTAL assists not only US farmers but [via USAID] also developing nations around the world with technology and training, materials and procedures for the advancement of biological nitrogen fixation from better quality inoculants and more efficient use of the technology.

As stated above, having achieved major improvements in and valuable development of biological nitrogen fixation in many parts of the world, NifTAL's major international funding via USAID has now ended. Extremely valuable published NifTAL material can be accessed at the website *ctahr.hawaii. edu/bnf*. Dr. Singleton's e-mail address may still be available at the University of Hawaii, email: niftal@hawaii.edu.

The 'Numbers Game'

In some of New Zealand's more isolated land areas where domestic livestock have not previously been grazed, there are few and even no *Rhizobium* in those soils at all. In those localities, responses to inoculation of oversown seed have been spectacular; even at just 100 rhizobia cells per white clover seed (*Trifolium repens*) at point of sowing, nodulation with the introduced élite strains has been very successful.

Fig. 8 New pasture from simple aerial oversowing onto never cultivated land.

This spectacular clover response in New Zealand's Te Anau Land Development Project, Southland, was obtained from aerial oversowing Prillcote™ inoculated and coated clover plus coated grass seed. This photo (showing the author's Triumph 2000 vehicle and a friend) is repeated in Chapter 3. p. 102 Fig. 29B below a classic nodulation failure also photographed by the author a few years earlier (while the local Farm Advisory Officer, Department of Agriculture) in a nearby location. That degree of contrast may never be seen again in New Zealand. Both photos taken by the author for his professional records were made available to CSL and published in its 'Technical Manual', edited by the author in the 1970's.

While this snapshot, taken by the author for his own records like others in this book, is reproduced from an old and 'rather tired' 35 mm transparency, it demonstrates that oversowing was used—not cultivation, because a few tussocks plus matagouri scrub (dark stalks), which would have been eliminated with cultivation, remain standing. Near this locality in previous years (see page 102), even after expensive cultivation, liming and topdressing, clover and grass seed totally failed! Why? Simply because the clover seed was not *Rhizobium* inoculated. As simple as that.

In all the research worldwide, the single most universally important factor in successful seed inoculation with sensitive rhizobia bacteria is that the numbers of rhizobia applied to seed prior to sowing is of supreme importance. Cell loadings must be high enough to compensate for extreme levels of mortality when applied directly to bare seed. Their survival to infection is paramount. It is therefore better to use a lower cell count with longevity protection as achieved in Prillcote—than a high cell loading but unprotected from hazards like seed toxins, dessication, sunlight/heat and delays in sowing, all of which cause rapid mortality.

Seed inoculation has to be economically worthwhile commercially of course and, as it is a relatively inexpensive process, it is not uncommon for farmers to apply double the rate of peat or lignite based cultures of *Rhizobium* as recommended by the manufacturer. It is important that farmers use only legume inoculants that are officially certified or meet official quality standards and which contain the appropriate strain or strains of *Rhizobium* recommended for their locality and the legume species to be inoculated. Packages of legume inoculants should also show an expiry date. In addition, inoculant storage conditions must be satisfactory (e.g., cool, dark, moist and toxin free, just as you would expect to find in good soils).

Liquid inoculant is increasing in popularity (as indicated by NifTAL's Dr. Paul Singleton) and can provide maximum numbers of rhizobia in situations where legume seed crops are sown with chemically pre-treated seed which would be toxic (as many insecticides and fungicides are, specially the heavy metal based ones such as copper or sulphur), if applied in close contact with rhizobia. With suitable precautions, seed can be chemically treated (as bare seed) then drilled into cropping soils where liquid inoculant is trickled into soil immediately below the chemically treated seed, which procedures growers can easily execute themselves. Where available from suppliers, liquid inoculant concentrate may be diluted on site with clean (unchlorinated) water for better distribution in the field.

Care with Inoculants

Rhizobia are not as robust as some other bacteria. They die readily when exposed to heat, sunlight, desiccation, mechanical damage and the toxic effects of natural and complex seed coat phenols and tannins (which are nature's protection mechanism, protecting seed itself from soil borne pathogens). *Rhizobium* also die when in contact with materials such as sulphur and sulphuric acid (found in superphosphate), most herbicides, insecticides and fungicides. When purchased, all solid inoculants, peat, earth or lignite based, should be stored in a cool place where ambient temperatures are below 20°C (in a refrigerator but not in the freeze box). Simple freezing of damp culture may significantly collapse and destroy the bacteria by cell rupture from the formation and thawing of ice crystals.

But, is all this fuss and pampering necessary? Of course it is; farmers take special care when they purchase a new bull, expensive machinery or select key employees to work on their farm. These are items that are rewarded by taking special care. Why would you not take special care with sensitive microbial inoculants when, by doing it well, it can earn more than $560/hectare and also because if it fails to work, your clover based grassland (or legume crop) will probably be reliant on resident rhizobia (if any) which may not be fully host-specific with your better quality introduced clovers or other legumes; either way it is likely to become the limiting plant food factor for growth no matter how well you provide all the other fertilizing elements—unless you buy artificial nitrogen. If there are absolutely no rhizobia in the soil (which is often the case with crops such as lucerne, soybean or cowpea) and seed inoculation fails, then you will have a pale green to yellow looking growth

Fig. 9 This Lucerne (alfalfa) field was drilled with inoculated and coated seed but the inoculant was subsequently found to be faulty: in laboratory tests it had passed the rhizobia numbers count but most of the cells were r. *trifolii* (for clovers) NOT *meliloti* (for lucerne). It had also passed a specificity test because there were SOME *meliloti* in the culture, sufficient to nodulate seeds in a laboratory agar test, but NOT enough to nodulate all of the seed in the field! This unforeseen and unexpected outcome resulted in a SUBSTANTIAL [but not total] nodulation failure as evidenced by the widespread yellowing and stunted plants in the picture. The tractor and drill are injecting dry 'Rhizocote' inoculant granules (*meliloti* culture coated onto inert calcite granules) which eventually saved this crop. The granules perform best when soil is pre-wetted as in rainfall or irrigation; the latter applied here in this border dyke irrigated lucerne crop. A standard seed drill can manage this job but a sod seeder (direct drill) is ideal. The granules only need to be placed 10 to 15 mm below soil surface.

result and will either need to buy and apply nitrogen fertilizer (regularly) or (as has been successful in New Zealand) apply rhizobia granules (freshly inoculate/coated onto an inert material such as marble chip and drilled into the crop) provided the soil has been moistened first by rain or irrigation. This product has been proven very successful.

So now that we know seed is protected from soil pathogens by nature's 'built-in' seed coat antibiotics, why do we put sensitive rhizobia bacteria there?

You would be justified in making the observation that if the surface of seed is such a hostile place for rhizobia, it would seem quite a bad idea to use seed to convey them into the seedling root zone (the rhizosphere).

However, microbes don't travel far in soil by their own mobility (but can be washed by stream or carried by livestock) so the closer we can get rhizobia into soil just below each germinating seed, the more chance there is for *Rhizobium* cells to locate and infect those tiny root hairs. When penetrated into the plant they are more safely protected, begin to form nodules and start converting nitrogen from soil air to plant food. If soils are saturated and pugged (compressed) there won't be much air for *Rhizobium* to absorb and this is partly why wet, pugged pastures turn yellow, grass dies and if badly saturated and compressed can even become barren.

FCC Ltd. technicians have commented as to just how toxic soluble seed surface poly-phenolic materials, including tannins, are, by washing them out of seed (which is not good in practice, of course, because it leaves seed unprotected from a multitude of soil pathogens) and then applying the washings onto cultures of rhizobia. The results have been spectacular; this liquid could be bottled as an anti-biotic, it destroys rhizobia promptly. Moisture mobilizes these water soluble compounds present on seed and the process of inoculation usually moistens seed. However, by use of protective peat (or similar) both as a carrier for rhizobia plus as a coating material which further separates rhizobia from contact with seed and adsorbs tannins, the natural toxicity of the seed is minimized but not necessarily eliminated.

Polyphenols comprise a wide group of compounds under the main classification of tannins, lignins and flavanoids. Each of these gives rise to more specific groups. The author/contributors are not experts in this field but suspect it is the tannins that are toxic to rhizobia. Among other effects, they precipitate proteins. Tannins are water soluble and well known to protect plants from attack by bacterial and fungal pathogens.

As mentioned above, there are other ways of applying *Rhizobium* for access to legumes, such as spraying cultures into soil in liquid form (as an aqueous suspension) which, while successful in cropping programmes, is not a universally practical solution for aerial oversowing of seed, nor for most seed drills which do not have liquid applicator equipment. The latter would require not only a large source of clean water available in the field but also extra time filling up application tanks; also in addition, much greater weights for tractors to pull and adversely compressing worked and aerated soil which is specially harmful if wet.

Nevertheless, it is used successfully and even preferred in some commercial cropping operations where the grower is set up for it.

However, using seed as the carrier is easily the most widely employed method of application at present levels of knowledge. That, of course, may not always be so.

There is research in progress, looking at the practicality of using solid inoculants, not just inoculant media materials, peat and lignite, but also inexpensive materials such as perlite, bentonite and other clays, cork compost, vesicular pumice, marble chip and coarse sand plus various protective polymeric materials; however, the practicality of their use remains under research study.

Selection, Certification and Distribution of Rhizobium Strains

As one would expect, due to their importance, legume seed inoculants are available commercially in many countries; some concentrated on temperate plant symbiosis and others on tropical plant symbiosis.

In Australia and New Zealand, commercially produced inoculants are manufactured (i.e., multiplied in favourable media) from mother cultures of host-specific strains of *Rhizobium*, each being distinctively homogenous strains of nitrogen fixing bacteria which form cross-inoculation groups (strains which are only effective within that host-specific group of legume plants). Strains of *Rhizobium* vary and are only effective with their specific hosts.

These strains have been selected in well developed agricultural systems (on behalf of the farming industry) by microbiologists, usually scientists of government agencies. This is a monitoring job suited to government control as it would be of prohibitive cost for manufacturers to attempt to select and isolate their own strains. They are also unlikely to be as successful. This work is of nationwide importance and really does need to be under strict, commercially independent, formal control by skilled and experienced people.

Fig. 10 Within an inoculant manufacturing laboratory.

The photograph here shows two steam sterilizing autoclaves, into one of which the laboratory technician is loading glass jars and equipment to be sterilized prior to use in *Rhizobium* multiplication procedure. If just one or two contaminant microscopic bacteria survive, finding their way into multiplication media, they will of course also multiply, some faster than *Rhizobium*, with negative results. Hospitals sometimes discard these autoclaves cheaply, but in Australia, sterilization is achieved by way of gamma irradiation.

Strains selection, testing, numbering and distribution are of national importance.

These strains are obtained from many field isolates (rhizobia from a single colony originating from a washed, crushed nodule on plant roots). They are taken from vigorous plants containing large pink nodules growing (ideally under stressed or difficult conditions) in known peak performing pasture and crop locations. If they are effective in slightly unfavourable conditions, one would assume they could be expected to fix even more nitrogen under favourable conditions (that however, is part of the mystery of *Rhizobium*—it is unwise to simply 'assume'; They may well be conditioned to the environment in which they were found, and not do quite so well in a microbiologist's concept of 'more favourable conditions').

Rhizobium strains growing in these significantly bright pink to red nodules, collected by the supply authority, are examined in the laboratory for stability of the strain, pot test performance in various soil types, survival on seed, strains purity, nitrogen fixing ability and ability to multiply quickly in (usually sterile) media such as agar or a broth later combined with modified peat, the peat having first been milled to a fine particle size and then sterilized for instance by gamma irradiation (Australia), or by autoclave, i.e., steam sterilizing (New Zealand).

A subject requiring more research is the infective ability of introduced strains compared to indigenous (or local native) strains. No matter how productive a strain is at fixing nitrogen, it would be quite useless if it could

Fig. 11 These pink clover root nodules indicate that they have been formed by an efficient host-specific strain of *Rhizobium* fixing a high level of atmospheric nitrogen to the benefit of the legume plant; more common grey to white nodules indicate a lower level of nitrogen fixation sometimes none at all [Illustration permission of Wikipedia Free Encyclopedia).

not compete for nodule formation with indigenous bacteria at various levels of efficiency. Studies of such competitiveness and perhaps selection of infective ability or the development of techniques that confer selective ability to an introduced strain are areas of research required which may take the technology another giant step forward.

International Inoculation Standards

There are however, good and not so good qualities of commercial legume seed inoculants.

Australia

In Australia, inoculant quality is checked by ALIRU (Australian Legume Inoculants Research Unit) which, after several changes motivated by their field and research experience, has more recently adopted a minimum standard of 500 viable rhizobia per seed at inoculation for small seed legumes such as white and alsike clover and 1,000 viable rhizobia per seed for larger seed such as lucerne or subterranean clover. This is quite a high standard as far as inoculant manufacture is concerned, but it may still fail, in certain situations, to effect successful nodulation of seedlings given the complexity of adverse factors which can mobilize against it—in particular, mortality on seed and competition from ineffective resident rhizobia of low nitrogen fixing ability. In many cases, the farmer does not know if his inoculation effort has been successful—and that problem 'cloaked in mystery' is unfortunately one of the major reasons legume seed inoculation is so tragically misunderstood and undervalued.

New Zealand

In New Zealand, the Department of Agriculture's Seed Testing Station at Palmerston North (which later became The National Seed Laboratory of 'Agriquality New Zealand Ltd.')—was headquarters for the ICSTS (Inoculant and Coated Seed Testing Service) which, like many other government services, was discontinued in the economically turbulent mid to late 1980's when the New Zealand Government sold all or part of its share in many of the publicly owned service organizations and privatized others as SOE's (State Owned Enterprises).

Some of these sales may have been justified but the complete closure of the ICSTS was a national blunder—a typical *error of understanding* by administrators with poor technical comprehension and an act of disloyalty to the New Zealand farmer who, sadly, was not well informed about the value of this service (the cloak of mystery at work again) and was largely unaware of his loss from this decision, let alone the merits of effective inoculation. ICSTS provided an important safeguard for farmers (and for the nation) employing just a single microbiologist/technician in the person of Ms. Dianna Johnston, maybe with an assistant at times. It cost 'peanuts' to run and saved farmers (and Government developers) millions of dollars, not only by prohibiting inferior microbiological products, i.e., inoculants, inoculated seed and inoculated and coated seed, but also by encouraging and, indeed, insisting on higher standards of all three.

Independent checks by scientists and others in New Zealand have subsequently identified inferior products arriving on farms since the ill-conceived closure of ICSTS, as might have been expected. The nodulation failure cost of poor quality product sown all over the nation is sadly much too complex to be able to be measured easily (by scientists or by farmers) and is the chief reason for lack of appreciation of its importance, including by senior administrators.

France

The standards for commercial legume inoculants in France are considered among the highest in the world which, in terms of soybean, is 10^6 per seed, a high number for a large seed. The industry is regulated and batches of commercial inoculants (supplied in sterilized containers) are tested in Dijon by INRA (Institut National de la Recherche Agronomique, The National Institute for Agricultural Research) where they are certified if they pass the standards set for each species of legume seed.

Canada

Because inoculants for sale must be registered and are prudently regulated under the Fertilizers Act administered by Agriculture Canada, current standards of rhizobia per seed are also quite high at 10^3 for small seeds, 10^4 for medium size and 10^5 for large seeds. However, experts there consider the standard needs to be raised by a further 1^{10} to be considered acceptable to microbiologists. There is no standard for contaminants in unsterilized carriers however and research has shown that a sterilized carrier, usually peat, will more readily attain high quality in terms of rhizobia per gram.

Thailand

Though not protected by legislation, the Thai Department of Agriculture conducts tests which require a national standard also at the level of 10^5 for soybean seed. Unsterilized carriers require the inoculants to be used promptly after manufacture. This however, is a further threat to effective application of *Rhizobium* and cause of poor nodulation—or outright failure because some unsterilized peat carriers have been found to contain virtually no *Rhizobium* present at all after even a short period of storage.

Many other nations produce commercial legume inoculants—including the United States of America (where Federal regulation other than field testing has been abandoned). At this time, incredibly, only one State authority retains minimum standards in protection of its farmers which, for the others, is at best casual, and at worst, careless. Inoculants are also used in Mexico, Argentina, Uruguay, Russia, Germany, Portugal, India, Papua New Guinea and elsewhere. Clearly, the value is becoming realized, the intentions mainly honourable, but is the practice successful? The problem is that farmers can't see and count rhizobia. Effectiveness in the field is also sometimes difficult to assess—both (frustratingly) when nodulation is very successful, or close to a failure.

Research literature (see 'Inoculant Quality', quoted below on this page) repeatedly makes the following two vital points which politicians around the world need to heed, one, a concerning observation and the other, a very important recommendation:

1/. Most commercial inoculant sold around the world has, sadly, been found to be of inferior quality, unable to achieve the boost in nitrogen production which farmers are both seeking and entitled to.

2/. Commercial inoculant quality standards need to be established in every manufacturing country by the government (or a trustworthy independent authority) and monitored by that authority whereby manufacturers who fail to meet the standard are initially supervised for a set period and if they still fail to meet the standard, are prohibited from either selling or giving the product away. All inferior product must be disposed of to avert failure.

Not every country need establish its own testing service. The governments in several countries might elect to have test samples selected at random by its own officials, then sent to a designated testing facility in the larger of those nations which also needs the service and has either established a laboratory for the same or arranged tests in a privately owned facility. The cost of running such a facility could be assisted by setting a fee for each sample sent. The fee collected by each local government would provide an equivalent fee to the testing authority (say, paid monthly) in the neighbouring test providing country. The originating authority can check on the efficacy of those tests, occasionally, by splitting some samples, sending part identical sample to another reputable laboratory for comparative analysis. This is extremely important on a world scale; without it, the nations of the world are missing out on this potentially giant 'nitrogen factory.

INOCULANT QUALITY

Australian rhizobiologist John Brockwell reported in ACIAR (Australian Centre for International Agricultural Research) Proceedings 109e (printed 2002) that he and others:

"were generally pessimistic about the prospects for the inoculants industry and its capacity for large scale production of high quality inoculants."

Photo [Left]: Australian Scientist John Brockwell
Photo permission of John Brockwell.

Scientist Brockwell continued:

"Inoculants of the highest quality tend to be those produced by the private sector under the umbrella of an independent Quality Control programme (as in France and Australia). The future of the inoculant industry, and its potential benefits for world agriculture, depend on improving inoculant quality, both numerically and in terms of strains effectiveness. New technologies may lead to improved inoculants in industrialised countries, but the fact remains that, in many countries, the 30 and 40 year old technology has yet to be properly mastered".

Photograph taken at a soils workshop permission of David Herridge.

Photo [Left]: Australian Scientist Dr. David Herridge

In this same research review paper, edited by Dr. David Herridge in addition to Greg Gemell and Elizabeth Hartley, both Scientific Technical Officers of ALIRU (Australian Legume Inoculants Research Unit), the Conclusion stated:

"The legume inoculant industry has made and continues to make an enormous contribution to the economies of individual countries. It is a paradox that, despite almost 100 years of research and experience, many of the inoculant produced in the World today is of poor quality. Even good quality inoculants are often not used to best advantage."

The above crucial observations remain as pertinent and viable as ever. This review importantly also stated: *"The whole question of inoculants and their use starts with quality. If the quality is poor, then everything else is irrelevant."*

In the USA the following extract from their expert rhizobia organization NifTAL (*Nitrogen Fixation by Tropical Agricultural Legumes*) which has expertly and generously assisted inoculant development not only in USA but around the world, is further evidence of the vital importance of commercial inoculant quality and the fact that such quality has, in general, been unsatisfactory.

Poor Inoculant Quality Limits Market Development

Nearly 50% of inoculants from developing countries tested by NifTAL were of unacceptable quality and would not have benefitted farmers.

> "A recent UNFAO consultant's report on Indian inoculant quality concluded fewer than 10% of the inoculant marketed was of useable quality (Thompson 1992[(2)]). According to this report, the Indian legume inoculant market has failed to develop primarily because of lack of farmer confidence. In most cases there was inadequate quality control in the production process and product quality was not monitored in the distribution chain."

So, here we have a brilliant opportunity to increase the wealth of the world which is sadly falling short of potential. Why? Not only because of lack of adequate care in manufacturing inoculants, but also because of lack of adequate understanding in their use and benefits—and, we must add, lack of further advanced research to overcome the overriding causes of inoculation failure, namely *Rhizobium* mortality on seed, and in general, the need for improved strains for both effectiveness and infectiveness. Additionally, and most importantly, there is a need to devise a method or methods for farmers to identify and recognize the added value of their inoculation efforts. The latter would revolutionize acceptance and understanding but may not happen without the development of a method for evaluation.

An Acceptable Standard for Legume Inoculants

In New Zealand, the standard set for acceptable inoculants by its former ICSTS was that it must:

- be a pure culture of *Rhizobium* species for the stated host
- be serologically identifiable with the initial culture supplied
- contain a minimum of 1.0×10^9 viable rhizobia per kilogram of the seed to be inoculated at any time up to the expiry date (six months or otherwise as determined by the manufacturer)
- be effective for nodulation and nitrogen fixation
- state clearly, the expiry date and batch number.

Primary selection of strains and the testing of their qualities, productivity and suitability for supply to manufacturers was, in New Zealand, quite a separate activity (conducted by DSIR in Auckland) from the testing of commercial inoculants and other product samples (by ICSTS at the Seed Testing Station in Palmerston North). Such commercial samples were obtained via random checks by officers of the Department (later, the 'Ministry') of Agriculture as well as those submitted directly by manufacturers. Samples were required to be submitted on a regular basis by manufacturers, failing which ICSTS could withdraw their name from the list of those eligible to submit tenders for supply of inoculated and coated seed to the government (e.g., Lands and Survey Department and Maori Affairs Department) for new land development, a substantial buyer in some years. These regularly submitted samples, both inoculants and inoculated coated seed, as well as samples from tenders were also required to meet the above standards.

After an intentionally stressed 28 day storage period at 20°C, the inoculated and coated seed was tested by MPN (most probable numbers) procedure (fully described in this book) whereby slurries of inoculant and coat removed from the seed were serially diluted from full strength through, 1/10th, 1/50th, 1/100th, 1/1000th (and more) dilutions which are then applied to bare legume seed of the same cross-inoculation type as the seed in the sample. Increased dilutions delivered a reducing level of successful nodulation of bare seed grown in laboratory sterile agar tube replicates from which data, through a standard mathematical formula (acceptable microbiologically), provided an assessment of *'the most probable numbers'* MPN of cells in the original slurry dilution. It remains the standard procedure for 'counting' rhizobia.

Even at the level of viability required by the New Zealand ICSTS, some of the more astute growers who inoculate their own seed are applying several times the recommended single rate of commercial inoculant per kg of seed due to their perceptive understanding of 'the numbers game].

Each of these farmers knew that if he found himself for instance, a Field Marshall, in charge of a commando raid through enemy territory to achieve an important goal, he could expect to lose some troops when running the gauntlet; so he had not only best start with as many as he sensibly could but also plan his approach well, maximize the optimum conditions of operation, minimize the risks, generally protecting and taking good care of the troops. Rhizobia are like those commandos.

Inoculated and Coated Seed Standards

The ICSTS (coated and inoculated seed) standard in New Zealand required at least 300 viable rhizobia per small seed after storage at 20°C for 28 days from manufacture.

These samples were also required to achieve at least 50% of the establishment of a laboratory prepared gum arabic/lime coated seed, inoculated at the recommended rate and sown one day after preparation in comparative field tests. The commercially bagged inoculated coated seed also had to be clearly stamped (on sacks of product) with both the date of manufacture as well as the recommended date by which the product should be sown.

These standards are somewhat arbitrary of course as in every field situation, the complexity of conditions, for example, concentration and competitiveness of indigenous strains in that soil (if any), length of delay between manufacture and sowing, post-sown soil moisture and temperature conditions, etc., can all contribute to the level of viable introduced rhizobia required to successfully infect sown plants in any given situation. At the minimal level of 'adequate numbers' must surely be the author's experience at the huge land development project at Te Anau in New Zealand, where it was found that just 100 viable cells per seed at sowing gave good nodulation results. There, we had the luxury of knowing nodulation had been achieved only by inoculation of seed with introduced rhizobia, simply because there was a total absence of indigenous rhizobia at that location when oversowing. Indeed the soil was ideal for demonstration and comparison. While sole Farm Advisory Officer (Department of Agriculture) for North/West Southland (which included Te Anau) in the 1960's, author Bennett sent several sacks of this rhizobia free soil to the DSIR in Auckland at request of Mr. Athol Hastings for some spectacular demonstrations of successful nodulation in soil in the Auckland laboratory.

Inoculant Carriers

The most widely used carrier to date worldwide has been natural peat which, normally being somewhat acid, needs to be pH modified for inoculant use. The most responsive and supportive pH modifier has been calcium carbonate powdered (by hammer mill or, in our experience, by Raymond Roller Mill) to a very fine particle size of which some 98% will pass through a 300# (mesh) British Standard sieve.

Because not all countries have access to peat deposits, other carriers investigated, and some used commercially, are sterile agar (about 1% agar, while not expensive, provides poor survival of rhizobia), selected clays, vermiculite, lucerne meal, cork compost, lignite, charcoal and some others, most of which are usually dampened with broth or pH acceptable clean water prior to sterilization. Some manufacturers use unsterilized peat; however, that requires maximum injection of rhizobia at the laboratory because if the conditions in which they are held or stored are conducive to multiplication of fast growing temperate rhizobia (i.e., lucerne and clover), such conditions will also allow contaminant bacteria, including pathogens, to multiply rapidly and competitively to the detriment of rhizobia. Research has shown that the most successful carriers are those first sterilized, which contain a high organic content, minimal salt content and are able to hold up to double their own weight of water in suspension.

Peat Characteristics

Peat is a biologically inert natural substance formed from dead plant material and often found in or near marshes and swampland. It is geologically the youngest stage in formation of coal, followed by brown coal, lignite and bituminous coal. It is generally believed that peat has proven to be such an effective carrier for sensitive rhizobia because it has a large surface area containing (microscopically) deep fissures, crevices and holes in which, when damp, resident rhizobia can 'hide' being protected from the most severe causes of death: desiccation, sunlight and heat. Even *Bradyrhizobium*, tropical legume *Rhizobium*, which is temperature safe stored below 25°C, expires increasingly at temperatures exceeding 35°C which is common in countries like India, Pakistan, Myanmar, Malaysia, Indonesia and Papua New Guinea and in the Middle East and Africa, creating difficulty in providing a cooling device, particularly in the field, and until more safely sown under soil.

Most peats also have adsorptive characteristics that enable them to lock up water soluble toxic compounds thus protecting rhizobia from that threat. Those polyphenols (especially tannins) which occur naturally on the surface of seed, protect it between sowing and germination from soil pathogens, being the reason seed can survive so long in damp, cold soils without growing moulds or becoming decayed. Perhaps this has been a classic example of selection of the species over the centuries whereby seed with genes for natural seed coat protection have survived and dominated to the present; as a result today, almost all seed is thus protected.

Peat Specifications

Peat deposits are generally confined to cold and temperate environments; they are not easily harvested neither is the product easily air-dried. Not all peats are suitable carriers for *Rhizobium* and there is no universal specification that defines the suitability of peat from one source over that from another. However, there are some general considerations that should be taken into account when deciding whether or not to evaluate peat from a given source. These include the size and consistency of the deposit, accessibility, ease and right of extraction and ease of drying and milling the particular peat to an appropriate particle size. Peats that have not aged and comprise considerable fibrous material are difficult to process. Some peats are excessively saline which positively rules them out.

Quality specifications for peat from a given source will need to be established once its physical and biological properties have been determined. The main characteristics requiring a specification are as follows:

- *Particle size* distribution as determined by wet sieve analysis, preferably in association with another measurement technique such as the use of a particle size distribution analyzer. The particle size of the peat or other carrier should be sufficiently small to enable the inoculant slurry to adhere to small seeds as well as pass through applicator nozzles.
- *Moisture content* as determined by loss on drying to a constant weight.
- *Absorptive capacity*. Various techniques are available to determine the oil or water holding capacity of absorptive powders.

- *Loss on ignition*. Standards tests are available which provide a reliable indication of the organic matter content of peat.
- *pH*. Most peats are acidic in nature and it is a common practice to incorporate finely divided calcium carbonate or dolomite to raise the pH to the desired level. A pH range for freshly processed peat and the formulated inoculant carrier mix will need to be established.
- The *quantity of water soluble materials present* in the peat will need to be determined and a maximum level established. High levels of naturally occurring soluble salts can be detrimental.
- *Biological characteristics*. This is a somewhat intangible specification but is an extremely important one and refers to the inherent ability of the peat to maintain high numbers of rhizobia in a viable state for long periods of time.

As mentioned above, not all peats are suitable and the fact that peats from different sources may have very similar physical and chemical characteristics is no guarantee that all will behave in the same way when used as a carrier for *Rhizobium*. Some peats simply lack the inherent characteristics required and others are rendered unsuitable by sterilizing.

A decision regarding the suitability of processed peat from a given source as a carrier for *Rhizobium* can only be made after it has been subjected to a full and proper evaluation. If the inoculant production method involves the multiplication of rhizobia in a nutrient rich sterile medium, then laboratory trials will need to be carried out to determine the optimum moisture content, pH level, nutrient levels. The numbers of rhizobia per gram of inoculant will need to be determined at point of production and after periods of storage at different temperatures. These trials might extend for twelve months or more. When the results show a favourable trend, it will be necessary to extend the evaluation to include seed inoculation studies and also trials to evaluate the peat as a carrier for other species of *Rhizobium*.

Managing Inoculants

Packets of commercial inoculum left lying in the sun, perhaps behind glass on the seat of a truck, for example, would be exposed to temperatures exceeding 40°C in many normal indoor and field situations and would, of course, be rendered useless in a short period of time.

It is important to understand that research papers based on field experimentation of inoculation procedures which 'gave no response' in certain treatments have no doubt been due, in some instances, to the misapplication of the technique which can occur in numerous ways, heat being one of them. *Rhizobium* numbers and their mortality are fundamental to any study or comparison. It follows therefore, that if we don't know how to count rhizobia, we have no business attempting to study their performance or compare treatments.

It is one of the tricky characteristics of seed inoculation contributing to the microbiological cloak of mystery which surrounds it (including significant human confusion) that failures of seed inoculation have often been failures of the technique, not failures of the technology. Exactly the same applies to research attempts.

Economics of Inoculant Use

It cannot be emphasized enough that the literature on the subject overwhelmingly supports an economic advantage from legume seed inoculation as opposed to no inoculation. In Australia, scientists have estimated that the cost/benefit ratio lies somewhere between 20:1 and 40:1, i.e., that farmers gain between $20 and $40 for each dollar they spend appropriately on inoculation.

Where soils are devoid of, or deficient in, an efficient host specific rhizobia strain, the ratio is likely to be even more favourable. Research literature also predominantly supports specific types of coating as having an additional advantage over bare seed inoculation; there are exceptions such as lime coating of *Bradyrhizobium*, acid tolerant (alkali producing) slow growing tropical *Rhizobium*, which have given depressed results from (high pH) lime coating compared to simple inoculation or other coatings.

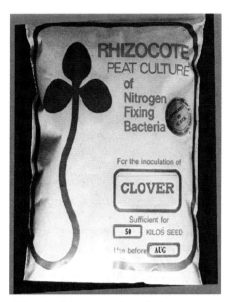

Fig. 12 A typical commercial pack of clover seed inoculant.

This substantial responsiveness of inoculation does not mean to say that all seed or soil inoculation is profitable, some has not been, or that all coating is additionally beneficial. Both can be unresponsive due to poor application and in some field situations as discussed in this book, at present knowledge, there may be no worthwhile potential or economic benefit to be gained from using either.

Inoculant Production

After sterilization in suitable, autoclavable polythene bags, these bags of modified peat (or where not available, other carriers such as, cork residue, charcoal, lignite or mineral soil could be used as all have been investigated and shown degrees of suitability) are injected by hypodermic needle with a measured aqueous (water based) suspension, a concentration of the appropriate *Rhizobium* strain, then promptly labeled (sealing the hypodermic syringe hole) clearly stating its host inoculation group, e.g., lucerne or white, red and alsike clovers (but in that case not subterranean clover for instance, as it has a different range of host-specific *Rhizobium*). It is then packaged into a light cardboard box appropriately labeled as to suitability for the range of legume seed, the quantity of seed the manufacturer considers it will successfully inoculate and most importantly a date of expiry. These packs have a shelf life, but when well assembled and kept in cool storage (i.e., temperate strains) they can maintain rhizobia viability safely for 6 to 9 months—up to a year is claimed by some manufacturers in the USA.

It is important for the manufacturer to supply culture packs fresh and directly to a user in the field or to a coating manufacturer. However, if and when commercial packs are supplied to retailers, seed companies, etc., the manufacturer needs to ensure that if stock is being held in anticipation of eventual sale, that s/he is satisfied both transport and storage conditions are acceptable in order to maintain the integrity of the product. In such cases, a useful rule might be, no refrigerator, no product. It is not in anyone's long term interest to sell or use ruined product.

The Numbers Game

The inoculant manufacturer will have calculated for each seed variety, an application level of viable rhizobia per seed considered necessary to provide an adequate number remaining alive at seedling root hair development. As stated, these numbers are arbitrary and it has become overwhelmingly evident in research literature and in practice that effective nodulation is increasingly likely to be successful with higher numbers of rhizobia applied per seed in the first place; it truly is 'a numbers game' at present levels of understanding. One day, scientists may discover or develop a robust and competitive strain which has low susceptibility to mortality plus high infective ability; as yet, to our knowledge, that has not been achieved.

Manufacturers cannot technically supply ever increasing concentrations of rhizobia in a given quantity of peat carrier or economically increase the volume of peat carrier several times the current level (with the same concentration of cells) at the same cost as the standard product. Neither can growers and farmers readily afford to apply many times the recommended single rate of inoculum culture per kilogram of seed; however, some do double it, and they will be the ones who are well informed as to 'the numbers game', the ones who realize their costs will marginally increase but who understand the benefits and cannot afford a failure.

In comparison with other farm treatments, seed inoculation and coating is not expensive (in fact if more processed seed survives than plain seed as in aerial oversowing demonstrated in this book, it can be a lot cheaper to use) so, where the benefits are able to return anything like $415 per hectare (of nitrogen, annually) it would be worth massively inoculating the legume seed, if it were required, and if it were practically possible. Most farmers do not know if it is required; it remains a mystery in too many cases. In practical terms, applications of peat culture will be less manageable at higher than current loadings on coated seed but doubling up is workable on bare seed if dried under shade prior to drilling.

Alternatives to Peat

As indicated above under 'Inoculant Carriers' other materials more readily available in some countries such as cork, bentonite, vermiculite and silica have been reported as equal or even in some cases superior to peat in recent studies which is important where suitable peat deposits are not available. The search will no doubt continue as long as rhizobia mortality in commercial culture remains significant.

Even where peat is available, there can be difficulty in exporting peat based inoculants to some countries; there may also be difficulty locating suitable deposits of peat in countries wishing to manufacture legume inoculants. Because of this, some years ago, the New Zealand company Mintech (New Zealand) Limited, through is agricultural division Mintech Agriculture, was successful in developing a highly effective and almost inorganic carrier. This was diatomaceous earth to replace the peat that had traditionally been used in the manufacture of 'Rhizocote' brand legume inoculants.

Diatomaceous Earth

Inoculants manufactured with this novel carrier proved very effective in independent laboratory and field trials carried out by New Zealand government scientists and became the subject of now expired patents. (More information about this alternative carrier can be found in expired Australian patent number AU1987069704 'Improvements in Carriers for Microorganisms'.)

Found around the world, often as a soft but solid rock called 'diatomite', it is used in various industrial functions, easily milled to fine particle size (increasing the surface area) thus is practically and structurally suitable for use as an inoculant carrier. In addition to this advancement, earlier tests had identified the importance of substantial rhizobial protection by incorporation of the chemical polyvinylpyrrolidone (pvp) into the inoculant. Together, these advancements produced a significantly improved and more effective legume inoculant.

Details of two studies are shown below:

Study 1/. 'Effect of Inoculant Formulation on Survival of *Rhizobium Trifolii* and the Establishment of Oversown White Clover {Trifolium repens}' a research study published in *'The New Zealand Journal of Experimental Agriculture'* 7 [1979] 311–14 by authors Dr. C.N. Hale, Dr. W.L. Lowther and J.M. Lloyd, of which scientists Hale and Lowther were completely independent of the inoculant manufacturer (being employed by DSIR and Department of Agriculture respectively) the following was reported:

'New Formula Rhizocote'® inoculant, produced to reduce the effect of seed coat toxins, increased the survival of rhizobia on the seed and increased the establishment of oversown white clover when compared with (standard) 'Rhizocote' peat inoculant. After imposing a 14 day storage stress on inoculated seed from date of inoculation, field experiments on hostile tussock grassland sites gave the following (statistically) highly significant results:

Table 1 Comparison of standard and improved inoculants.

Treatment	*% of vigorously growing seedlings*
®Rhizocote [peat carrier]	*31*
New Formula Rhizocote	*60*

Note: In the research paper, the new carrier (diatomaceous earth) was not named as it was then commercially sensitive to the manufacturer.

* 'Rhizocote' is the trade name for legume seed inoculants manufactured by:
AgBioResearch Ltd. P.O. Box 3414 Richmond, Nelson, New Zealand.
Phone (+ 64 3 544-2379)

And in study 2/. Being a paper presented by Dr. C.N. Hale, Head of Bacteriology Section, PDD, DSIR (New Zealand) to the 10th North American *Rhizobium* Conference (Hawaii, Aug. 1985) titled *'Legume Inoculant Production and Development of Alternative Carriers to Peat'*, the results of several independent studies carried out by government research scientists were reported.

Results of various experiments were consistent and as an example of them, the following test, *'Survival of Rhizobium Trifolii on White Clover Seed Using Peat and Alternative Carrier Inoculants'* illustrates the degree of significance where, in fact, the 'alternative carrier' was diatomaceous earth and the 'additive' pvp (polyvinylpyrrolidone), known to suppress seed surface toxins, i.e., toxins which protect seed but are lethal to rhizobia.

Table 2 Comparison of three carriers in sustaining viability of rhizobia.

Results were :_

CARRIER	RHIZOBIA/GRAM	RHIZOBIA/SEED (% count at x hours)	
		24 hours	48 hours
Peat	5.0×10^9	7.5	4.8
"Diatomite"	4.2×10^9	7.2	3.8
"Diatomite " + PVP	5.6×10^9	97.7	76.9

Such a spectacular result is not often demonstrated so emphatically & positively in microbiological research.

Lignite

While peat is the most juvenile ranked of the series of coals found in nature, widely used in isolated localities for cooking, for heat and (in Russia) for the production of steam, in New Zealand, it is used for horticultural and toxin absorption purposes. Its limited availability and ranking among the coals series has led to the investigation of a successor.

Lignite, as a replacement for peat, as a carrier for rhizobia, has shown promise though not all lignites are suitable as carriers for rhizobia. These studies preceded knowledge of the virtues of diatomaceous earth, but remain important for countries with suitable lignite but no peat.

Fig. 13 Three legumes from left, Lucerne, Clover and Lotus inoculated in each right tube (successful nodulation showing green growth) and not inoculated in each left tube (failure, no nitrogen) sown in sterile nutrient agar gel. The three failures are due absolutely ONLY to nitrogen starvation because all 6 tubes are identical—except for inoculation.

With modification, lignite could become an established carrier for legume seed inoculants. Lignite is of similar composition to peat and while vast deposits of it exist in many countries, it is more readily accessed, handled and transported in some than others. It has been reported as being more consistent in quality but this may be in reference to rhizobia survival because it is in fact of variable structure and has diverse physical characteristics including a high content of volatile matter according to each deposit. It is more easily sterilized than peat and while not as clean to work with, has recently been found a satisfactory legume inoculant carrier in replacement for peat, specially when milled to a fine particle size. Milling of coals in general to fine particle size can be hazardous because coal dust is usually explosive when concentrated and activated by static electricity. Most such facilities, however, are equipped with explosion suppression systems or purpose built blast walls which readily blow out releasing the energy.

Strains and Cross Inoculation Groups

Commercial inoculant manufacturers in UK, USA, France, Canada and elsewhere will have developed their own materials and techniques; however, it is important that an independent register of cross inoculation strains, each with its own number and its history of performance be maintained (probably by an appropriate government agency) as well as a nucleus store of viable strains for periodic supply to commercial manufacturers and for research.

To ensure effectiveness, a supply of mother culture to a manufacturer for multiplication may contain two or three strains for a single host-specific inoculant. Because of their long term environment, some strains for (say) clover (*r. trifolii*) may be more suited to and effective in dry climates, or more acid or alkaline soils, more suited to white clover, or alsike, or red clovers and as these are in the same cross-inoculation group (i.e., all such strains are capable of infecting all seed varieties in that group) then it may be wise to include a range of rhizobia which will cover those host specifics or potential environmental variations in any locality. Neither the mother culture selectors or inoculant manufacturers know where their inoculant will finish up or on which variety of seed.

While this broad spectrum of strains specificity is a strength, it can also be a weakness in the system because it means that if any one strain of rhizobia is dominantly successful with any one seed species in that cross-inoculation group, or is dominantly successful in that new environment—then, if it is one of (say) three strains of rhizobia in that mother culture—and later of course in that commercial culture, then, in best effect, we only have one third of the ideal numbers of rhizobia for that specific inoculation job.

There goes our "numbers game"!—because we are aiming at maximum numbers of effective rhizobia from the outset.

In an ideal world, it would be better to be able to identify THE best *Rhizobium* strain for each specific sowing of seed—maybe even inoculate and coat seed varieties separately with the best strain for each, then mix them for sowing—or identify the most suitable strain of rhizobia for a particular location and soil type and inoculate all the seed for that situation. However, the logistics of achieving that degree of precision over many different seed mixtures, soil types and environmental variations, even in a small country, would be practically and economically unachievable—particularly given the present lack of specificity knowledge.

Commercial Packs

In New Zealand, commercial packs of ®Rhizocote legume inoculants are made in pack sizes to suit the customer's requirements. Depending on species and standards required, 500 grams of 'Rhizocote' culture will successfully inoculate 500 kilograms of seed. In the past, off the shelf packs of inoculant were sufficient to inoculate 40 lbs of seed, either clovers or lucerne. Manufacturers will usually adapt to a specific size pack to suit regular customers. 'Rhizocote' inoculant for small seed white clover provides approximately 5×10^9 viable rhizobia per gram of carrier at manufacture and guarantees not less than 2.5×10^9. The aim is to apply at least 500 viable rhizobia per white clover or lucerne seed at inoculation and 300 viable cells after 28 days storage at 20°C.

As we shall see in a later chapter, survival can also be enhanced by coating the seed at the time of, or prior to inoculation. Also, additives to the seed such as polyvinylpyrollidone (PVP) or finely milled charcoal both act as adsorbents of the natural toxins in seed coats and can dramatically protect *Rhizobium* from those polyphenols and tannins which so effectively destroy them, as well as other soil bacteria.

Polyphenols and Tannins of Seed

Tannins are polyphenolic compounds. There are 516 different types of polyphenols that are divided into five classes. Tannins fall into the phenolic acid category of polyphenols. It is important to note, not all polyphenolic compounds are tannins but all tannins are polyphenols. The following is a description of tannin from Wikipedia Free Encyclopedia:

A tannin (also known as vegetable tannin, natural organic tannin or sometimes tannoid, i.e., a type of biomolecule, as opposed to modern synthetic tannin) is an astringent, bitter plant polyphenolic compound that binds to and precipitates proteins and various other organic compounds including amino acids and alkaloids.

The term tannin (from tanna, an Old High German word for oak or fir tree, as in Tannenbaum) refers to the use of wood tannins from oak in tanning animal hides into leather; hence the words "tan" and "tanning" for the treatment of leather. However, the term "tannin" by extension is widely applied to any large polyphenolic compound containing sufficient hydroxyls and other suitable groups (such as carboxyls) to form strong complexes with proteins and other macromolecules.

The tannin compounds are widely distributed in many species of plants, where they play a role in protection from predation, and perhaps also as pesticides, and in plant growth regulation. The astringency from the tannins is what causes the dry and puckery feeling in the mouth following the consumption of unripened fruit or red wine. Likewise, the destruction or modification of tannins with time plays an important role in the ripening of fruit and the aging of wine.

Tannins have molecular weights ranging from 500 to over 3,000 (gallic acid esters) and up to 20,000 (proanthocyanidins). Tannins are incompatible with alkalis, gelatin, heavy metals, iron, limewater, metallic salts, strong oxidizing agents and zinc sulfate, since they form complexes and precipitate in aqueous solution.

Manufacture of Legume Plant Inoculants

This book is intended to be a guide to the whole industry of seed inoculation, coating and precision pelleting; therefore, the author/contributor have included this section on the manufacture of legume seed inoculants for those able to employ one or more microbiologists to prepare this extremely important accessory product for supply into various international agricultural economies.

We wish to make clear however, that there are other important publications available describing legume inoculant production written by some of the world's most experienced and expert microbiological scientists, including (but not necessarily limited to):

1/. 'THE PREPARATION AND USE OF LEGUME SEED
     INOCULANTS' by R.J. Roughley
     Division Science Services, Department of Agriculture, Rydalmere
     NSW, AUSTRALIA.
     Published in "Plant & Soil" 32, 675–701 [1970]

2/. 'LEGUME INOCULANT PRODUCTION' by R.A. Date
     CSIRO, Division of Tropical Agronomy,
     Cunningham Laboratory, St. Lucia, 4067,
     AUSTRALIA

3/. 'LEGUME INOCULANT PRODUCTION MANUAL'
     JOE C. BURTON, Ph.D.
     Inoculant Production Specialist 1984

     NifTAL Center—MIRCEN; University of Hawaii
     Department of Agronomy & Soil Science
     College of Tropical Agriculture and Human Resources
     1000 Holomua Road, Paia, Maui, Hawaii, 96779, USA

4/. And in NifTAL's website references (quoted in this Chapter above) at: www.ctahr.hawaii.edu/bnf

The above publications provide in-depth technology associated with the production of legume inoculants which scientists and students will continue to study. In this book, however, the author/contributor have compiled a practical Manual of Procedure for those wishing to develop a more complete range of products, not just inoculants, but also coatings, pellets, granules, film coating and more, for which basic techniques are described; these can be expanded or modified with the assistance of various technical publications including those listed above. Neither author, contributor nor publisher accept liability whatsoever for the interpretation or use of these techniques, over which they have no control.

ICMP

In New Zealand, the ICMP (*International Collection of Microorganisms from Plants*), now a very valuable resource, includes a wide range of isolates of microorganisms containing both fungi and bacteria from which the most élite strains of rhizobia identified to date in New Zealand are available, in addition to some from overseas.

The strains acquired from overseas cannot be released in New Zealand unless first officially approved by the New Zealand Department of Agriculture, a biosecurity measure.

Landcare Research (New Zealand) now holds this very valuable collection of more than 12,000 individual microorganisms obtained over many years including (on 1/10/10) some 793 strains of rhizobia of which 373 were collected locally and from which a carefully selected subset of 43 strains comprise the recommended list (see Table of recommendations below) for optimum *Rhizobium* inoculant performance as at July 1992, when last published.

This collection is now overseen by Science Team Leader (Biosystematics) Dr. Peter Buchanan and managed by Research Scientist (Biosystematics) Dr. Bevan Weir, both Scientific Officers of Landcare Research. They may be reached at Landcare Research, Private Bag 92170, Auckland Mail Centre, Auckland 1142, New Zealand. Street Address 231 Morrin Road, St. Johns, Auckland 1072. Phone +64 9 574 4100, Fax +64 9 574 4101.

The whole ICMP collection has been moved from care of the former New Zealand Department of Scientific and Industrial Research of which the Plant[3] Diseases Division located at Mt. Albert Road laboratory was the original collector (and supplier of mother culture to manufacturers).

To the dismay of the author, but no doubt in compliance with the mandate of the current keepers of ICMP, they place no greater importance on *Rhizobium,* which make up just 4.4% of the entire collection, than any other organism they hold. While we realize the enormous (actual and potential) value of a wide range of fungi and other bacteria, apart from rhizobia due to their already established huge value to mankind, we nevertheless consider rhizobia deserve re-instatement of independent and specific facilities which indeed were operative in the 1960's to 1980's. These were unwisely disestablished during the political 'cut and thrust' of economic revision in the recessionary 1980's prompted by the perceived oil crisis of 1979, followed by a worldwide share market crash in 1987. Economic revision in New Zealand was, of course, necessary as it eventually achieved a drop in the unemployment rate from 10% in the 1990's to just 3.4% (and a fiscal net surplus in excess of $2 billion) in 2007.

However, it was almost inevitable that in their enthusiasm for economic conservation, politicians and administrators would unconsciously stumble, thus destroying vital components of a healthy agriculture; partly perhaps because the benefit of a particular service may not have been clear to them but also because the strongest 'lobby' in support of rhizobia needed to be the users themselves, i.e., New Zealand farmers (specially developing and expanding grassland farmers). They, sadly, did not do enough to save this agricultural gem, simply because they were not then sufficiently aware of the huge importance of rhizobia. And, as stated elsewhere in this book, lack of awareness is probably due to the fact that farmers have no clear way of measuring the value of inoculation of legume seed, except in those negatively extreme situations where legume crops do not do well, or clover seed does not grow at all, as in the first sowings of New Zealand's Lands and Survey Department grassland development at Te Anau. This was also evident when lucerne stands turned yellow and failed to thrive as in the 1970's in New Zealand, when contaminated lucerne inoculants insidiously changed from *meliloti* to *trifolii*, undetected by (then) current DSIR testing procedure[5] and thus probably multiplied under commercial manufacture.

Fortunately, on that occasion, the lucerne crops were saved by application of 'Rhizocote' lucerne inoculant granules, developed by G.G. Taylor and J.M. Lloyd of Fruitgrowers Chemical Co. Ltd. for Coated Seed Ltd., which manufactured the granules on an urgent basis, thus assisting the recovery of several thousand acres of lucerne crop during that season, mainly in Canterbury Province. The lightweight granule was an improvement on heavy water truck overspraying which, however, also worked in saving crops.

New Zealand needs an Independent INOCULANT SUPPLY and Testing Service as does every Nation using Legume Inoculant

Though there may be only one manufacturer of commercial legume inoculants in any small country, there needs to be an independent mother culture testing authority plus an equally independent viability testing authority for commercially manufactured inoculants, inoculated seed, inoculated coated seed, inoculated granules, liquid inoculants and any other formulation of rhizobia that is used in agriculture and horticulture.

There is some slim research evidence that *Rhizobium* transformants may be able to fix nitrogen on non-leguminous plants such as wheat, barley, sorghum, brassicas and rice. If that ever becomes commercially viable and standard practice, then the above supply and testing facilities will certainly be required and will become big business worldwide.

Because the nature of biological nitrogen fixation by legumes is complex and its importance to agriculture and food production enormous, it becomes vital to ensure that the quality and performance of commercial legume inoculants and inoculated legume seed is monitored by competent independent authorities. As stated, actual cost per hectare of inoculants or inoculation is generally very low compared to the substantial benefits gained and ensuring that these benefits are achieved through a rigorous quality testing service far outweighs the cost of maintaining the same. The author firmly believes that it is in the best interests of the farming community and the economy of New Zealand for government agencies to become more pro-active once again in this area.

Periodic quality tests need to be conducted by random compulsory sampling plus storage time stress to emulate commercial delays until sowing. Such sampling, a simple job, can be carried out regularly by officers of one or more government departments, i.e., Agriculture, Consumer Affairs, Conservation, and/or one or more state owned enterprise, i.e., AgResearch, Landcorp or AgriQuality New Zealand Ltd.

The maintenance of a dedicated élite *Rhizobium* suite with an ongoing service both in selection and testing of strains purity, stability, viability, survival on seed, infectiveness and productiveness in nodules is a professionally demanding task requiring highly skilled staff. However, it could be partially funded by fees for product supplied (with higher fees for better product) and with the reduced cost which would accrue from that activity if it were a part time function of an existing microbiological laboratory, in New Zealand's case, possibly Landcare Research.

For an existing laboratory, the added cost would be reasonably modest, and certainly justified.

Mother Culture Supply in New Zealand

Landcare Research advises cultures are available for overseas research—and possibly for commercial inoculant production as well, subject to conditions.

Pioneers of the current collection of *Rhizobium* in New Zealand, all from within DSIR, were Senior Technical Officer Mr. Athol Hastings and his assistant Mr. A.D. Drake. Their work was supervised by Scientist Dr. Doug Dye and later Dr. Chris Hale.

These are some of the far sighted New Zealand visionaries of the gigantic N^2 world, their contribution sparked a keen interest in legume seed inoculation in New Zealand formalized by the publication of several imperative research papers and most importantly, the very practical 1966 DSIR Information booklet Series No. 58 "Legume Inoculation in New Zealand" which positively introduced seed inoculation and coating to New Zealand farmers and generally into New Zealand agricultural practice.

While the others are now deceased, Dr. Hale remains active but is now partially or fully retired.

The Current New Zealand Rhizobia Strains Recommendation

NB. As the ICMP is now a research collection, commercial use of strains needs to be discussed with Landcare Research, Auckland, New Zealand to obtain approval for release which will include a condition that there can be no guarantees as to the effectiveness of any strain supplied by them (which prudently and probably remains unchanged from the days of supply by the former DSIR).

The strains number (right hand column) is the ICMP number for the strains currently recommended for the legumes named in the left hand column.

Legume name	Botanical Name	Ploidy	Rhizobium strains
White clover	*trifolium repens*	2668; 2666; 2163 [cc 275e]	
Red Clover	*trifolium pratense*	"	"
Alsike clover	*trifolium hybridum*	"	"
Subterranean clover	*trifolium subterraneum*	5938 [WU 95]	
Kura clover	*trifolium ambiguum*	Diploid	4071 [cc 227] 4072 [cc 231a]
"	"	Tetraploid	4074 [cc 286a]
"	"	Hexaploid	4073 [cc 283b]
Lucerne [alf alfa]	*Medicago sativa*	2751 [u 45] 2752 [su 47]	
Lotus	*pedunculatus* Diploid & Tetraploid [Bradyrhizobium] {5798 [cc 814s]		
"	" [incl. cv. Maku]	{5942 [nzp 2021f]	
Lotus	*corniculatus*	3663 [su 343]	
Lotus hybrid G4712	*pedunculatus* [tetrap] x corniculatus	1326 [nzp 2037] 5798; 3663	
Soybean	*glycine max*	2860; 2862; 2863; 2864; 2882 [wb 61]	
Note: It is recommended that a combination of strains be used to cover varieties requiring specific strains for effective nodulation.			

Legume name	Botanical Name	Ploidy	Rhizobium strains
Pea	*pisum sativum*		5943 [nzp 5225]
Bean [Faba]	*vicia faba*		5943
Vetch	*vicia* spp.		5943
Bean	*phaseolus* spp.		3305 [cc 511]
Lupin [Crop]	*Lupinus* spp.		4771 [nzp 2076]
			3155 [nzp 2243]
			8377 [wu 425]
Serradella	*ornithopus sativus*		4771; 3155; 8377
Vetch [Crown]	*coronilla varia*		5085 [nzp 5462]
			5944 [nzp 5357/1]
Sainfoin	*Onobrychis viciifolia*		3157 [nzp 5301]
Lupin species for High Country vegetation :-			
	Lupinus polyphyllus		5376 [nzp 2141]
	" *nanus*		5376 [nzp 2141]
	" *vallicola*		4683 [nzp 2257] ; 4771
Desmodium spp.			
Winged bean	*Psophocarpus tetragonolobus*		8467 [cb 756]
Sulla	*Hedysarum coronarium*		6932 [nzp 5410]
Chick pea	**Cicer arietinum**		6937 [cc 1192]
Cowpea	*vigna* unguiculata [Chinese longbean]		8467 [cb 756]

ICMP Current Cost of *Rhizobium* strains supply (in 2014)

Culture isolated from New Zealand ....…......NZ $120.00
Culture of non-New Zealand origin ....…......NZ $150.00
Difficult to grow cultures (e.g., Nitrobacter) ...…..........…....NZ $250.00
In addition to the above fee, costs of packaging, shipping (postage, courier, air freight) GS tax (i.e., 'goods and services' within New Zealand only), Ministry of Agriculture Transfer/Export Permit (overseas only) and any other essential expense—are to be added.

There is a 10% discount for 10 or more cultures. The above prices are subject to change without notice. Further details are available from:

Landcare Research Attention: Maureen Fletcher (or associate)
Private Bag 92170 Phone: 09 574 4100
Auckland Mail Centre DDI: 09 574 4157
Auckland 1142 New Zealand Fax: 09 574 4101

The information to order cultures is on the ICMP website, and the main culture database is searchable at *http://www.landcareresearch.co.nz/resources/collections/icmp/depositingordering-strains*

Legume Inoculant Manufacture in New Zealand
(Contributed by J.M. Lloyd)

Biological Laboratories Limited of Auckland was the first company to produce legume inoculants in New Zealand and the history of this interesting initiative was given by J.R. Callaghan in a paper delivered to a New Zealand Grasslands Association conference. Prior to 1955, nearly all legume inoculants used in New Zealand were supplied by the Plant Diseases Division of the Department of Scientific and Industrial Research. This service, which was mainly for lucerne, was introduced and controlled by Dr. W.D. Reid. It was due to the retirement of Dr. Reid and also because the large demand for culture was proving an embarrassment to the Division that Biological Laboratories Limited decided to provide this service. Although their laboratory was primarily concerned with medical bacteriology, it was felt that their facilities and staff were suitable for the production of inoculants and on his retirement Dr. Reid agreed to act as a technical advisor to the company, being a great help over subsequent years.

Initially, only agar cultures were supplied, but as several merchants in New Zealand were importing moisturized powdered peat inoculants from Australia and from the United States, it was felt that this type of inoculant should also be produced. Considerable research on methods, types of peat, and suitable strains of rhizobia was undertaken before this product was finally marketed. At about the same time, the Plant Diseases Division commenced a certification scheme whereby all inoculants sold in New Zealand were subjected to laboratory and field trials for effectiveness and keeping qualities. At that time, no work on rhizobial strains was carried out by the inoculant manufacturer and master cultures were obtained from either Professor J.M. Vincent, Sydney University, or the Plant Diseases Division of the DSIR Auckland. These strains had been selected over the years as being the best for commercial use, having remained stable and not given rise to avirulent mutants during fermentation and mixing. It was considered important for them to compete in the rhizosphere with other rhizobia.

For lucerne inoculants a mixture of three strains was used:

1. The standard Plant Diseases strain, which has been used for many years with most satisfactory results, known simply as P.D.D.
2. A strain originally isolated in Western Australia known as WA16/1.
3. A strain isolated by Sydney University known as SU277/1.

This combination of strains was able to live in harmony and gave very satisfactory results with all lucerne types. Over time, further strain improvements have been made.

In the selection of clover rhizobia, great care was taken to select strains which were effective for the three major clovers: subterranean, red, and white. The commonly used strains for this inoculant were obtained from cultures originally isolated in Australia. Those used commercially at the time (1960) were designated NA30 and T.A.1. Several New Zealand strains had been used previously but at some stage, gave rise to avirulent mutants. All the strains used in New Zealand legume inoculants at the time had been fully tested in the field and had demonstrated very good results under New Zealand conditions. Individual strains became available for peas, beans, cowpeas, lupins, soya beans, lotus, and peanuts, but the demand for these inoculants was very small.

To ensure that the bacteria remained stable and virulent, they were grown on yeast mannitol agar covered with a layer of sterile paraffin and stored in the refrigerator until required. Under these conditions, they remained in an excellent state of preservation for up to 12 months. A further method used was to grow the host plant in tubes of seedling agar, infect with the rhizobial strain, then deep freeze the plant and re-isolate from the nodules when required. Freeze dried cultures were later used.

Methods of Manufacture (Propagation)

The fermentation and propagation of rhizobial strains was carried out in fermentors under sterile conditions using a procedure very similar to that used in the fermentation of antibiotics such as penicillin. The pure master culture was inoculated into the fermenting vessel containing the appropriate culture medium. This comprised of peptone, mannitol and yeast extract at a pH of 6.8 incubated for three days at a temperature of 27°C during which time the culture was continuously aerated and agitated. At the end of this time, considerable multiplication of the rhizobia occurred with counts of up to 5×10^9 bacteria per ml achieved. Each batch of broth was subjected to the following tests before addition to the peat carrier:

1. Microscopic tests by phase contrast microscopy and by stained smear for the presence of contaminating bacteria other than rhizobia and for fungal spores
2. An estimation of total number of viable rhizobia cells
3. The identification or confirmation of the strains by means of antigen-antibody tests using specific anti-sera prepared in rabbits
4. Routine checks for nodule formation on plants growing in a nitrogen-free nutrient solution

When more than one strain was being used in an inoculant, these were then pooled, having due regard to total viable numbers in each fermentor, then added to the powdered peat. The peat itself was, surprisingly, the most expensive component used, as it is necessary to ensure the raw peat did not contain large numbers of contaminating microorganisms, particularly actinomycetes, as these can produce antibiotics harmful to rhizobia.

Peat was dried and ground to a fineness of less than 53 microns (mesh) and during the process, the fibrous portion of the peat was separated from the silica. Only the fibrous material was used. This was then sterilized in revolving heated drums at 120°C for one hour. Temperatures above this produce charring and were considered to be unsuitable for the maintenance of the cultures. The peat is usually acid and this has to be brought up to pH 6.8, at which stage the bacterial broth was added. Moisture content has been found to be of great importance in the maintenance of viable rhizobia and a moisture content of 45% aimed at. When packed in sealed polyethylene film bags no difficulty is experienced in maintaining this water content.

Peat cultures were issued with a shelf life of six months during which time the total number of organisms must not fall below 1×10^8 per gram. Agar cultures which are normally dispatched direct to the farmer have a shelf life of only one month, but in effect can remain potent for three months provided they are stored in a cool, shady place. In South Australia and Queensland where large quantities of agar cultures were still used from laboratories, there was a guaranteed shelf life of three months.

Methods of Testing

In addition to the New Zealand Plant Diseases Certification scheme in which all cultures are fully tested, rigorous quality control measures are also carried out in the manufacturer's laboratories. These include total numbers of viable rhizobia, and a minimum figure of 5×10^8 viable rhizobia per gram was aimed for. Dilution tests in seedling agar with the host plant were carried out and a satisfactory culture at time of sale must still be able to nodulate its host plant effectively when diluted 100,000 times. Pot trials using soil in which it is known that rhizobia do not exist were also used, and these, in conjunction with the field trials conducted by Plant Diseases Division, gave a good estimate of how the culture would perform under field conditions. Regular monthly checks incorporating all these procedures are carried out during the life of the inoculant.

The sole New Zealand manufacturer of legume inoculants at that time was also the only one in Australasia who had implemented such a rigorous quality testing programme and it is believed this was the reason their products never failed to reach certification standards.

At that time, there had already been a great deal of research throughout the world in an endeavour to develop methods whereby legume seed could be inoculated by the merchant and sold to farmers as pre-inoculated seed. This proved to be quite a challenge. It was believed that one of the difficulties arose from the fact that inoculant manufacturers were dealing with a living organism which has been taken from its natural environment, subjected to all sorts of artificial procedures, then put back into soil via the seed in the hope that it would still perform its natural functions. Some believed that after living 'on the fat of the land' as it were, under pampered conditions, having nothing but the best of food and conditions during several life cycles, it was perhaps asking a bit much of the bacteria to compete in the rhizosphere for available food with the other naturally occurring, more hardy, soil organisms.

For this reason some form of protection should be given, it was believed, to ensure that the bacteria would survive until the seed germinated and the root hairs reached a stage conducive for infection. It was also recognized at the time that a further complication existed from the toxic water soluble compounds present in clover seed shells which had demonstrated considerable antagonism toward rhizobia bacteria. The use of milk and sugar solutions has been found to be of great assistance in maintaining the life of the bacteria on seed, and tests had shown that when peat was used in conjunction with molasses syrup, adequate numbers of bacteria remained alive for at least two weeks after treatment of the seed. The survival rate of rhizobia, when dusted as a peat inoculant on the outside of lime coated seed, was very good. Lime coated seed treated in such a manner, with a good quality peat inoculant, gave satisfactory nodulation for up to one month from time of treatment.

At the time and in light of these good results, it was suggested that the answer to the problem of achieving good rhizobia survival on inoculated clover seed would be to formulate the inoculant as a very finely divided dry powder for direct application to seed. It was generally believed at the time that in order to survive, rhizobia required an environment with a moisture content of at least 25%. However, some observers found it difficult to reconcile this with the fact that rhizobia seemed to survive quite well in some soils during times of significant drought. More recently, one New Zealand company has been successful in developing a dry powder formulation of rhizobia with good shelf life which was formulated in a manner that enabled it to adhere quite well to legume seed as a dry powder coating.

Legume inoculant production methods vary between manufacturers, those basic procedures have been well documented elsewhere. 'Rhizocote' brand legume inoculants were produced for many years in New Zealand by growing rhizobia within a sterilized mixture of finely ground pH adjusted peat, water and nutrients. While the specially selected peat used proved an excellent carrier, it was difficult to harvest and process and it came from a relatively small deposit on private land that the owner was going to develop for other purposes.

Alternative carriers

The manufacturer of 'Rhizocote' commenced a major development programme to identify a suitable replacement for peat that could be obtained to a consistent specification from a reliable long term supply source. A wide range of predominantly inorganic absorptive fine powder carriers from various suppliers was evaluated and most rejected on the grounds that they did not have the inherent ability to act as a suitable medium for the multiplication and long term survival of rhizobia. However, the efforts of the manufacturer were eventually rewarded with the discovery of a particular grade of diatomaceous earth that possessed all the desired characteristics and proved an excellent replacement for peat for all the strains of *Rhizobium* used commercially. The manufacturer's findings were supported by independent laboratory and field tests carried out by scientists of the New Zealand Ministry of Agriculture and Department of Scientific and Industrial Research[4].

Legume inoculants based on this new carrier were the subject of successful patent applications in New Zealand and Australia and subsequently became the property of ICI Australia Operations Pty Ltd. Although the patents were primarily directed to the new legume inoculant formulation, they also covered other microorganisms such as *Agrobacterium radiobacter, Serratia* spp., *Pseudomonas* spp., ericoid mycorrhizal fungi, *Metarhizium anisopliae, Trichoderma harzianum* as well as *Rhizobium* spp. The main claim was for a microorganism culture comprising a sterilized inorganic material of fine particle size, with a surface area ranging from 5 to 20 m^2/g and a water absorption capability of at least 120% together with the microorganism. The inorganic carrier medium was described as processed diatomites, synthetic magnesium or calcium silicates, fumed silicas, precipitated silicas and expanded siliceous materials.

The carriers of this invention were exclusively inorganic materials of fine particle size, preferably between 2 and 50 microns, having a bulk density preferably between 80 and 200 kg/rn^3, a large surface area preferably between 5 and 20 m^2/g and a high water absorption of at least 120%. A desirable feature of the carrier was for it to be substantially free of soluble salts (preferably less than 0.1% by weight) and have a pH as a suspension in water of preferably between 5.5 and 6.0. Without being bound by any theory, the inventors envisaged that the water holding capacity of the carrier medium of this patent enables transfer of essential elements throughout the substrate through absorbed and retained moisture while a large surface area enables sufficient aeration to occur for rhizobia to survive and grow. Preferred materials were processed diatomites, particularly those of fine particle size supplied by Johns Manville Corp. of USA and sold under the trademarks 'Celite', 'Super Cel' and 'Filter-Cel'. Synthetic magnesium and calcium silicates, fumed silicas, e.g., 'Aerosil 200' from Degussa of Germany, precipitated silicas, e.g., 'Sipernat 228' as supplied by Degussa, and expanded siliceous volcanic glass, e.g., Perlite 432 as supplied by Grefco Inc. of the USA were also referred to.

The carrier medium can be prepared in any suitable form but the preferred method was to mix the carrier together with the nutrient material and pH adjusting agents (which comprised a buffer of the required type) preferably dissolved or suspended in water. The amount of water is chosen to provide desirably a damp product on completion of the mixing rather than a dry powder or slurry. On completion of mixing, the carrier medium is packed into high density polyethylene bags, an aeration tube inserted and sealed tightly around the mouth of the bag and a gauze type filter placed over the top of the tube. The packs are then sterilized by steam autoclaving or gamma radiation. A suspension of the strains of *Rhizobium* selected for their high performance is then added to the carrier under aseptic conditions, the contents of the bag blended by external manipulation before storing the packs in a temperature controlled incubation room for 7 days or more to allow multiplication of the rhizobia to occur.

Packs from each batch produced are subjected to plate counts and contamination testing; suspensions are applied to host plants growing in seedling tubes in a plant growth chamber for effectiveness and infectiveness testing. 'Rhizocote' legume inoculants have been tested by independent scientists and found to consistently meet official New Zealand, Australian and Canadian standards for quality. The manufacturer's standards are: Number of viable rhizobia per gram at manufacture, 5×10^9 minimum; Contaminants, Nil.

In New Zealand, as stated, the strains of *Rhizobium* used for the manufacture of 'Rhizocote' legume inoculants are obtained from *The International Collection of Microorganisms from Plants* (ICMP) of Landcare Research, one of eight Crown Research Institutes. CRIs function as independent companies but are owned by, and accountable to, the New Zealand Government.

ICMP is an international collection of live cultures of fungi and plant-associated bacteria available to researchers which may be available commercially on application to the Director. It retains major, worldwide collections of plant pathogenic bacteria, along with rhizobia. Most strains are stored using liquid nitrogen or in vacuum–dried ampoules. The ICMP laboratory is a MAF approved PC2[6] Containment Facility and a Transitional Facility for Microorganisms, located in Auckland, New Zealand.

ICMP had its origin in 1952 as the personal collection of plant pathogenic bacteria and rhizobia of Dr. Douglas W. Dye[7]. It expanded as the culture collection (PDDCC) of Plant Diseases Division, and later, the Plant Protection Division, of the New Zealand Department of Scientific and Industrial Research (DSIR). Initially, cultures were maintained by serial transfer of slope cultures, but in 1962, the process of vacuum-drying was introduced. Following the reorganization of science services in New Zealand in 1992, the collection was transferred to Landcare Research.

On receipt of a pure 'mother' culture from the ICMP of Landcare Research, the manufacturer of 'Rhizocote' prepares a specific number of agar 'stock' cultures from it which are maintained at 3–5 deg. C until used for the production of a commercial batch of legume inoculant.

Mother cultures are normally supplied in a freeze dried state within a sterile sealed glass vial and the bacteria in it are hydrated and transferred to the surface of sterile nutrient agar in 'stock' culture bottles which are incubated at 25°C until good growth of *Rhizobium* occurs. Because the survival of *Rhizobium* on nutrient agar is relatively poor and to avoid the risk of the strain of *Rhizobium* developing aberrant characteristics through repeated sub-culturing, fresh mother cultures are obtained at regular intervals.

Fig. 14 Vial of Freeze Dried Rhizobium.

Fig. 15 Agar Stock Culture of Rhizobium.

Quality Control (measuring *Rhizobium* viability)

Having manufactured quality inoculants, we need to be able to identify how they fare in commercial use. Their viability can only be measured by counting the number of survivors after applying them in certain treatments (such as coating seed) or after storage delays such as those which happen in real life.

There are essentially three standard methods of measuring viability: the first is the seedling tube method which simply shows successful nodulation or degrees of failure and the second, the MPN (Most Probable Numbers) test which, by series dilution, estimates approximate actual numbers since rhizobia cells are invisible to the naked eye. (Both these methods are described below, but after the explanation of some important microbiological principles.) The third method involves plate counting of colonies (of rhizobia grown from a single cell) which are visible to the naked eye. All three can be further supported by field testing.

Due to the difficulty of obtaining predominantly single cell suspensions, the result of any quantitative test can only be taken as an indication of the 'nodulating capacity' of a given sample of inoculant or coated inoculated seed.

Depending on the method used to manufacture the inoculant and prepare it for application to the seed and the actual seed coating composition, any specific method used to prepare samples for testing may not accurately reflect the actual number of rhizobia present. It is inappropriate, therefore, to compare the results obtained by MPN or plate count from inoculants or coated inoculated seeds prepared by different methods. However this does not detract from the importance of having official standards for the minimum number of rhizobia per gram of inoculant or per coated seed provided these can be linked to the percentage of plants that form nodules in a typical field sowing situation.

In New Zealand, field testing provided an official and extremely valuable part of the legume inoculants and coated seed testing programme formerly operated by the ICSTS.

Enumeration of Viable *rhizobia* in Legume Inoculants on Inoculated Coated Seed

As earlier emphasized, the number of effective viable rhizobia present on inoculated coated seed at the time of sowing is of major importance in achieving successful nodulation. However this aim can be severely modified by the effects of sunlight, heat, fertilizers, water soluble toxic seed exudates and other adverse conditions or influences from which rhizobia need protection.

There are various methods by which the presence of rhizobia can be qualitatively detected on seed and some of these can be used to provide a reliable indication of the survival characteristics of *Rhizobium* on inoculated legume seed prepared with different adhesive and coating compositions, thus providing a relatively quick means of screening different experimental coating formulations. The coating may be dispersed in sterile water and a sample of the suspension spread on sterile nutrient agar in petri dishes. Colonies of rhizobia can usually be detected after incubation although, at times, the presence of other organisms, especially if they are fast growing, can make accurate detection difficult. The accuracy of this method can be enhanced through the use of a selective growth medium that contains additives antagonist toward bacterial contaminants.

A simple method is to plant the seeds in sterile, nitrogen free nutrient media in test tubes (seedling tubes) and observe the development of nodulation.

A variation of this technique is to disperse the coating from the coated seed in sterile water and deliver aliquots of the dispersion to legume seedlings previously sown in the nitrogen free media in seedling tubes. For qualitative assessment, these methods are preferred over the previously described technique of spreading an aqueous suspension of the coating material on nutrient agar in petri dishes. The qualitative method described below has proven to be useful in formulation screening studies where a large number of samples need evaluation but it does not replace the quantitative Most Probable Number (MPN) test which will be described later.

Assessment of Nodulation Potential: Seedling Tube Method

To be able to conduct these tests, the following materials and equipment (major items) are necessary:

- Laboratory balance, two decimal places
- Autoclave
- 3 litre stainless steel open top jug (or other suitable capacity vessel)
- Supply of 150 mm x 19 mm rimless test tubes
- Stainless steel baskets or similar for holding tubes during autoclaving
- Supply of suitably designed drilled wooden blocks or similar for holding seedling tubes
- Mechanical laboratory stirrer
- Temperature controllable hot plate
- Peristaltic pump
- Temperature controlled plant growth chamber (light room) equipped with a time controlled illumination system to provide simulated sunlight
- Tweezers
- Stainless steel probe (rod with tapered end)
- Bunsen or spirit burner
- Ethanol in secured short cylindrical container
- Laminar air-flow cabinet
- Distilled water
- Laboratory grade chemicals as listed under nutrient agar formula below

The Method

Prepare the nitrogen free nutrient agar medium according to the formula below:

Note: **%w/w means** the percent by weight of solute in the total weight of solution.

Ingredient	% w/w	Grams/litre
Sodium chloride	0.02	0.2
Ferric chloride hexahydrate	0.01	0.1
Magnesium sulphate heptahydrate	0.02	0.2
Di potassium hydrogen phosphate anhydrous	0.02	0.2
Calcium hydrogen orthophosphate	0.01	0.1
Agar (dried)	0.80	1.0
Distilled water	100	1litre

Now, proceed as follows:

1. Add 1 litre distilled water to the stainless steel jug; place on a hot plate under a mechanical stirrer.
2. Activate the stirrer, add the ingredients to the water and bring to the boil.
3. Simmer the mixture for approximately 10 minutes until the agar has dissolved.
4. Test the pH of a drawn and cooled sample and if between 6.5–7, no adjustment is required. If below 6.5 use a 0.1 M solution of sodium hydroxide to correct. If above pH 7, use dilute hydrochloric acid to correct.
5. Before the solution has cooled, use a peristaltic pump to transfer 10 ml to each of the desired number of test tubes (10 per sample to be tested) and plug each with fresh cotton wool. The medium should be agitated with the mechanical stirrer during the tube filling process.
6. Place the tubes upright in the stainless steel basket and autoclave 10 minutes at 138 KPa.
7. Allow the tubes to cool in vertical position and when the agar has formed a gel, transfer the tubes to a laminar air flow cabinet and, using the tweezers gently transfer one inoculated coated seed from the sample to be tested to the surface of the nutrient agar in the tube.
8. Using the stainless steel probe, make an indentation in the surface of the agar in the tube and manipulate the seed to sit in this. Manipulate the seed further to moisten and remove most of the coating which should remain present as a slurry. Note: The tips of the tweezers and probe are flame sterilized before each transfer by dipping in ethanol, flaming off under a burner and allowing to cool before use.

9. Repeat the process until 10 seeds of the sample have been 'planted', replace the cotton wool plugs and transfer tubes to the wooden racks marked by rows for identification.
10. Transfer the racks of prepared tubes to the light room maintained at 25°C with a 14 hour illumination and 10 hour darkness cycle. Seedlings are assessed visually for the presence or absence of root nodules after 30 days.

If this method is to be used for screening experimental formulations of coated inoculated seeds, the samples should be stored in sealed containers at a constant temperature and tested at monthly intervals. Accelerated storage at an elevated temperature is useful in sorting out the very good from the very bad more quickly as indicated by the period of time 100% nodulation continues to be achieved.

Most Probable Number (MPN) Testing Method

This qualitative method is particularly suitable for determining an approximation of the number of rhizobia present on inoculated coated legume seed, soil samples and legume inoculants containing microbiological contaminants.

This and other methods such as the plate counting techniques are based on the assumption that *Rhizobium* are present as single cells in the suspensions prepared but it has been demonstrated that this is rarely the case. The ratio of single cells to clusters of many cells will vary depending on the type of inoculant used (manufacturing method), method of preparing the same for inoculation and the dispersion method used to prepare samples of inoculant or coated inoculated seed for testing.

It is also assumed that the *Rhizobium* cells are dispersed in water evenly and that a small aliquot will contain a proportionate number of cells relative to the whole sample. Even under ideal conditions, the variation in distribution is such that sampling errors are unavoidable. Whether counting is done from colonies growing on a nutrient agar plate or from seedlings that have nodulated, the basic theory is that one colony or one nodulated plant arises from a single rhizobia cell. With some inoculants and coated inoculated seed such ideal dispersion cannot be achieved and consequently, clumps of rhizobia in the peat or coating material are counted as single cells. This was adequately demonstrated in a 1975 unpublished paper, *'Enumeration of Viable Rhizobium Cells in Commercial Peat Inoculants'* by J.M. Lloyd and K.G. Walker of Fruitgrowers Chemical Co. Ltd. which is reproduced in part below:

Introduction

Amongst the standards set for defining the quality of commercial legume inoculants the number of viable *Rhizobium* cells per unit weight of inoculum is considered to be of major importance. There are many factors which influence the successful infection of legume seedlings but great emphasis is placed on the use of inoculants containing a large number of rhizobia.

The number of viable rhizobia in commercial inoculants is usually obtained by means of a plate count or plant infection (Most Probable Number) count. Both methods rely upon dispersion of the test inoculum in sterile water, serial dilutions of same with aliquots from the dilutions used for seedling inoculation or applied to nutrient agar in Petri dishes. In the calculation of 'numbers' it is assumed that rhizobia are dispersed as single cells thus inevitably clumps of two or more cells are counted as one.

In laboratory research where the methods of culturing and counting rhizobia are similar, the 'numbers' derived from a standard method of counting are relative and the actual number is less important. A different situation arises where comparisons are made between different commercial inoculants where the manufacturer's methods and materials vary considerably. These methods range from culturing rhizobia on nutrient agar or in liquid broths for use alone or (in the case of broths) mixed with peat for multiplication of rhizobia in a sterilized mixture of peat and nutrients. This paper reports on laboratory studies carried out on commercial legume inoculants produced in accordance with the latter method where opportunities occur for rhizobia to multiply on and within the peat particles, and where large amounts of gummy substances (exocellular polysaccharides) are formed, all of which inhibit the dispersion of rhizobia as single cells.

Examination of Commercial 'Rhizocote' Peat Inoculants

Peat used for the manufacture of 'Rhizocote' legume inoculants is milled to a particle size whereby 95% w/w will pass a 53 µm aperture test sieve. The particle size distribution of the milled peat ranges from less than 1.0 µm to 150 µm. *Rhizobium* cells are generally 2–5 µm in size therefore many peat particles are able to carry more than one bacterium. The finely divided particles of peat tend to agglomerate when moisture and nutrients are added and these agglomerates are not readily dispersed by conventional laboratory shaking or stirring methods; refer to Table 3 below.

Table 3. Wet sieve analyses of aqueous suspensions of milled peat and 'Rhizocote' inoculant prepared from same after periods of time in a mechanical flask shaker and after recirculation through a rotary disc Mill (Percent w/w retained, cumulative, dry basis).

Test Sieve	Raw Milled Peat	'Rhizocote' Shaken 1 Hour	'Rhizocote' Shaken 5 Hours	'Rhizocote' Rotary Mill 5 Hrs
300 µm	Trace	15.3	10.9	Trace
150 µm	Trace	22.6	16.0	Trace
53 µm	0.45	51.8	42.8	1.6

Fig. 16 A typical field of very narrow focus showing single rhizobium cells and clusters of same as green objects; peat particles appear orange in colour.

'Rhizocote' legume inoculants are dispersed for routine plate counts by adding the inoculum to sterile water in a flask and shaking same mechanically for one hour which is similar to the method used by many research organizations.

Suspensions of 'Rhizocote' inoculant prepared in this manner were examined by acridine orange fluorescent microscopy (courtesy the late Dr. Ben Bolhool of the Cawthron Institute, Nelson) and many clusters of *Rhizobium* cells were found. Photographs of four representative fields were examined and the number of single and multiple cells counted; refer to Table 4 below. The greatest number of cells estimated on one peat particle was 128.

Table 4. *Rhizobium* Cell Counts.

An approximate distribution of single cells and cell clusters in 'Rhizocote' legume inoculate after dispersion of an aqueous suspension by shaking for one hour in a mechanical flask shaker. Results are expressed as a percentage of peat particles (units) by count carrying one or more *Rhizobium* cells within the different cell count categories shown. They cover counts of four representative fields during examination by acridene orange fluorescent microscopy.

Field Number	Single Cells	Units of 2–5 cells	Units of 6–10 cells	Units of 11–15 cells	Units of 16–20 cells	Units of >21 cells
1	38.2	41.2	11.8	2.9	2.9	2.9
2	28.4	59.4	8.1	2.7	Nil	1.4
3	41.0	44.3	13.1	1.6	Nil	Nil
4	34.4	62.3	3.3	Nil	Nil	Nil
Mean	35 5	51.8	9.1	1.8	0.7	1.1

Improving Dispersion

In order to determine whether a higher plate count could be obtained by improved dispersion of the peat inoculum slurry, laboratory equipment was constructed for the purpose of better disintegrating the clumps of peat particles, gum and rhizobia by a gentle grinding motion; refer to Figure 17 below.

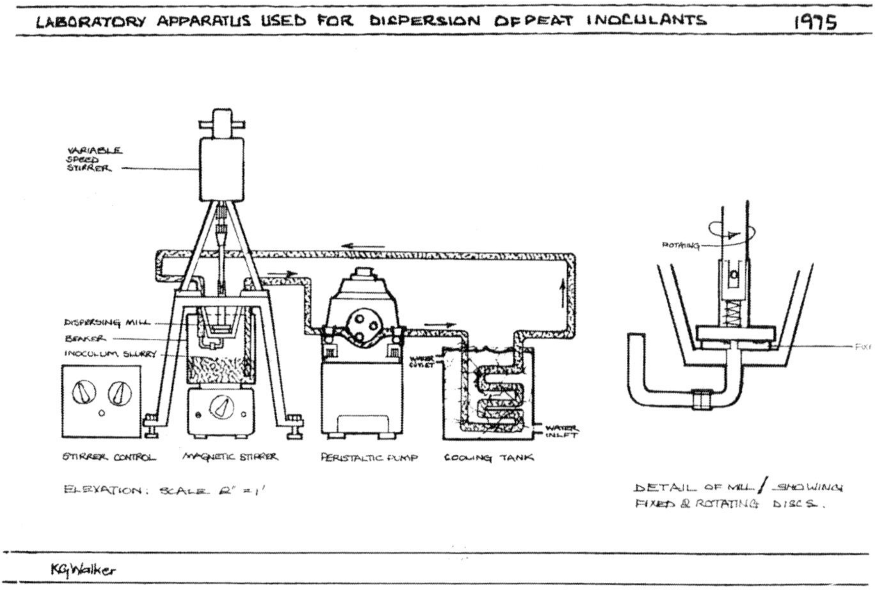

Fig. 17 Peat Inoculant Dispersion Apparatus. [Walker/Lloyd/Axis]

Fig. 18 Photo of Disc Mill Apparatus Used.

The initial slurry was recirculated from a cooled reservoir by means of a peristaltic pump through a hole in the centre of a stainless steel fixed disc. This disc supported a similar size solid rotating disc above, that was attached to a spring loaded shaft which turned at approximately 750 rpm. The contact surfaces of both discs were closely machined and lapped with the aid of a fine abrasive.

Inoculum slurry was pumped through the discs at approximately 27 litre per hour and maintained at 20°C by a heat exchanger and cooling tank. All apparatus was pre-sterilized by autoclave and all experiments were carried out in aseptic conditions in a sterile room. The results of one wet milling trial are reported as follows:

Method

1. One 160 g pack of 'Rhizocote' peat based inoculant for white clover was transferred to a 3 litre flask containing 1 litre sterile water and mechanically shaken for one hour. Serial dilutions of the dispersed inoculum were prepared to the 10^{-7} level and 0.1 ml aliquots spread over the surface of nutrient agar in each of 6 petri dishes as a control.

2. A representative 450 ml sample of the original slurry from 1 above was recirculated through the apparatus shown in Fig. 17 and 10 ml samples drawn at 2, 3 and 5 hour intervals. Each sample was diluted to the 10^{-7} level and aliquots plated out as described under 1. The petri dishes were held in an incubator for 70 hours at 27°C and the colonies of *Rhizobium* counted. Results are given in Table 5 below.

3. Another pack of 'Rhizocote' inoculant from the same batch was shaken, diluted and plated as for 1 above. This original slurry was then held for 5 hours at 20°C, shaken by hand for 2 minutes and set out for a repeat plate count to determine whether any change in the number of rhizobia had occurred. An increase of 2.4% on the first count was recorded but this was well within the margin of error in testing and not significant.

Table 5. Rhizocote Inoculant Dispersion Trial
Plate Count Results.

Measurement Details	Shaking 1 Hour	Plus 2 Hours Disc Milling	Plus 3 Hours Disc Milling	Plus 5 Hours Disc Milling
Mean no. colonies per plate	117	281	408	127
Mean no. colonies (control as base 100)	100	240	349	109
Rhizobia/gram inoculant	8.5×10^9	2.0×10^{10}	2.9×10^{10}	9.2×10^9

Discussion

It is evident from the results above that adequate dispersion of the peat based inoculant is not achieved by a conventional shaking method.

The rotating disc mill used in the experiments improved the dispersion of the microbes considerably but colony counts after 5 hours milling indicated that Rhizobium cells were being damaged in the process.

In order to obtain a more reliable indication of the number of rhizobia present in commercial peat inoculants it will be necessary to improve the dispersion of inoculum slurry by methods which do not adversely affect rhizobia.

The author/contributor gratefully acknowledge Dr. Ben Bohlool of the Cawthron Institute[8] Nelson for the use of his microscope and Mr. Geoff Taylor M. Agr. Sc., Technical Consultant, Fruitgrowers Chemical Co. Ltd. for his advice and interest.

Because of the difficulty of obtaining predominantly single cell suspensions, the result of any quantitative test can only be taken as an indication of the 'nodulating capacity' of a given sample of inoculant or coated inoculated seed.

As indicated above, depending on the method used to manufacture the inoculant and prepare it for application to the seed and the actual seed coating composition, a given method used to prepare samples for testing may not accurately reflect the actual number of rhizobia present. It is inappropriate therefore to compare the results obtained by MPN or plate count from inoculants or coated inoculated seeds prepared by different methods. However, this does not detract from the importance of having official standards for the minimum number of rhizobia per gram of inoculant or per coated seed provided these can be linked to the percentage of plants that form nodules in a typical field sowing situation.

In New Zealand, field testing comprised an official and extremely valuable part of the legume inoculants and coated seed testing programme operated by the ICSTS.

Any official standards established should take into account the enormous differences in the number of seeds per gram of different legume species. A much lower inoculation rate per kilogram of seed can be applied to seeds of subterranean clover where there are only around 250 seeds per gram compared with, say, white clover having typically around 750 seeds per gram.

The 'Most Probable Number' testing technique involves suspending rhizobia from a quantity of inoculated seed, soil or an actual inoculant, serially diluting that suspension, adding aliquots of the serial dilution to host seedlings and determining the number of rhizobia in the original sample by determining the number of successfully nodulated seedlings. Accuracy of result is influenced by the dilution level at each step and the number of seedling tubes used at each dilution level. Using more tubes and reduced dilution levels will provide greater accuracy but the time and equipment required per test will be greater. A good compromise between accuracy and time requirements is to use 10 fold dilutions and 4 seedling tubes per dilution in accordance with Table 3.5A from '*A Manual for the Practical Study of Root Nodule Bacteria*' by J.M. Vincent IBP[9] Handbook No. 15. This manual was first published in 1970 and is still regarded as the 'Rhizobiologist's Bible' by those involved in legume inoculant research and production today.

Material and Equipment Required for MPN (Plant Infection) Testing:

In addition to the materials listed for assessing the nodulation potential of inoculated coated legume seed listed on page 41 the following will be required:

- Surface sterilized legume seed (for the *Rhizobium* species to be tested)
- Supply of heat resistant glass bottles, e.g., 35 ml 'Universal' culture bottles with metal screw cap and rubber wad
- Variable 'Pipetter' 0.1–1.0 ml adjustment and supply of sterile disposable plastic tips for same
- 250 ml measuring cylinder
- Squeeze bottle
- Glass funnel with a 500 µm aperture stainless steel sieve to fit
- 250 ml conical flask
- 250 ml beaker
- Supply of distilled water previously sterilized by autoclaving for 10 minutes at 138 kPa

Method

A) **Surface Sterilizing of Seed**

Prepare a stock of surface sterilized legume seed from a high interim germinating line. This is done at 6 monthly intervals as follows:

1) Immerse up to 50 g of seed for 5 minutes in a mixture comprising 70% ethanol, 30% distilled water by volume.
2) Strain the seed using a suitable sieve and rinse with 500 ml sterile distilled water.
3) Dry thoroughly at 40°C overnight and store in a sealed container.

B) **Pre-germination of Seed**

1) Cut clean sterile gauze wads to fit sterile petri dishes and moisten with 7–9 ml distilled water.
2) Sprinkle about 200 seeds on the surface of the gauze in each plate, place the lid on the dish and store within a sealed container (to maintain humidity) in an incubator at 25°C (or in a light room).
3) When the radicle is 0.25–0.5 mm in length, the seedlings are ready to be transferred to the seedling tubes.

C) **Seedling Tube Preparation**

1) Prepare a supply of seedling tubes in accordance with the procedures set out on page 44 but instead of allowing the tubes to cool in an upright position as described in step 7 (p.65), allow the tubes to cool at an angle of 60° so that the surface of the nutrient agar forms a slope when cooled.

2) Using a stainless steel probe, transfer healthy seedlings to the lowest point of the agar slope in the seedling tube, flaming the probe before each use.

3) Plant two seedlings about 5 mm apart in each tube by manipulating with the probe to insert the radicle into the agar surface and replace cotton wool plug.

4) Place the tubes in numbered wooden racks in discrete rows of 4 tubes (24 tubes per rack) and transfer to the light room where they are grown on for 7–10 days prior to inoculation.

5) Maintain the light room temperature at 25°C with 14 hours artificial illumination and 10 hours darkness per 24 hour cycle.

D) Water for Serial Dilutions

Add 9 ml distilled water to each of the required number of 'Universal' culture bottles, lightly replace the metal caps and autoclave for 10 minutes at 138 kPa. Twist metal caps tight when cool.

E) Sample Preparation (example for coated inoculated seed)

1) Weigh a representative 50 gram sample of inoculated coated seed into a 250 ml beaker and add 100 ml sterile distilled water.

2) Place the beaker under a laboratory mechanical stirrer and stir at a speed that forms a moderate vortex for 20 minutes to remove the coating and disperse the rhizobia.

3) Place a glass funnel into a 250 ml measuring cylinder fitted with a 500 μm aperture stainless steel sieve and strain the contents of the beaker into the cylinder.

4) Wash beaker and seed clean into the cylinder with sterile distilled water from a squeeze bottle using a glass rod to stir the seed for maximum rinsing.

5) The filtrate is made up to 250 ml with sterile distilled water and transferred to a 250 ml conical flask.

Note: All glassware should be thoroughly washed in boiling water and then rinsed in sterile distilled water before use.

F) Preparing Serial Dilutions and Inoculating Seedling Tubes

1) Prepare a serial dilution from the above 250 ml sample as follows:

50 g seed diluted to 250 ml	A
1 ml A + 9 ml sterile dist. water	B
1 ml B + 9 ml sterile dist. water	C
1 ml C + 9 ml sterile dist. water	D
1 ml D + 9 ml sterile dist. water	E
1 ml E + 9 ml sterile dist. water	F

Additional dilutions can be made if required. Results can be read over any six dilution series.

2) At each dilution level use a mechanical 'Pipetter' to transfer 0.2 ml diluent to each tube on the base of the agar slope.
3) Aspirate the tip of the 'Pipetter' twice before making a transfer to 'saturate' the inner surface with rhizobia.

G) Incubation and Recording

1) Each numbered rack of inoculated seedlings is returned to the light room.
2) The roots of the seedlings are examined for nodules after 28 days.
3) The presence of nodules on one or both plants is recorded as positive.
4) Observations for nodules are made in bright light; plants can be removed from agar for closer observation if in doubt.

H) Calculations

1) Use the table below calculated from Table 3.5A 'A Manual for the Practical Study of Root Nodule Bacteria' by J.M. Vincent.

Table 6 Calculation of the Most Probable Number of rhizobia per gram of coated seed.

Positive Tubes	MPN Per Gram	95% Fiducial Limits	Positive Tubes	MPN Per Gram	95% Fiducial Limits
1	1.5×10^1		15	2.5×10^4	6.6×10^3–9.5×10^4
2	1.5×10^1	4.0×10^0–5.7×10^1	16	4.3×10^4	1.1×10^4–1.6×10^5
3	2.5×10^1	6.6×10^0–9.5×10^1	17	7.8×10^4	2.1×10^4–3.0×10^5
4	4.3×10^1	1.1×10^1–1.6×10^2	18	1.5×10^5	4.0×10^4–5.7×10^5
5	7.8×10^1	2.1×10^1–3.0×10^2	19	2.5×10^5	6.6×10^4–9.5×10^5
6	1.5×10^2	4.0×10^1–5.7×10^2	20	4.5×10^5	1.2×10^5–1.7×10^6
7	2.5×10^2	6.6×10^1–9.5×10^2	21	8.5×10^5	2.2×10^5–3.2×10^6
8	4.3×10^2	1.1×10^2–1.6×10^3	22	1.7×10^6	4.5×10^5–6.5×10^6
9	7.8×10^2	2.1×10^2–3.0×10^3	23	2.5×10^6	6.6×10^5–9.5×10^6
10	1.5×10^3	4.0×10^2–5.7×10^3	24	4.5×10^6	1.2×10^6–1.7×10^7
11	2.5×10^3	6.6×10^2–9.5×10^3	25	8.5×10^6	
12	4.3×10^3	1.1×10^3–1.6×10^4	26	1.5×10^7	
13	7.8×10^3	2.1×10^3–3.0×10^4	27	2.5×10^7	
14	1.5×10^4	4.0×10^3–5.7×10^4			

2) To obtain the most probable number of rhizobia per coated seed, divide the MPN per gram by the number of coated seeds per gram.

E.g.: A + + + + B + + + + C + + + + D + + + − E + − − − F − − − −

Number of positive tubes = 16, MPN of rhizobia per gram of coated seed = 4.3 x 10^4 (from table); Number of coated seeds per gram = 750, therefore the MPN of rhizobia per seed = 57 (14–213 per seed, 95% fiducial limits).

Note: Initial tests on samples of coated seed often require a further tenfold dilution at the start of the series. MPN value in the table is therefore multiplied by this factor.

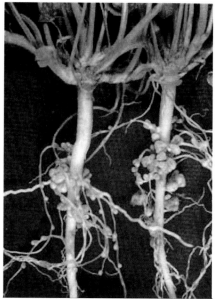

Fig. 19 The Goal. Healthy effective pink nodules which have so successfully colonized the rhizosphere of these nitrogen rich, deep green white clover plants.

Photo from 'Inoculating and Pelleting Pasture Legume Seed NSW Department of Primary Industries, Australia.

An excellent publication as is this photograph.

Modifying the Formal Technology

There is a major practical problem with this technology: in developing economies, it may not always be possible to assemble and use the laboratory techniques and technical equipment described in this chapter. However, the contributing author's first duty is to describe formalized procedure of a well established legume inoculant manufacturing and/or quality testing organization.

While these quite time consuming and complex techniques have been essential to the development and establishment of highly successful inoculant and seed coating formulae, as in New Zealand's 'Rhizocote' inoculant and Prillcote™ coated seed, they also have the capacity to slow down practical use of legume seed inoculation, escalate cost (which may not be sustainable initially in a small and undeveloped market) and may have potential to create problems for all concerned if and where such formalized procedures might be mishandled, giving false information.

There is, therefore, a need for more simplified methods of inoculant manufacture and subsequent testing of inoculated and coated product even if such product and their tests are less precise (actually the MPN counting procedure widely adopted commercially and in research which is described in this chapter is itself, simply a guide, it is not precise). A 'shortcut' procedure would have huge benefits for undeveloped agriculture and horticulture which the contribution author recognizes but has not adopted because there has been no need for it in New Zealand.

There are recent developments in simplifying production and testing which may be available from Axis Associates Ltd, a privately owned New Zealand Research and Development Company, which does have an export policy and is able to provide 'hands on' commercial experience to any new facility, not only of inoculant manufacture and viability testing of product, but also (somewhat uniquely), of practical aspects of large scale coated seed manufacture, which expertise is not readily available.

In addition, advancements have been made available by the NifTal Centre at the University of Hawaii by the development of a Micro-Production Unit [MPU] which, when fully refined, should become the key to global then more localized intensive use, eventually allowing economically viable, more sophisticated and accurate methods of production and testing.

NifTAL was instrumental in the establishment of inoculant production facilities in many tropical locations and has developed tools for more efficient and economically viable inoculant production and quality control.

Worldwide research needs to be gathered, filtered and combined into several essential documents covering strains collection and evaluation, inoculant manufacture, coating, testing and production. Some of the required methodology probably exists already. For instance, a simple method of evaluating the viability of lucerne (alfalfa) and clover seed inoculation was publicized in the American Society of Agronomy's *'Agronomy Journal'* 74; 921–923 [1982] more than 30 years ago, whereby inoculated seeds were planted in plastic growth pouches and successfully appraised. Such simple procedures, probably carefully refined as to detail, could well be sufficient to support a fledgling 'cottage industry'.

The author/contributor would have included in this chapter, details of the tremendously important work which was carried out around the world by the University of Hawaii's NifTAL organisation through the USA's foreign aid programme USAID[10]; however, we consider it better to convey the following information from NifTAL itself (by permission of Director Dr. Paul Singleton).

"Appropriate inoculant production systems can exploit small markets in developing countries.
Although inoculant production can be made fairly scale-neutral, cost-effective systems with low throughput volumes have not been specified.

The startup costs and difficulty developing inoculant markets among small farmers in developing countries require that firms recover their investment and operate profitably at low volumes. NifTAL's earlier conceptual design of a Micro-Production Unit (MPU) addresses this issue and has received favorable reviews from several producers. Precise design and specifications, protocols and materials now need to be formalized, tested and made available to prospective producers."

(This was followed by a paragraph titled, 'Poor inoculant quality limits market development' which is reported in an earlier part of this chapter.) NifTAL went on to reveal:

"Part of the quality-assurance problem can be addressed by training; however, there is a need in both developed and developing countries for improved quality-assurance methods. Quality-assurance methods must be more rapid, precise and convenient if producers and external quality control agencies are to rigorously monitor inoculant quality from production to application in the field. An example of the technical inadequacies of an existing standard method is the plant infection method (MPN). This method is labor intensive, requires three weeks, and is precise to within only one log (order of magnitude) of the actual count. This method cannot help the industry as a tool to make production decisions. Existing quality assurance methods also create inefficiencies in some markets by delaying product distribution until lengthy assays can be completed by regulatory authorities."

To address those constraints to improving BNF (Biological Nitrogen Fixation) in developing countries, NifTAL developed information on inoculant production and quality control, details which have been and hopefully still are available on their website at www. ctahr. hawaii.edu/bnf.

Potential (by author)

Many underdeveloped nations are short of food yet there are huge areas, literally millions of hectare, around the globe lying semi-barren, producing nothing edible because that land is considered 'hostile', too hot or too cold or lacks one or more of the essentials of plant life, e.g., water which however can be found underground even in arid locations accessed by deep bores, or channeled from mountains and rivers, also stored in natural or manmade reservoirs. In addition, nutrients such as calcium, phosphate, potassium and sulphur are all obtainable from deposits in many locations but not nitrogen; yet it, too, is available in abundance from that universal location, the atmosphere, worldwide via *Rhizobium*.

While some land is considered too hot and hostile in its natural state to successfully grow plant material edible by humans or animals, the use of increasingly available heat tolerant species, combined with planting with spaced protective shady trees under which both temperature and heat are modified, by using *Rhizobium* more tolerant of warm soil conditions [which are also cooled by surface growth itself] are all accessible to planned programmes being 'holes in the barrel of productivity' which can be plugged more readily today using investigation, hard work and current technology paid for, one would hope, from natural resources being discovered around the world—in Africa, South America, Asia, Indonesia, Soviet States, Mongolia the Middle East, and many more. Specially those nations which can benefit from the vast resources of the seas and of the rivers of the world.

Other land is considered too cold for plant growth but by growing fast maturing legumes (which may need to be selectively or genetically enhanced or specifically developed for each location) during the short summer thaw in cold locations and storing that product over winter where refrigeration is a readily available natural resource, some of these locations could be more productive. Temperate strains of *Rhizobium* are tolerant of cold, even periodic freezing conditions, particularly when safely inside the host plant.

Yet other areas, including part of central Australia, are too saline for both plants and their symbionts. However, as man has found a way to de-salinate water, so may there be prospects for economically neutralizing salt in soil with effective research, as well as growing salt tolerant grass species. Salt removed from that land over time, by means of cutting and selling hay, silage or via livestock sales, will very gradually de-salinate the soil under otherwise normal conditions (i.e., where there is no ongoing re-salination). The former Ahuriri lagoon north of Napier, New Zealand, is now growing grass successfully; it was a 1 or 2 metre deep tidal lagoon but miraculously raised from the sea by a major earthquake. It became a Department of Agriculture 'laboratory' for successful reclamation of saline soils, proving it can be done. In the case of saline soils it may be necessary to supply nitrogen artificially (at first) because the *Rhizobium* genus is generally not salt tolerant, another avenue awaiting research for a salt tolerant strains selection.

All these more challenging prospects require serious localised initial experimentation, probably by governmental organizations (e.g., small research teams) followed by pilot development and taxpayer investment. When parameters and possibilities are established, the technology can be passed on to farming people who are willing and able to pursue new opportunities to enrich their lives and the environment in which they live, thus provide more food for their hungry families and often growing populations. They also pay tax on profits earned to repay that initial outlay. Much neglected 'barren' land can grow palatable grass but the investigation must be done, then the more important holes in the barrel of productivity can be systematically plugged by supply of essential deficiencies, not least of which will be nitrogen.

References

(1) Gill, A.M., R.H. Groves and I.R. Noble. 1981. [AUSECO] Publication Fire in Australia Extract from an Internet Article.
"Fire has some direct impacts on the soil. Heating to temperatures of 3000C–4250C can cause chemical transformations in the top few centimetres of soil and at lower temperatures sterilization of the soil, including death of leaf litter flora and fauna occurs. Loss of nitrogen also occurs (up to 1.7 million tonnes per year in northern Australia) and this is important considering the naturally low levels of nitrogen in Australian soils".
AUSECO is a privately owned and run field study company, set up as a resource for K-12 teachers. Above extract acknowledges Fire and the Australian Biota, Australian Academy of Science: Canberra.

(2) Thompson, J.A. 1992. Consultant Report to UNFAO [IND/86/003. FAO, Rome], Italy 48 p.

(3) Hastings, A. 1970. Plant Diseases Division New Zealand DSIR (Department of Scientific and Industrial Research). Rhizobia infection of plants is technically categorised 'a disease', a beneficial one.

(4) Hale, C.N., W.L. Lowther and J.M. Lloyd. 1979. New Zealand Journal of Experimental Agriculture 7: 311–14. Effect of Inoculant Formulation on Survival of *Rhizobium Trifolii* and the Establishment of Oversown White Clover {*Trifolium repens*}. A research study in which Dr. Hale and Dr. Lowther were completely independent of the inoculant manufacturer (being DSIR and Department of Agriculture scientists, respectively).

(5) Taylor, G.G. et al. 1970's [Personal comment]. Plate counts of colonies may show adequate numbers of *Rhizobium*, and seedling tube samples may show lucerne plants successfully nodulated, but neither would reveal that 90% of the rhizobia cells were in fact *trifolii*—not *meliloti*—a trap for unwary microbiologists! Lucerne (alfalfa) will only nodulate with *Rhizobium meliloti.*

(6) [ANZS] Australian and New Zealand Standards. Refers to Physical Containment levels. Indoor containment facilities such as laboratories, glasshouses and animal facilities are approved to specific Physical Containment (PC) levels referred to as PC1, PC2, PC3 or PC4. These levels are arranged in order of increasing stringency of operational and structural requirements. The requirements are described in ANZS 2243.3 (with any exemptions listed in the MAF/ERMA New Zealand Standards [Ministry of Agriculture and Fisheries and Environmental Risk Management Authority]. PC1 is the least stringent level, with PC4 being the most. The 2243.3 can be purchased from: The Standards New Zealand website: www.standards.co.nz.

(7) Dye. D. Circa late 1960s, Dr. Doug Dye was head of the expanding Bacteriology Section of Plant Diseases Division of DSIR, Mt. Albert Rd. Auckland, from which he retired in 1983.

(8) Cawthron Institute. [Established in 1919]. The last will and testament of Nelson philanthropist, Thomas Cawthron created this Institute. He had a vision of science contributing to the growth of a young New Zealand (which it certainly has done).
Following his death in 1915, £231,000 was bequeathed by Cawthron, the largest single bequest in New Zealand at the time, to establish and maintain a technical school, institute and museum to be known as the Cawthron Institute.
Cawthron Institute was officially opened in Nelson in 1921 following the establishment of the Cawthron Institute Trust Board.

(9) Vincent, J.M. 1970. A Manual for the practical study of root nodule bacteria published for the International Biological Programme [IBP] Professor, Sydney University, Australia.

Contents:
The cultivation, isolation and maintenance of rhizobia; The qualitative characteristics of rhizobia; Examination with the light microscope; Cultural and metabolic characteristics; Antigenic properties; Bacteriophage and lysogeny; Recognition of rhizobia; Enumeration; Determination of total growth; Counts of viable rhizobia; The assessment of nodulation and nitrogen fixation; Methods for greenhouse and light room; Field trials; Assessment of the need for legume inoculation; The production, control and use of legume inoculants.

(10)   [USAID] The lead USA Government agency that works to end extreme global poverty and enable resilient, democratic societies to realize their potential. It represents the tremendous generosity of both the people and Government of the United States of America.

3

Legume Seed Inoculation and Coating

The Technology Leap....... and Sequence of Development

As early as the 18th Century and probably by lucky chance, observant arable farmers began to realize that soil spread from a legume crop field onto another about to be sown to the same crop, sometimes produced healthier or greener looking plants.

They did not know why, so it is no wonder such mysterious phenomena have historically fed the superstitions of mankind. More likely, the benefit observed was a straightforward result of both fertility and bacteria transfer.

This practice actually transferred as yet undiscovered rhizobia nitrogen fixing bacteria from the old crop to the new. As a principle of agriculture, it was not a sound practice because pathogens [other bacteria, harmful to crops], plus fungi, both of which increasingly multiply and colonise existing crops as they mature [which is often the reason for eventual failure of that crop], were also transferred to the newly sown field of highly vulnerable young seedlings which of course would conspire to shorten the useful life of that new host—be it lucerne, soybean, chickpea, lupin or beans, by giving the pathogens a "jump start" to-ward infecting that newly sown crop.

The illustrations below are posted on the website of Wikipedia free Encyclopaedia with photo [B] submitted by The Plant Pathology Department of the University of Nebraska at Lincoln, USA. Alfalfa is considered the largest forage crop in the world—excluding grasslands of course. It should be drilled on clean soil not previously in an alfalfa crop for several years. Rejuvenation drilling into a thinning stand is also at risk of early infection.

Technical understanding advanced when microbiologists later identified the marvel of nitrogen fixing ability of rhizobia bacteria[1], and so, marginally, began the on farm practice [around the early 1930's] of applying cultures of valuable rhizobia root nodulebacteria to the new crop—but without the soil

[A] Healthy disease free alfalfa.   [B] Phytophthora root rot infection in alfalfa.

Fig. 20

and without the diseases. And, what better practical way to achieve that than applying the cultures of rhizobia to the seed itself which was about to be sown—thus wherever that seed became deposited in the soil, its rhizobia bacteria would be right there with it, ready to infect root hairs of the newly germinated plants which then start manufacturing valuable nitrogen from the enormous N^2 resource in the atmosphere.

Work by Rhizobiologists identified problems with this technique however, because they recognized very high levels of *Rhizobium* mortality when these relatively sensitive bacteria were applied onto bare seed—sometimes leaving no viable cells at all after several hours or days, which of course led to failure of the technique we now call *legume seed inoculation*. Because rhizobia are such sensitive organisms adapted to cool dark moist soils and the interior of plant tissue itself, their high rates of mortality were caused by various antagonisms including—heat, desiccation [rapid drying in wind or sun], mixing with any fertilizers which have an acid content such as the sulphuric and phosphoric acids in superphosphate, or even simple delay itself in an incompatible environment—all caused death—but the most potent killer of all was found to be a natural but most formidable toxin—unhappily [for rhizobia] from the seed itself.

Many seed varieties carry water soluble phenolic compounds which are anti-biotic protecting that seed from soil pathogens when lying on the ground— believed to be nature's way of ensuring seed survives until germination, those natural seed surface exudates [polyphenols] which are bactericidal [including hydroxybenzoic acid for example] are mobilized when seed becomes wet [which unfortunately occurs when applying *Rhizobium* inoculant aqueous suspensions] and these natural substances of course destroy rhizobia as well as other bacteria, quickly and in large numbers.

Several remedies have been tried to overcome this phenomenon— coating seed with protective materials such as finely ground adsorbent and microscopically pitted peat [in which microbes can escape from toxins], or applying alkaline or pH/neutral finely ground minerals—lime, rock phosphates and dolomite directly to the seed—but also simply applying very high numbers of rhizobia to counter high levels of mortality. Also by applying rhizobia cultures to the soil, not seed, as a liquid "inoculant"—a method which would only work in arable agriculture where machinery drills the seed but can also transport the huge extra weight of water required [mixed with culture] to satisfactorily deliver the suspension and "inoculate" drill seed rows through a mounted tank and calibrated hose directed at those seed rows. On a small scale, this can also be achieved manually.

Fig. 21 The three stages of original Prillcote inoculated and coated legume seed: [lucerne above] Left: Plain seed plus pre-treatment, Centre: Inoculated and peat coated, Right: Plus Lime coated.

Despite the degree of difficulty, a promising new method was developed.

That breakthrough in technology was achieved by coating the seed and including the inoculant simultaneously. Firstly by a method using either methyl cellulose or gum arabic applied in solution as adhesives followed by finely milled lime—a procedure promoted by the New Zealand Department of Scientific and Industrial Research. That method was later substantially improved on by commercially developed coatings—notably Prillcote™ coated seed produced by the leading manufacturer of the time, a company called Coated Seed Ltd.

A major advance was then achieved with the professional development of seed coatings which do not interfere with those polyphenol natural seed surface protectants [in fact encloses them with a film over each seed], providing a more safe environment for sensitive rhizobia even when mixed with non-acid fertilizers. It also allows a small but vital quantity of finely milled, pure and readily available nutrient such as phosphate, magnesium or lime, plus trace elements such as molybdenum and in the case of lucerne for arable drilling, fungicides—all able to be precision placed wherever the seed may deposit, pending germination and then nodulation.

In New Zealand, large areas of hill and high country, most of it stabilized by tussock, fern, matagouri, manuka and kanuka "scrub" and native grasses of low nutrient value, too rugged or remote and too expensive to cultivate simply to grow better grassland for foraging sheep and cattle [more recently also domestic deer]. Nevertheless were found [in the 1950's] to be receptive to surface sown clovers [for nitrogen] and more productive but hardy grasses [for better dry matter production than legumes] particularly where certain rules, identified by grassland scientists, were followed.

This revolutionized the development of oversown grassland which for a century earlier had been less successfully seeded by hand, often on foot—or from a "pikau" [Maori sack bag] slung over the horses neck with seed tossed directly into the cold ashes of a "burn off" of scrub or native bush [usually after a "cutover" and extraction of valuable native timber].

Fig. 22 At 500 to 600 metre a.s.l. [above sea level] in New Zealand hill country (where there is snow in winter and temps over 30°C in summer), white and subterranean clovers are well established among the natural shelter of tussocks, rock and native foliage. When well managed, the native grasses are gradually replaced by these introduced clovers and grasses as fertility increases (see nitrogen green clover in the photo).

The New Zealand [and worldwide] surface sown pastoral revolution had begun. Its progress and viability however, always dictated by market values of wool, lamb, mutton, venison and beef.

Not only could clover and grass seed be more reliably sown directly on the surface of native land, but it could be sown much faster covering vast areas of steep and remote country which had never before been accessible physically and seldom financially, but could now be oversown by aircraft also able to top dress the land with fertilizer—usually superphosphate which supplied both phosphate and sulphur, deficient in many of the world's soils and most of those in New Zealand. Sulphurised "super" is now applied where sulphur is more positively deficient—but also sometimes lime on very acid soils although raw limestone just finely crushed has proven difficult to manage by aerial application—this is less of a problem today where granulated.

Production of Coated Seed in New Zealand

From the introduction of coating in 1966 by the DSIR, whose Director later collected and published the value of Inoculants, Coatings and Seed sown in New Zealand during 1973 to 75 being:

Inoculant	$115,235
Coatings	$1,230,720
Seed	$2,783,425

the industry has grown significantly and in the 1976/77 year, one manufacturer alone produced 842,000 kg of [coated weight] seed with an estimated value of about $1,350,000 followed by an output exceeding 1 million kg for the 1978/79 year. Today, more likely well over a million tonne are processed worldwide.

While inflationary effect alters subsequent results it is reasonably clear the industry continued to expanded in New Zealand where turnover from coated seed activities by larger seed merchant companies in the 1990's is estimated to have reached about NZ$5 million a year [and possibly approaching $10 million in value by 2012].

The situation which developed in New Zealand may serve as a "role model" for some aspects and an avoidance warning for others, to the seed industries of other countries. It was certainly not all "plain sailing" however and "modification" would happily avoid some of the problems which surrounded the industry in New Zealand in its earlier years.

Former Coated Seed Ltd. [CSL] was assembled and supported by expertise from two sources: one company from the seed industry and the other from the chemical and mineral processing industry. This combination was sufficiently substantial to employ specialized factory staff, technical and administrative support plus a qualified agriculturist as development manager. It was nevertheless restricted in its aim to provide specialized seed processing services to the whole New Zealand seed trade by the fact that its part ownership [the seed supplier] being just one Firm [Wright Stephenson & Co. Ltd.] was opposed to the development manager's proposal that all seed firms in New Zealand be invited into joint half-ownership, the Processing company to become the other half. In author Bennett's opinion, Wrightson™ may have earned more and saved a great deal of the expense it invested in development of coated seed if it had allowed the whole seed industry to share production and costs while earning for itself a royalty fee on all production.

The aim of Wrightson's™ original three employee/directors on the Coated Seed Ltd Board was to offer all New Zealand seed companies the processing expertise of Coated Seed Ltd. {CSL} on a wholesale processing basis using only Wrightson's™ seed. The development manager advised the three directors who represented the other half owner of Coated Seed Ltd, the Processor [Fruitgrowers Chemical Co. Ltd. {FCC Ltd.}—that in his opinion other seed firms which have their own seed at the core of their business *would not buy another Firm's seed* unless that were the only source of a particular line or type of seed, whether it was expertly inoculated, coated or not—they would not agree to any scheme which would limit the selling of their own seed, processed or not. While that opinion proved to be correct [i.e., the other seed trade did not agree to purchasing a competitor's seed] it placed an element of tension between Wrightson™ staff/directors and the development manager (appointed by FCC Ltd) which persisted and which FCC Ltd directors did not fully understand.

A compromise was reached whereby any member of the Grain and Seed Industry in New Zealand could send their seed to the Coated Seed Ltd. factory for processing at wholesale rates, but if they did not have the appropriate variety or quality of seed required by a farmer client [and indeed often traded among each other to replenish stock] then Coated Seed Ltd would draw on Wrightsons™ first for supply of that seed which it would process and dispatch to the requisitioning firm. Most New Zealand firms accepted this and most used the Prillcote™ process of CSL. Periodically, CSL bought seed from those other Firms to satisfy an order.

However, even this arrangement was not fully acceptable competitively as one or two firms refused to participate where a seed competitor had a ½ share in CSL and certainly would not buy its seed. Notably of those, the original Hodder & Tolley Ltd, a large seed merchant company supplied its own seed processed at a Palmerston North facility and sold it mainly throughout the North Island of New Zealand. This firm was unable however to apply the research proven manufacturing and microbiological expertise which CSL & FCC Ltd were able to do during the development phase and did not manufacture commercial inoculants. A couple of other seed firms too manufactured their own inoculated and coated seed however according to comments from Govt. technicians who tested competing products during the 1960's and 70's they seldom if ever matched the higher standards of Coated Seed Ltd's Prillcote™ processing—specially as to retaining rhizobia viability which was the chief aim.

Commercial Considerations

In a paper presented to a Legume Nodulation Conference in Australia in the late 1970's, two New Zealand scientists[2] reported that as a result of their investigations, while commercially coated seed "from one firm" which had been used extensively in New Zealand had given satisfactory establishment results with just 100 to 200 viable rhizobia per seed at point of sowing, the majority of other firms attempting to produce competitive product frequently had difficulty meeting that minimum standard of 100 viable rhizobia per seed after a stress delay—being a minimum essential to qualify for supply to Government contract purchases, and these products had also given poor results in oversown trials.

In 1977 however, those same scientists realised that the standard of 100 cells/seed was not a reliable indicator of subsequent field performance because the Firm who's product was dominant in the marketplace and which at that time carried only 100 to 200 cells/seed at sowing [after a delay stress] outperformed inoculated and coated seed from other firms—even where their coated seed product carried up to 400 cells/seed at sowing—yet gave much lower eventual establishment—in fact gave results no better and sometimes even less than plain inoculated seed. The investigators were of course puzzled by this result and concluded that the reason might be either the use of different strains of *R. trifolii* between manufacturers samples—or perhaps the more successful coated product with its modest [but high survival] loading of cells employed a *superior quality coating*. It is much more clear today that the latter reason was correct as emphasized in the results of further research Tabled under CLOVER in this Chapter.

CSL then, easily the major producer in New Zealand, became this somewhat uniquely co-operative and successful venture—amalgamating most of the seed industry into co-operative support, its agriculturist delivering technical and colour illustrated lectures to the Seed Staff of all participant Firms, quite remarkably so, in an industry which in New Zealand is strongly sales-competitive.

While CSL success arose from a commonsense approach—including through careful promotion [the development manager visited all firms throughout New Zealand—frequently, offering technical support and for promotion, but only representing CSL—tactfully not mentioning Wrightsons™ when visiting other seed firms—or vice versa when visiting Wrightson™ Branches]—but also because leading seedsmen with profound common sense such as John Paterson, a Director of Pyne Gould Guinness' Ltd. and Manager of its Seed Department plus those in Donald Reid & Co. The Farmers Co-ops and others, all realized their clients would need this product and that it would also help sell their own seed.

However, from this experience, the ideal model for any group would be a centrally located processing facility owned by a consortium of all compliant seed suppliers in that given location, not necessarily within one Nation, but possibly from and for supply to several adjoining countries such as may be found in Africa or South America. These firms would select three experienced directors [in seed merchandising, in human resources and in finance at the very least] who would meet periodically in conjunction with another three specialist directors [with skills in agricultural science, in microbiology and in the manufacturing industry, also at the very least] appointed by the other half owner being one or more chemical/mineral processing companies thus 6 directors for formulating joint policy. These directors would need to avoid the distraction of self interest, both as contributor, and personally. If these directors are themselves simply employees of part owners then there is a danger they have no authority to effect change or improvement, are afraid to challenge the others—which leaves policy being dictated by a minority—or even maybe just one persuasive or authoritarian Director. Worse, an absentee, ill-informed, financially driven "godfather" controlling all directors—a situation to be avoided by any company.

The ideal model would combine seed and processing skills and hopefully, unlike the New Zealand experience, avoid all seed competitor jealousies, thus more likely produce a consistently high quality product for farmers and to have the financial strength and skilled expertise to survive threatening fluctuations in trade caused as is so often the case in agriculture, by climatic change, floods, drought, economic and man-made crises.

Despite a common New Zealand practice of appointing independent "directors" who have no knowledge of that industry [or the expertise of its staff], appointing favoured associates or ex-employees in retirement simply to "make up the numbers" or to keep costs down or as a reward for past services, being people who may have little skill or experience in advanced business procedure or human relations, the importance of appointment of executive[3] directors skilled and knowledgeable in that industry, cannot be overemphasized.

At CSL too, both contributors were obviously unaware of this importance.

One CSL director, in a letter*[1] to the development manager wrote *"I have always been concerned at the lack of organization when we have our Board meetings"*.

This dysfunction became another threat to survival of CSL [and eventually CSL as such did not survive, but, the work of its technical team—the technology, certainly has done—it has spread around the world].

Uncultivated Pastoral Agriculture

Inoculation of legume seed with nitrogen fixing bacteria has now been an adopted pre-sowing practice for many years in well developed agricultural economies—originally more commonly applied with field crops on cultivated seed beds—such as lucerne [alfalfa] where the host-specific rhizobia {r. *meliloti*}

*[1]That letter is available today.

or soybean {r. *japonicum*} do not usually survive in soil for more than two or three years in the absence of the host plant. That fact meant inoculation of the seed [or soil by liquid inoculants more practical in countries where peat carrier is not available] became essential for healthy nodulation of the crop.

More recently it has become an important advance in world agriculture [unfortunately not yet widespread], that not only crops drilled into cultivated seedbeds, but *really productive new pastures are also able to be established by simple surface sowing of seed on uncultivated seedbeds.*

Fig. 23 A contrast in colour. On this rugged high country in New Zealand at 2,500 to 3,000 ft [a.s.l.] inoculated and phosphate coated seed was aerially oversown without other fertilizer (foreground) whereas the distant native hill was not oversown or topdressed.

This result at the 90,000 acre "Muller" Station in Marlborough was achieved on soils of above average natural fertility, nevertheless shows what good inoculation and a small dose of fertilizer coating on both clovers and grass seed can do. Forage value almost doubled within a year in terms of sheep and cattle numbers carried, and in pot tests root development of grasses increased 120% compared to plain seed both sown in Muller soil without topdressing. The green is due to nitrogen fixation.

This has opened up Land areas around the world previously too steep, too boulder strewn, too wet, too broken [humps and hollows] and sometimes too inaccessible, for wheeled access including cultivation machinery.

Oversowing unproductive native grassland is not necessarily cheaper than a cultivated seed bed sowing, but it is faster, more versatile of ground

Fig. 24 The center hill in this photograph was oversown with Prillcote™ inoculated and coated clover seed and phosphate coated grass seed and topdressed with superphosphate—whereas the lower left foreground and distant hills were not. This is not just a contrast in colour via nitrogen, topdressing and good seed, it is also growing a lot more non leguminous grass than previously and, usually prone to clay slips under heavy rain, these former slips and slumps are being well healed and returned to production as well.

Fig. 25 Another colour contrast. The foreground sown in Prillcote™ inoculated and coated clover and grass seed on the property of Mr. Eric Nicolson, Alexandra, Otago, New Zealand. The distant easier but sun bleached country had not been oversown. The green colour is due to rhizobia nitrogen fixing bacteria from topdressed and oversown seed, not significantly from more rainfall which, limited in this locality anyway, was similar over both the green foreground and the distant brown.

conditions, less prone to windblown or waterborne erosion, is not long out of grazing production, but requires careful planning to reliably succeed.

New Zealand has hundreds of examples of these successful oversowings which are now taken for granted as a runholder would take for granted the mobility and power of modern machinery.

Planning Required

While oversowing onto the surface of existing vegetation is actually a simulation of nature where plants [or through birds and animals] simply drop seed onto the surface of the soil—nature has no "bean counter", is prolific in reproductive volume including seed manufacture, can afford high levels of failure yet still achieve a few successfully established new seedlings most years and seems happy with that. Whereas farmers are accountable to time, often to a Bank or lender and cannot afford high rates of failure of purchased seed and fertilizer [the latter wasted if new grassland fails]. However, overcoming the reasons for natural establishment failure and achieving a high percentage of success at every sowing, is not a simple matter, indeed quite challenging where good results of both seedling establishment—and legume seed nodulation are required. The risks must be minimized.

Consistently good results require careful planning via both careful thought and usually quite a lot of preparatory work—all of which will reap rewards.

Thought and planning will include selection of the area to be sown. A sound principle of farm management is that the best and easiest land should be developed first to return maximum profitability with which to tackle more difficult less rewarding areas. It will also require working out the economically optimum seed species mix, the optimum quantity [see below sowing rates related to % of successful establishment], what the best seed treatment available is, an arrangement if possible to test a sample of that product [selected at random during sowing and if available, laboratory tested for rhizobia survival] a date for manufacture [if it is to be inoculated and, coated] and a date for prompt supply after inoculation—to coincide with seasonal and application requirements.

The work required may include items such as fencing into grazing manageable areas related to stock numbers available, i.e., to use livestock as a "tool" to force hard grazing to "open up" [expose bare soil] native growth prior to seeding, to compel stock to eat competitive weed species and later to provide a natural fertiliser "injection" from livestock. Also spelling an established area {no grazing for a month or two to allow additional growth and re-seeding}—all good management "tools"—just like a good tractor which is also a valuable management tool. Such subdivided areas need to be provided with natural, piped or bore water for livestock. Shelter should be provided either by inclusion of natural cover, or by planting of shelter—ideally, some years before it is required.

Ground preparation by removal of dense scrub or weed, hard grazing just prior to oversowing, identification of the optimum sowing time in relation to subsequent rainfall, danger of post-emergence frost lift but also by overcoming traditionally dry periods when seedlings need to be established to survive are all important. Concentration of livestock on low quality native grass at pre-sowing can be hard on animals but is an essential and formidable tool of development provided care of animal welfare is also exercised.

Timing of these measures—particularly sowing, is critical.

In New Zealand's dry, winter cold, summer hot Central Otago region for instance, late August to early September {early spring} is optimum for oversowing, but March/April {autumn} may be more successful in higher rainfall, milder winter, but summer drought prone North Island East Coast hill country. Every region will have its own climatic, vegetative, physical and soil combinations to deal with.

Because seasons do vary, there is always *some* risk—that of course is "farming".

We can nevertheless reduce the risk by following the principles of the formula as closely as possible.

Germination of Coated Seed

As will be shown in this Chapter, best results from oversowing have been achieved with seed not only inoculated, but also coated with various finely milled nutrient materials—lime, rock phosphate [which is not acidic], dolomite [magnesium], etc.

But what does enclosing seed in these materials do to its subsequent germinating ability? The illustrations below best describe what happens.

From these tests it is clear that coating [as formulated by CSL] not only did no harm to seed viability, in some species and conditions [surface sown] it quite clearly enhances speed of germination—and sometimes total %, not surprisingly because coating partially simulates covering the seed with soil. This advantage is unlikely to be found with seed drilled into soil, but these tests prove that professionally formulated coatings do not harm seed germination in general—and enhance it when seed is surface sown.

Similar tests have been carried out with legume seed not only coated with these nutrients, but with peat based inoculants [and other carriers]— also without damage to germination—and with an improvement when surface sown. Whatever the treatment, seed must not be left in damp storage conditions.

To date hundreds of thousands of samples of inoculated and coated legume seed and of coated grass seed have been grown in laboratories and in the field throughout the world, we have yet to hear of a single adverse effect on germination from correctly processed and dried product.

Fig. 26 All surface sown at the same time, using identical seed, soil and moisture with above result at 14 days.

Early Coated Seed Ltd germination tests with identical seed (taken from one seed line [i.e., a consistent quality] in each case at random, half being inoculated and Prillcote™coated)— [top left] subterranean clover [top right], lucerne (alfalfa),and [bottom left] white clover then red clover. Coating the seed often hastens germination as shown here with the Prillcote™ sample on the left and the plain seed on the right—in all four treatments. Although bare seed catches up in time, coated treatments sometimes reach a higher germination % probably due to activating more "hard seed" earlier. These treatments were all surface sown simulating oversowing.

Having established that we have developed the best oversowing product possible [shown to be the case later in this Chapter] and because that product's greatest value is in surface seeding—more so than when drilled into soil, the next most important factor, is to determine the optimum overall procedure for maximum success from surface seeding [oversowing].

Fertilisers, Mixing and Avoiding Damage

Achieving satisfactory soil fertility in an oversowing situation will usually require supplementary correction of major and/or minor elements so organizing a soil test, subsequent fertilizer transport and then topdressing— sometimes combining fertilizer with the seed, is advisable if not essential.

Thus, on-farm mixing of seed and fertilizer right at time of sowing needs to be arranged. Mechanical mixing must not damage the coated seed [steel augers employed to mix plain seed often do knock the coatings off coated seed, creating much dust, all of which is detrimental] however, recently developed more stable coatings are better able to withstand mechanical mixing.

Coatings resembling a firm, shiny pharmacy pill will better withstand mechanical handling than a more loosely attached unstable and powdery coated seed—but better survival of rhizobia is more important than stability of coat as mixing can be adjusted to cope. An ordinary concrete mixer with fertilizer applied into it first with a very short mixing time after coated seed is added will often suffice. For smaller quantities, a wide mouth shovel for mixing on a concrete or wooden floor can also succeed without significant damage to seed or coat.

The most expensive elemental fertilizer—nitrogen—should be substituted by inoculation with rhizobia. Indeed, artificial nitrogen [e.g., sulphate of ammonia] is not found naturally in deposits like the other 5 major plant food elements—phosphate, magnesium, calcium, sulphur and potassium. If it were, and we applied it as topdressing onto native grassland when oversowing, it would do quite a lot of harm, i.e.:

- It could not be added to the coat of coated seed due to inoculant damage and as fertilizer would simply boost ALL plant growth including the competitors of our delicate newly introduced oversown species.
- It would suppress clover growth which happens when artificial nitrogen is introduced—because rhizobia activity is switched off by the presence of adequate alternative nitrogen in plant tissue.
- It would add considerably to the cost of development being the most expensive fertilizer of those usually required for grassland development.

The goal is to have an inoculated clover seed product which can maintain *Rhizobium* viability despite delays between inoculation and sowing, a product which can also hold viability of both inoculant and seed itself even when mixed with materials such as acidic superphosphate.

Fertiliser can be applied a week or two prior to application of oversown coated seed but that may [depending on growth conditions] also tend to stimulate existing grass growth in competition with the seed to be oversown.

Neither do we wish to pay for an aircraft [or spinner, etc.] to pass twice over our development area just to apply inoculated seed and superphosphate separately—to keep inoculant separate from "super"; ideally we need a product which can be held >3 weeks [for delays of transport, weather, etc.] then be mixed with fertilizer briefly [preferably granulated fertilizer] and still deliver a healthy dose of both inoculant and seed viability to the oversown land development investment.

Fig. 27 Superphosphate and coated seed are dispersed to a known width and known rate of spread but is subject to wind direction and strength, height and speed of aircraft—all under control of an experienced pilot. The fertilizer seen leaving the aircraft here is dusty, not granulated, the "fines" can travel long distances, undesirably onto neighbours land, into riverbeds, the sea or scrub and bushland. Accuracy is essential.

Aerial Considerations

While good progress has been made in the manufacture of higher quality coated seed and specialised fertilizer products in the latter 60 years, other problems have developed. The declining number of fixed wing aircraft remaining in aerial oversowing and topdressing in New Zealand, the increasing average age of pilots with very few new trainees coming on are large concerns. It has historically earned the reputation of being a dangerous occupation where cost cutting has encouraged unsafe practices such as overloading, inadequate aircraft instrumentation, flying in both substandard weather and daylight conditions and shortcutting the level of and provisions for training young pilots.

Helicopters are becoming less costly and have advantages over fixed wing—which may become the immediate future of the application industry— and who knows what else? It is an industry which must continue, but pilot deaths to date in New Zealand are deeply regrettable and will be largely avoidable if all participants closely observe all of the rules at all times.

So these qualities of 2–3 weeks inoculant-safe retention and safe fertilizer mixing sounds easy? It has been the subject of intensive research for over 50 years. Hopefully this book will save developing Nations from having to spend another 50 years [and a great deal of money] on reaching the same goal.

We do know how to make such product now—that is a chief subject of this book.

Not surprisingly, the full narrative account of seed, inoculation and precision pelleting has not been published elsewhere for two main reasons, past commercial secrecy [now terminated by High Court Order obtained by author Bennett] but also because both the author and contributor of this book have "grown up" with the full and somewhat unique experience of the complete technical, commercial and agricultural progress to recent date in this field. We know of no one else still living who has done so or who could write all of this book. The world will benefit from the sharing of this technology.

"The Wizard" of Aerial Oversowing[*1]

In temperate climates such as in New Zealand [NZ] where seasons are reasonably positive, rainfall usually reliable and temperatures [below 1,000 m altitude] not extreme, there has nevertheless been the need for a carefully researched formula for ensuring optimum results from oversown grassland establishment. That field research has been very successfully achieved by New Zealand Department of Agriculture Scientists—chiefly Dr. Bill Lowther [retd.] formerly of Invermay Research Centre, Mosgiel, Otago, and his technicians. Dr. Lowther's work has been absolutely essential for successful development of consistently reliable aerial seedling establishment into hostile native grassland environments in New Zealand. He can appropriately be titled *"The Wizard of Aerial Oversowing"*—on a world wide basis. His work, different to, but of similar National magnitude to that of his eminent predecessor Sir E. Bruce Levy [former Director of the former Grasslands Division, Department of Scientific and Industrial Research] who is reported in his famous book "Grasslands of New Zealand" [3rd edition 1970] to have stated:

> *"The New Zealand farming Industry still needs more and better grass, more and more stock ......"*
>
> [published by Govt. Printer 1951]

to which Dr. Lowther has certainly made a greater contribution to our knowledge than any other New Zealand grassland establishment expert. His work has piloted the way for very successful commercial development benefitting New Zealand farmers Nationwide. More importantly, his meticulous investigations have been of an extremely high and searching

[*1]The "wizard" being our description of Dr. William Lowther [Ph.D, University of Western Australia], Scientist, New Zealand Department of Agriculture {later "AgResearch"} Mosgiel, Otago, New Zealand, (retired).

professional standard which New Zealand has been able to totally rely on, earning multi-millions of dollars for farmers and runholders—much of it unrealized by those farmers, runholders—or by successive Governments. Public formal recognition is well overdue.

He wrote classic work on oversowing. By grasping an understanding of reasons for adopting key items of Dr. Lowther's "formula", we can modify it to suit other climates, other soils, other altitudes, provided conditions of soil, climate and freedom from predators exist which are not prohibitively extreme. Hill and high country which can be successfully sown with good seed able to establish when accompanied by effective pre and post sowing management— and of course some rainfall, ideally around 750 mm a year and reasonably spread—but as low as 450 mm a year and seasonal, hopefully reliable and not a total drought [which New Zealand has experienced regionally as elsewhere in the world].

Dr. Bill Lowther was not the only New Zealand scientist who worked successfully on legume seed inoculation and surface establishment of clover and grass seed, there were others such as P.M. Bonish, Dr. David Scott, Ewan Vartha, Peter Clifford, J.A. Douglas, Prof. Jim White, Nelson Cullen, Peter During, Dr. Ron Close, R.J.M. Hay, Heather Patrick and several others including University researchers.

However, there is no doubt that Dr. Lowther was pre-eminent in this field—particularly concerning inoculation of oversown clover which he has recently claimed is only necessary in virgin or barren situations in New Zealand now, due to previous and highly successful widespread use of inoculation in New Zealand.

Fig. 28 Oversown white clover nicely established among sheltering tussock. Dr. Lowther has shown how important shelter is to establishing seedlings—see Tables 8A and 8B in this book.

That situation may well be true at least until infectively proven new strains of *Rhizobium* become available commercially, but does not apply where newly introduced cultivars require specific strains of *Rhizobium* such as hexaploid Caucasian clover (*Trifolium ambiguum*).

Formula for Successful Surface Sowing of Seed

On New Zealand's hill and high country, oversowing is carried out up to about 1,000 metre [3,280 ft] altitude, where even at 300 m winters are cold and native grassland occasionally covered in snow during 3 months of winter.

Cold moist winters however, are followed by drying warm winds in spring and long hot dry spells over summer [annual rainfall ranges from 450 mm in the traditionally driest localities, and is poorly spread up to about 700 mm average, winter wet and summer dry, and 1,200 mm on higher and more Westerly facing hill and mountainsides with more even distribution but colder climate]. Overall, this is a significantly hostile and challenging environment for establishment of new grass by aeroplane—it is nothing like the more favourable lowland with its arable grassland, much of it in dairy production for which New Zealand is well known, and of which, substantial areas of that easy country are irrigated. In addition to the hostility in general, for establishment of inoculated seedlings, shady hill faces with fern, tussock or other cover usually give better results from oversowing than exposed and open sunny faces which get very dry and very hot in summer.

The formula for optimum oversowing success in such hostile hill country conditions is:

Firstly, Prepare a Plan: In addition to the need for clean water for livestock, secure fencing, hopefully livestock shelter, predator control -
- it should also include these
5 essentials:-
1 ☑ Seedbed Preparation
2 ☑ Soil Test. Use Clean, Dry Flowable [or granulated] Fertiliser.
3 ☑ Accurately Calculate and Order [Proven] Seed Treatments.
4 ☑ Ensure optimum Time of Sowing.
5 ☑ Use freshly inoculated and quality proven coated seed. If mixed with fertilizer or lime, mix immediately prior to use.

1. Seedbed Preparation

Removal of dense and unproductive cover is necessary, such as thick manuka and native scrub, gorse, lupin, sweet briar, blackberry and other aggressive, dominant and hostile weeds which not only survive but indeed thrive on low to medium fertility soils where there is little competition.

Worse, *most weeds respond to fertiliser.*

In particular, legume weeds such as gorse and lupin which "fix" nitrogen—and others like rhizomatous sweet briar, also vigorously respond to fertilizer intended for grass. Their eradication may wisely have been carried out over several years prior to a long term development plan which would also, as mentioned above, include fencing into manageable size blocks for good stock control.

It does not include removal of helpful harmless native ground cover/shelter plants such as tussock or bracken fern which generally improve survival of oversown seed by protection from hot sun and drying wind as well as retention of surface moisture and providing frost protection. It also does not include total removal of stabilizing woody plants such as manuka or trees on steep potentially unstable slopes which are vital in controlling erosion—specially in higher rainfall areas with unstable clay soils widely found in the North Island of New Zealand. Such stabilizing cover also offers shelter and shade to livestock. Oversown grassland must work in harmonious association with natural phenomena—not against it.

When "cleaning up" a native area to be oversown using livestock, they are being forced to eat everything remotely palatable to survive, so a careful watch needs to be kept as to the maximum period they can be confined to the area to be oversown without inflicting health damage such as significant weightloss, wool growth "checks" or vulnerability to disease and cold.

Make certain there is adequate clean drinking water available for such livestock and remove them to a sheltered better feed situation in the event of a forecast storm because you are using the good condition of livestock—usually sheep and/or goats as "an implement" to control and improve the condition of the area to be oversown—but in so doing will inflict hard conditions on them which will become marginal in bad weather.

Chemical desiccation of existing foliage by spraying a wide spectrum weedkiller such as glyphosate, even at a reduced concentration to effect dormancy [not death] would be expensive, not just because of the cost of the chemical, but also due to the weight of water dispersant required—a prohibitive cost for most engaged in aerial application. That may also kill sensitive beneficial native grasses or shelter plants and is unnecessary if the use of livestock, for control of competition to establishing seedlings—is managed effectively. But where thick concentrated infestation such as gorse (*Ulex europaeus*) is the problem, there may be no alternative to spraying, then crushing by concentrated heavy cattle—or a roller run downhill off a bulldozer winch and/or burnt where fire control of fences and shelter allow and where environmental concerns can be alleviated. Removal of thick debris by fire is usually helpful prior to oversowing, but not only is good fire control necessary [often a permit required from the local authority so Firefighters {if any} are made aware of the burn], but the fire will sometimes massively germinate subsoil seed [for which gorse is notorious], thus burning must be followed up immediately by oversowing and topdressing to stimulate seedling grass growth [because fertilizer will also stimulate gorse growth]—thus allow early grazing to clean up gorse seedlings as well, while they remain soft and grazeable [for which goats are valuable]—with care for the new grass. Re-spraying of thick gorse re-growth after initial clearing and burning—but prior to oversowing, while expensive, will secure oversowing more successfully.

The whole idea is to introduce better grasses and nitrogen fixing legumes into native foraging as cheaply and effectively as possible to begin a gradual improvement in the quality of feed and in the quality and numbers of livestock which graze it.

2. Soil Test [Order Clean, Dry Flowable [or granulated] Fertiliser]

Unless the developer is already familiar with soil fertility of the area to be sown, a random representative soil test should be arranged whereby the top 5 cm is plugged and bagged and then another auger placed in the same hole calibrated to remove the next [deeper] 5 cm—also carefully bagged separately in pre-labelled clean plastic bags. A comprehensive test will reveal soil pH [acidity or alkalinity] and levels of phosphate, potassium, calcium and maybe magnesium at both depths of soil. The top level is more important for fertility correction but it is important to know what the fertility level is lower down. It can take several years for surface sown elements—such as broadcast lime [calcium carbonate] for instance, to move down naturally through soil from the surface to just 10 cm deep, even under average rainfall. There are many laboratories worldwide which offer soil testing services and from about 2004 kitset testing has been offered Internationally with details shown on the Internet simply by Googling "Soil testing". It is important to ensure the technical competence of the tester.

Aerial liming is often considered uneconomic, however, measurement of the benefit needs to be related to time—at pH 5.6 or lower, liming will eventually be beneficial to grassland—we just don't know when "eventually" will be. Application of molybdenum as molybdic superphosphate [or as molybdic coated seed] can [in some situations[4]] also replace some or all of the value of liming—and at a fraction of the weight. Other trace elements, boron, manganese—for plants, or cobalt, selenium and copper—for animal health, are usually applied with fertilizer or separately as granules and would need to be rhizobia-harm tested before being included as part of inoculated coated seed [copper for instance is very likely to be lethal to rhizobia].

Both seed and particularly rhizobia have been shown to benefit from the small but precisely placed finely milled calcium carbonate contained on an inoculated lime coated seed. In relation to establishment, the tiny quantity of lime in the seed coat has extraordinarily been found equal [during establishment] to standard broadcast lime on moderately acid soils [Lowther 1975].

A Timely Note About Soil and Fertilisers

Nitrogen is not normally tested by foliage analysis or soil plugs in undeveloped native soils where we can be reasonably certain nitrogen levels will be low, however, if the legume seed is effectively inoculated, nitrogen will be supplied from the atmosphere—an enormous resource equally available right around the World—but sadly, not understood in many parts and therefore a neglected asset.

Due to cost, fertilizer [and occasionally lime] are traditionally only sparingly applied to surface sown or oversown hill country grassland, but even a small quantity of a deficient element can change failure into success. The lime coat on legume seed mentioned above—particularly where sown on an acid soil [which at a normal seeding rate would supply only 10 to 12 kg per hectare of lime] has an impact on nodulation and subsequent growth [at 2 years] equal to the traditional broadcast weight [usually not less than 1 tonne/hectare] being many times the coated seed weight, *simply due to fine particle size and precision placement with seed as shown in field research.*

The authors would expect that longer term, the tonne of lime per hectare would catch up and exceed the liming value of the lime coated seed—but what use is that if there is sparse or no grass there to benefit? If the lime coating achieves establishment and early growth—then it will be more affordable to apply bulk lime [calcium carbonate], or dolomite [magnesium carbonate] subsequently.

Some soils have the disadvantage of high levels of "neutralization" of major plant food, specially *phosphate* where ionic bonding can make elemental phosphate unavailable to plants. In such situations, as much as 600 kg/hectare of superphosphate can be bonded to iron or aluminium oxides making it initially plant food unavailable thus requiring further levels of "Super" to be applied to achieve availability to plants as plant food.

Such soils are expensive to develop, in New Zealand a prime example being the yellow-brown loams of the Te Anau/Manapouri region of North/West Southland comprising glaciated debris containing greywacke and granite—the latter essentially a feldspar made up of various elements including aluminium silicate which bonds with phosphate rendering it nutrient-inert.

In field research carried out on these soils by Scientists Peter During, Nelson Cullen and local Farm Advisory Officer Gerald Bennett [all of the New Zealand Department of Agriculture] published in *"The New Zealand Journal of Agricultural Research"*, Vol 5 No's 3 and 4, June and August 1962, a phenomenal and almost lineal increase in dry matter production was discovered on these new soils [above 4 cwt per acre] by increasing applications of up to 9 cwt of "capital superphosphate" per acre applied in the first year [with other trial evidence of *responsiveness to increasing quantities all the way up to an unaffordable 15 cwt super/acre in the first year!*].

A

Fig. 29 A. The top picture shows expensively cultivated and drilled new grass [all dead] despite application of superphosphate and lime—A TOTAL FAILURE. Drought played no part in this result (see water lying in puddles in this high rainfall area).
Photograph by G.M. Bennett, Farm Advisory Officer, Department of Agriculture, Northern Southland.

B

Fig. 29 B. The bottom picture shows inexpensively aerial oversown new grass (no cultivation) also topdressed with superphosphate but less lime (than above sowing)—located within 5 km of above picture: A TOTAL SUCCESS!

Note: The author pursued and took photograph B for his own records specifically to complement photograph A—knowing we may never again see such a contrast under identical conditions now that the importance of inoculation is known in New Zealand and quality products now available to deal with it. These two illustrations are classic and unlikely to be repeated—with photograph B identifying the event, its authenticity and its location by inclusion of the author's vehicle in the picture. Both A & B were made available to CSL for its Technical manual published in 1974, edited by this author.

The importance of inoculation

The two illustrations on the previous page could not be more convincingly graphic:

Both, A & B are closely located to each other at the former Lands and Survey Department's 1960/70's land development scheme at Te Anau, Southland, New Zealand.

And the reason for difference?

Apart from expensive heavy cultivation out of native bracken fern and tussock in A and simple oversowing in B (see the tussock and some matagouri sticks remain undisturbed)—the difference was simply this----
THE CLOVER SEED IN "B" was: *INOCULATED & COATED*.
This was a lesson well learned !.

The only developer who could afford such heavy superphosphate applications was Govt's Lands and Survey Department which, guided by above research and using well developed skills of its own Staff from previous projects, particularly that of District Field Officer [later Superintendent of Land Development, David R. Marshall] transformed that native wilderness of many thousands of acres into productive clover rich grassland creating around 56 excellent productive new farms.

This Te Anau/Manapouri land development scheme, after suffering a large grassland establishment failure at its first sowing,*[1] was subsequently guided by urgent Department of Agriculture field trials, became transformed into a huge success, a credit to all concerned with it, including the Government of the day endorsing the fact that the most successful developer of land in difficult environments is likely to be [adequately resourced] Government.

In both A & B photographs (above) grass seed was included with clover, but in (photo A) ALL the new grass is dead because the clovers did not become nodulated (there were no rhizobia in those soils) so the clovers became starved of nitrogen—an essential nutrient in plant growth. Because the clovers did not supply them with nitrogen, the grasses also died. This whole area was subsequently re-sown with Prillcote™ coated seed—an expensive lesson. In photo B only Prillcote™ coated grass and inoculated/coated clover were used when research identified the problem.

Responsibilities of Government

This fact seriously challenges recent political leanings in New Zealand toward private enterprise as being inherently "the most economically efficient

---

*[1]Local farmers were dismissive—"We told you so!"—their earlier attempts to grow grass there had also failed.

manager" of large enterprises. In New Zealand, Electricity generation and distribution, Tele-communications, Railways [sold then purchased back by Govt.), Forestry and Mining ownership and development, Agricultural advisory and financial services to farmers and growers, Agricultural and Industrial Research, National Apprenticeship and Industry training and many others—all used to be provided solely by Government—but are now wholly or partly privatized. Either amalgamated with a profit not service modus, some now minimized, non-existent, neglected or, as in the case of Railways, re-instated after unsuccessful privatization.

Some Governments fail to see that the much heralded profit goal can be enhanced by Public Service support of wide spectrum private enterprise. Many people will succeed financially given a helping hand to get started. Take a look around the globe. Which nation is bounding ahead in growth despite these days of fragile International economics?—China of course with its cleverly combined socialist and private enterprise policies. State owned enterprises (SOE's) which are functioning as private enterprises but remain largely Government owned do fulfill a dual role enterprise and have the added protection that where they are operating in specialized fields (such as AgResearch) any financial deficit can be alleviated by Govt. support in the National interest.

In New Zealand, private enterprise contractors have tended to adopt "the cheapest" way to maximize profitability, not long term best practice (which resulted in thousands of recently built "leaky homes" in New Zealand for example). There is a constant commercial temptation to take shortcuts, and with those savings, instead of improving the service, pay itself inappropriately handsome rewards, specially to shareholders, ceo's and directors. These levels of remuneration are frequently neither necessary in a small politically placid country like New Zealand, or justified because they have the ultimate effect of escalating costs for everybody else thus fuelling inflation.

Government can afford adequately but sensibly salaried, qualified and skilled staff who adopt professional standards, initiate, interpret and follow local research as their guide. Such State managers are usually more able to deploy large financial resources from which to develop the asset, for proven results.

Govt. can also sustain delayed returns—not easy for private enterprise where there are shareholders focused on early profitability, but in land development for instance, can eventually provide a wonderful permanent asset to be divided sensibly into individual, economically viable privately owned farms sold to adequately experienced private farming families [selected by ballot if necessary] under suitable loan schemes, which then begin to return profitability via income tax [for evermore] to "the taxpayer" which funded the project initially.

Such land development has been hugely successful in New Zealand and there is no good reason that it might not perform equally well in many other countries. The main ingredient being engagement of proven skilled and qualified staff with the integrity, energy and vision to succeed, backed by sensible adequate financing with freedom from negative bureaucracy and from any form of corruption. Further training of adequately experienced future farmer occupiers could and where necessary should be run parallel with development of the land. Modest minimal production levels should be established for each State developed farm which if not met each year should in fairness lead to enquiry and if not remedied, lead to compulsory recovery of the asset—compensating the failed occupant the value of any improvements paid for by that occupier.

While huge quantities of superphosphate topdressing gave spectacularly good results at Te Anau, the Lands Department sensibly capped subsequent development to a maximum of 12 cwt/acre of capital superphosphate in the first year with a follow up maintenance topdressing of 2 to 4 cwt per acre in the subsequent year.

Today, some of those pastures are the greenest and most productive in New Zealand—not only are the clovers exclusively nodulated with introduced elite strains[1] of *Rhizobium* [because there were virtually none in the soil originally] but, in addition, while these pastures are regionally located where they will always be winter cold, sometimes –10 or –12°C {heavy frost} and 50 to 100 mm of snow on low levels for a week or two—they nevertheless also benefit from a reasonably evenly distributed rainfall—a relative rarity in New Zealand with quite hot summers.

Ionisation (Covalent bonding) and Phosphate "Fixation"

Other soils around the world are likely to suffer the expensive chemistry of covalent bonding through ionization presenting a cost barrier to production.

An "obvious solution" would be to apply some less expensive element than P {phosphate} to "feed" the bonding capability of soils notoriously laced with aluminium and iron oxides—however that is not possible under present knowledge. It is also a different process from "phosphate fixation"—an exchange of phosphate and hydroxyl ions.

The following explanation of covalent bonding of phosphate in New Zealand soils, and a note about "phosphate fixation" has been kindly provided to the author of this book at his request by Dr. A.H.C. Roberts who is Chief Scientific Officer of "Ravensdown", New Zealand's largest manufacturer and supplier of Superphosphate. [*www.ravensdown.co.nz*]

> *"New Zealand soils are naturally poor in available phosphorus (P) for supporting intensive pastoral or crop production. In order to reach a soil's productive potential nearly 100% of New Zealand soils (there are areas of recent soils in Northland and the Waikato which are exceptions) need continuing annual applications of P fertilizer to maintain productivity.*

---

[1]Elite strains, because they were selected for quality by the Department of Scientific and Industrial Research and supplied to Inoculant Manufacturers.

During initial soil development (from the unimproved state) large amounts of P (capital applications) are required to build the pool of plant available P in the soil, once this has reached the optimum level, then smaller annual additions of P are required to replace the P that is sold in product, transferred to non-productive areas of the farm (by animal dung) and some inevitable soil losses.

The amount of capital P required varies greatly between the major soil orders in New Zealand. This is due to the differing amounts of clay minerals containing iron and aluminium oxides which bind strongly with P. We can differentiate between soils using a laboratory test called the Anion Storage Capacity (ASC), formerly the phosphate retention test. Simply, soils with high ASC will require more capital P to achieve the optimum soil P status than a soil with low ASC. The volcanic soils of Taranaki, Northland and Waikato typically have ASC ranging between 70–99%, while the sedimentary soils of the South Island will typically be 20–40%. The ASC should not be interpreted to mean that 70–99% of P in volcanic soils, or 20–40% P in sedimentary soils is 'locked up' or 'fixed' and permanently unavailable to plants. The P added specifically binds to the iron and aluminium oxides but is exchangeable with P that is in the water between soil particles, which is where plants absorb their P from.

It is not possible to avoid this retention of added fertiliser P in soils by additions of some other nutrient or mineral. The demand by the soil for P (in these specific adsorption reactions) must be satisfied before much P becomes plant available.

Phosphate retention must not be confused with 'phosphate fixation' which is a separate process whereby small amounts of P does become permanently 'lost' to plants as it becomes part of the soil particle. Recommendations for both capital and maintenance P take the small quantities of P fixed into account.

Possibilities and Limitations of "Pasture Pellets"

Yet another possible solution may be the development of "pasture pellets" [p/pellets] in which the soil–deficient element [say phosphate] is applied at several times the weight of seed in the form of a large pellet. Work to date on pasture pellets using reverted superphosphate at 5 times, 10 times, 15 times and 20 times the weight of the seed attached [which would be either inoc./clover or grass seed] has surprisingly not yet been found sufficiently beneficial to cover the extra cost of production.

More research is required—particularly out in the field and over an extended period of time to find not only the cheapest and most effective pellet formulation—variable according to local conditions, but a really effective means of application, then to compare the cost/benefit ratio—not only of establishment, but also longer term [e.g., >5 years or more].

Fig. 30 Surface sown grass seed pasture pellets under laboratory trials.

Because they are much larger than coated seed, p/pellets may not be capable of reaching the same terminal velocity when aerially spread but be more likely to sit up on top of existing foliage if surface sown into native cover.

However, in granulated fertilizer research, larger particles around 4 mm diameter have shown better velocities and spreadability than micro granules of less than 1mm diameter. This situation will need to be further investigated. Perhaps also by shaping the pellet, higher terminal velocities may be possible.

In New Zealand, a M.Agr.Sc. thesis[5] [summary] found that while coated seed penetrated through a chemically killed sward twice as successfully as bare seed in calm sowing conditions, this difference was evened out when foliage was disturbed by hand [in research plots]—or by wind. Unwelcome levitation might be overcome for instance by use of power projection—i.e., a large [say 20 mm barrel] compressor powered air gun which fires pellets from a vehicle or aircraft at high velocity into seasonally selected soft wet soil {ideally intensively broken up by stock hooves} penetrating both existing foliage and placing pellets into and onto the surface. More work is required because these pellets and the principle of precision placement have a lot to offer potentially once optimum formulae have been developed.

Guess work? Yes indeed, however, because it has been proven that a tiny amount of finely milled lime coated onto seed can produce the same effect as conventional broadcast lime on both establishment and dry matter production of white clover for more than two years from oversowing [on some soils] as reported by scientist Dr. Lowther, then it would seem reasonable to anticipate that a relatively small quantity of finely milled reverted superphosphate [or other nutrient] precision placed with seed on any soil deficient in that element—[indeed much more so when those soils also display a capacity to inactivate a substantial proportion of traditionally applied fertilizer]—might give an establishment and subsequent growth response well above the response level from randomly topdressed fertilizer.

It is a question of cost in relation to benefit.

Fig. 31 A similar concept to pasture pellets is the granulation of crop seed which may also be sown on the soil surface. These are wheat granules on the right and bare grain on the left.

Pasture Pellet Limitations

The principle here is that we are not attempting to fertilise the whole area [say 100 hectare] to 50 mm depth, that would contain 50,000 m^3 of soil which at [say] 75% the weight of water would weigh about 37,500 Tonne [@ 1,000 kg per tonne = 37,500,000 kg of soil].

As we would normally apply about 500 kg/hectare of fertilizer with a new grass oversowing [about 4 cwt/acre] that would apply 500 kg x 100 ha = 50,000 kg which is just 1 kg of fertilizer to permeate each 750 kg of soil—or 1 gram of fertilizer [which would easily fit on a pinhead] to enrich each 750 grams of soil [which is the weight of just over 1½ pounds of butter—or about 4½ cups of flour], a ratio of 750:1 being a huge dilution of an essential element. In addition, because topdressing is randomly spread, some of it feeds associated weeds, some is leached below root access or washed away on hillsides and some is lost when massively exposed to ionization and "locking up".

But if we concentrate fertilizer with the seed by creating a pasture pellet weighing [say] 20 grams of elemental fertilizer plus at least 1 seed [clover or grass] who's roots initially colonise an area of [say] 10 cm square, i.e., 3.16 cm x 3.16 cm = [approx.) 10 cm^2 by 5 cm effective depth [in the first year], that will be 50 cm^3 of soil (.00005 m^3) weighing [at say 750 kg/m^3] about .037 kg [or 37 g soil] which with 20 g of fertilizer provides a soil/fertilizer ratio more like 2:1, e.g., 375 times the effective fertilizer ratio to soil [as 750:1 above] in conventional topdressing. More importantly, if we sowed 4 kg/hectare of white clover that will apply about 5.28 million seeds/ha [e.g., per 100 million square centimeter] which theoretically provides one seed for about 20 cm^2 [about 8 inches square] which is roughly the area a good white clover plant and its root system should occupy at maturity.

That 5.28 million seeds/hectare x 20 g per pasture pellet with [say 2 seeds attached to ensure at least one grows] = 2.64 million pasture pellets each of 20 g of fertilizer = 52.8 million grams [52,800 kg] of fertilizer to be spread which would provide for—a whopping 52.8 Tonne per hectare!.

Obviously this is a grossly excessive quantity of fertilizer, so lets see it from the seed perspective, how many 20 g pasture pellets would be required to apply just 500 kg of fertilizer per hectare? The answer is 25,000 which at 2 seeds per pellet would only apply 50,000 white clover seeds/ha—or [at 600,000 white clover seeds per lb or 1.32 million/kg] only about 1/3rd of a kg of seed per ha and that has traditionally been quite inadequate.

If however, we sowed a much smaller pasture pellet than 20 g—but designed so that it also "force fed" the clover, still excluding some of the mass exposure to covalent bonding or "locking up", each p/pellet comprising a core of [say] 4 plain white clover seeds—that product may find some middle ground between the [above] two extremes.

These could be manufactured various ways, perhaps "snap" suction lifted off a flat moving bed of seed by a tube complex, then [reverse air] injected into the tubes comprising putty like pellets—then extruded and chopped off to size. These pellets made of nutrients [hammer milled and screened] such as reverted super fertilizer, lime [or whatever is required] dried with warm air and stored pending sale. When sold this product would be freshly, promptly and easily oversprayed with a large loading of elite rhizobia culture without necessity for suppressing poly-phenols using expensive biological protectants such as polyvinyl-pyrrolidone because the seeds, inside the pellet, are not in contact with rhizobia. The overspray would include very little moisture or, use a volatilising medium or alternative dispersant such as a bio oil with no moisture; the product not wet, not requiring drying, seed inside the pellet protected from any moisture outside it, bagged and shipped off for oversowing.

This would be sown at about 100 kg/hectare providing everything required, e.g., total topdressing [OF THE SEED—not the soil] for initial establishment, adequate seed, concentrated fertilizer against "locking up" in difficult soils, lower total handling weight, faster and safer aerial [or land] application, better ballistic properties [with added pasture/pellet weight], much reduced overall cost and above all, very high numbers of elite rhizobia (where required) injected into that new grassland right alongside that seed. Two possible refinements may be 1/. That the centre of the pellet is fertilizer but an outer layer on each pellet comprises a [yet to be discovered] inexpensive "ionizing anti-dote" which would slough off the pellet first and "fix" onto bonding and ionizing elements in the soil. And 2/. The pasture pellets may be [by extrusion] shaped for optimum ballistic effect permitting a faster terminal velocity, presumably shaped like a double nosed bullet and maybe requiring weighted balance.

At say 200 kg per hectare, this would apply about ½ the normal fertilizer topdressing but through precision placement, a much higher rate per plant [after all, we are topdressing the plants, not the hectares] and at [say] 5 g per pellet unit [plus seed and inoculant] would apply 40,000 p/pellets per hectare [ha = 100,000,000 cm^2] = 2,500 cm^2 to each capsule with [say] 4 seeds which would apply 160,000 white clover seeds per hectare [about 20 grams of seed per hectare, not the 2 kg to 4 kg normally sown], so, is well below requirements. In addition these seeds are clustered not spread and some will not succeed. Assuming say two of those seeds became successfully established per pellet on average, that still remains a minimal seeding rate of 1 successful seedling per 1,250 cm^2 [125 m^2 or about 36 cm x 35 cm] which is even more inadequate—and still "clustered" with one plant on top of another—a further impracticality. From these analyzing calculations however, there do appear to be some possibilities.

Pasture Pellet Possibilities

A/. That an inexpensive "pioneer pasture pellet" for totally undeveloped land or "renovation pasture pellet" for basic grassland improvement might be used to introduce an initial light seeding of [say] white or subterranean clover into a situation whereby this "superpellet" contains everything required to successfully establish miniature nitrogen factories each containing an elite strain of *Rhizobium*, an important dose of fertiliser, lime if required, two good quality seeds, ballistic properties via the pellet weight—with lower cost [and risk] requiring no fertilizer other than as in the p/pellet.

At this lower cost, with precision placement of both rhizobia and fertilizer [critical in low fertility soils] these p/pellets would identify the degree of success of such legume establishment in nitrogen starved situations and where minimal but successful introduction would create a more certain grassland environment and basis for subsequent pasture establishment and improvement—either using normal coated seed and fertilizer, or repeated sowing of p/pellets.

And B/. That a much superior sward of introduced clovers and grasses than has been achieved in the past be established via a "blitz" surface sowing of a range of clovers and grasses as p/pellets where an application rate of [say]

1,000 kg/ha [approx. 8 cwt an acre] or more is acceptable [because the pellets may also contain the added weight of lime or dolomite in acid soil situations] and where a rapid transformation of grassland quality is required. This "Big blitz" approach may possibly be justified for instance in urgently upgrading hill country for dairy "run-off" [off season grazing for 500 to 2000 dairy cow herds which saves damage to home pastures in wet seasons or where flooding has occurred]. Innovators will blaze this trail when the product becomes available.

There are many more possibilities and opportunities in addition to pasture pellets which need close examination, albeit at a cost. In relation to enormous research cost, agricultural systems in undeveloped farming countries have benefited from a slow start because they are now able to adopt published scientific core research such as recorded here avoiding the high cost of intensive laboratory, factory and field investigations, most of which in fact fail to produce economically worthwhile benefits.

Customarily, only a very small % of research is ultimately rewarding, but the rewards can be huge when successful.

On some soils, trace elements like molybdenum, when deficient and applied at very low rates with seed, can make an important contribution. This is certainly true where it has been found that growth response to traditional liming of some soils has in fact been simply a response to the small quantity of molybdenum contained in that lime [calcium carbonate] and/or its steadily greater availability under increasing pH. In such situations, provided pH is satisfactory, liming can be discontinued and small quantities of sodium molybdate applied with the seed coat or as an additive to other fertilizer, i.e., molybdic superphosphate with equally satisfactory results. However it is the seed which needs the molybdenum—not the soil.

Materials and Test Services Essential for Development

Human populations around the world need higher food output—but if the needy cannot pay for it, an option for some is to grow it themselves.

Soil tests and the local commercial availability of fertilizers or lime [calcium] or magnesium [dolomite] to supplement soil deficiencies plus availability of effective strains of *Rhizobium* as commercial inoculants are absolutely fundamental items of National importance in any Society in improving undeveloped agricultural systems which, to help "kick start" more intensive and successful farming, may need to be provided initially by Government and Aid Agencies in financially impoverished nations.

Soil tests plus inoculant product tests are both well publicized manageable services already available in various locations around the World. Education in their effective use will need to accompany availability; field trials using locally preferred crops or grasses comparing • time of sowing [specially for surface sowing of grassland] • rates of fertilizer • inoculation and coating of legume seed • inoculation of drill row in arable crops with granulated or liquid rhizobia cultures, always using bare seed [or the locally adopted method of sowing seed]—as a "control"—or measuring base, are necessary not only to find successful local formulae, but also to convince local farmers that there is a better way. Small scale trials do not cost much and provided they avoid errors and truly represent the "real world"—can provide accurate information and avoid waste.

Good QualityInoculant Manufacturers Need Encouragement

There are now restrictions in some countries as to transfer of soil and organisms across international boundaries, however, a controlled activity such as soil and rhizobia tests where samples are securely packaged and handled by authorized and professional technicians should be acceptable to most authorities. In some cases where International boundaries are contiguous across the same land mass [such as in Africa] where animals, birds, insects and microbes migrate across-boundaries regularly, exclusion or even rigid control of a vastly profitable microbe like rhizobia when supplied in a sterile media, would seem pointless.

Regrettably, some well developed Nations including Australia have imposed such demanding restrictions on import of *Rhizobium* culture that export to those countries has become uneconomic. One can only speculate that these restrictions may be more focussed on protecting local production—than protection from imported national environmental issues. In tests conducted by New Zealand's Dr. W. Lowther when FCC Ltd. produced "New Formula Rhizocote" inoculant, he found little difference in survival of rhizobia on white clover seed after a stress period between the only New Zealand commercial culture ["New Rhizocote"] and the two Australian cultures available at that time, having found the Australian cultures superior in earlier years. In MPN tests conducted by FCC Ltd., it was clear the New Zealand "New Rhizocote" commercial culture gave higher cell survival counts than either of the two Australian cultures. The carrier for the two Australian cultures was almost certainly irradiated peat, whereas the carrier in the New Zealand culture was autoclaved diatomite [Rhizocote manufacturers now use only diatomite having abandoned peat]. However, peat may still be eligible as an added coat in seed coating. Both Australian and New Zealand commercial cultures have been of generally good quality which hopefully will continue and indeed improve.

Protection of a local manufacturer might be justified if that manufacturer's commercial viability were "fragile" in relation to sales volume, because it is important that each nation have access to its own reliable, strains selected legume inoculant cultures, where some local control over quality may be exercised. But to exercise protectionist activities of a continuously poor product—or simply to block import of a superior product to protect internal [specially robust] production would be an official abdication of responsibility to that nation's farmers—indeed that nation itself.

For spring sowing [which will be preferable in most temperate climates] one should, during the previous autumn, hard graze [the area to be oversown] with maximum concentrations of livestock [inexpensive temporary electric fencing using a car battery can assist this] which is aimed at controlling most edible plant growth prior to winter and reducing that native sward to an open, hard grazed ground cover—but not bare, which hopefully will also be well trampled and even lightly "pugged" [hoof penetrated holes] to create a better more sheltered, moisture retaining seed bed environment for oversown seed. Then, if possible, immediately prior to sowing, again concentrate livestock at least for a short period to hard graze, remove native competition and further open up the ground cover with hooves. Surface sowing into snow is acceptable rather than wait for clear ground which may be too late in the season for good seedling establishment—but snow will prevent coated seed from exercising its initial ballistic advantage in penetrating native foliage to achieve ground contact for later root penetration into soil. Higher density coated seed will penetrate snow more effectively than bare seed and be less attractive to winter hungry birds. Melting snow will settle seed eventually.

3. Accurately Calculate & Order Proven Seed Treatments

Arrange for the supply of both seed to be oversown and fertilizer to be topdressed specially sourcing good quality seed. A representative random sample of such seed should have ideally been examined [hopefully at a qualified seed testing station] for viability [specially for high interim germination] and for noxious or potentially harmful weed seed contamination. Such seed lot may be "certified" [usually by Govt. agencies such as Department of Agriculture] which provides a test certificate showing results of the examination.

As stated, the analysis should indicate no noxious weed seed in the sample, whether certified or not. If there is such contamination, it must either be "dressed" out of that line of seed—which is not always possible, and re-examined otherwise it should most definitely NOT be used to sow on land difficult to access. A very small quantity of other weed seed may be acceptable—but a high interim germination count is vital for legume seed because slower germinating "hard seed" which, while also classed as viable seed, is however, not suitable for such oversowing particularly where clovers are concerned because rhizobia, vulnerable until safely inside plant tissue, will be depleted by delay and may all be dead by the time hard seed germinates some weeks or months after sowing.

In some samples, hard seed can comprise as much as 50% of the total which is not suitable for this work but can be used for any subsequent sowing where legumes are already established—and where rhizobia have also by then become established in the soil [nodules slough off clover roots—i.e., decay— releasing both *Rhizobium* and available nitrogen into the soil—the latter can then be absorbed by non-legumes [hopefully by the associated grasses—not by unpalatable weeds].

While slower germinating seed may become beneficially infected from rhizobia introduced into that soil on adjoining seed—a high interim germination count is always more desirable for seed which is to be inoculated and coated for first time sowing.

Grasses too can be beneficially coated [but they are not legumes, do not carry nodules on their roots and are not inoculated like legumes]; both grasses and clovers should be coated with the most protective and supportive coating for oversowing, i.e.—only inoculated coated seed which can be backed up with proven results as to a high and sustained level of *Rhizobium* viability after a stress delay—should be used.

Coated Seed Use

Research evidence throughout the literature and in subsequent commercial experience is overwhelmingly [but not exclusively] in favour of coating inoculated legume seed when it is to be surface sown. Whereas, surface sowing of inoculated bare seed clover at the recommended and affordable rhizobia loading, has been overwhelmingly unsuccessful.

The evidence for grass seed [more fully discussed in Chapter 4] is not as positive though it favours coating for surface sowing into existing native vegetation and if % establishment is increased by coating as it has been in some well documented field research, it is only sensible that seeding rates may be reduced for seed when coated, which will usually result in similar cost. That of course is just % establishment—if subsequent growth is better with coated—then that feature is an additional bonus leaving little room for argument.

But is the coated seed *you* can buy—unproven [e.g., no independent research results], unavailable or too expensive?

This book carefully examines the relative merits of plain versus coated seed [see below in relation to oversown clovers "Which costs less?"]

However the author is conscious of the fact that coating of seed in general is preferred [about which the author is somewhat uneasy as there are as many different types of coatings as there are mushrooms and toadstools in the field—edibly, some good, some indifferent and some very bad]. Also, that in remote locations and under-developed parts of the world, good quality inoculation and coating may simply not be available, or available but not supported by qualified independent test results, or may just be too expensive in relation to the alternative of simply sowing a lot more seed.

In these events, farmers can make up their own inoculated and coated legume seed or coated grass seed by following the methods related in this book [see Chapter 6—formulae]. These are proven formulae.

It is the duty of commercial manufacturers to prove their product—wise manufacturers don't attempt to sell a brand of motor vehicle until the model is well tested, modified and proven, following which they can confidently turn out thousands more produced to exactly the same specification.

Though somewhat different (but their use more expensive than a motor vehicle in many development situations] inoculated and coated seed carries exactly the same principle—e.g., formulae needs to be field proven and manufacturers need to find independent grassland scientists [such as Landcare Research in New Zealand] to conduct tests, and yet more tests if the formulae need improvement.

If the soil to be sown is acid [say pH 5.5 or less] then a coating of finely milled high grade lime will be likely to effect the best response as shown in research, by moderating soil acidity in the rhizosphere [root zone] of plant development below each germinating seed [also with legume seed primarily for support of the inoculum culture]. But other coatings such as rock phosphate [free of that acid content which can kill rhizobia and possibly damage seed itself as can superphosphate], and dolomite [magnesium carbonate], are valuable as a finely milled seed coat as they also offer some protection of rhizobia and can add a small but precisely placed beneficial pinch of nutrient and alkali to the seedling which element may be found deficient in the soil to be oversown.

Less alkaline coatings may be preferable for some of the slower growing, acid tolerant, alkali producing {Bradyrhizobia} associated with tropical legumes where lime coating has actually depressed nodulation results in some Australian field research. However, other publications recommend lime coating of Bradyrhizobia, e.g., Daniel Real and Geoff Moore Department of Agriculture and Food, Western Australia Titled "Siratro, atro" (a tropical legume).

Seed should be sown as soon as possible after inoculation and coating, however, the best coatings are able to maintain survival for several weeks of cool storage—one extraordinary sample examined recently, imported from USA, held a good high *Rhizobium* count after 12 months—but that is not likely repetitious and certainly not a guide.

In most situations, good quality inoculated coated seed should be oversown within 2 weeks of fresh inoculation unless ongoing tests show otherwise.

The same seed purity and high interim germination qualities should also be sought for grasses as for legumes.

4. Mixing Just Prior to Sowing

For ground or aerial oversowing the safest procedure with the most potentially successful seed and inoculant viability, best fertilizer spread and minimal risk to pilot and aircraft where aerial application is used, will be obtained from transporting freshly inoculated and coated seed plus *clean, dry, flowable* fertilizer onto the site for sowing— be it an airstrip, a shed or bin, immediately prior to use, both mixed [where required] on-site just prior to application and applied on target without delay.

That is a tall order one might say—however there are many jobs on a farm which require careful planning, dry weather, an early start, maximum effort on the job and determination to get the job done as soon as possible and certainly prior to onset of bad weather—this is one of them.

Fertiliser manufacturer, seed coating manufacturer, transporter and aerial or other operator need to be advised that your product must not be allowed to get wet or even damp until sown—there can be no exceptions. Product should be returned at senders cost if found to be wet. This condition of supply should be written into the original order for product and service—if a supplier cannot guarantee that condition—the remedy is simple, cancel the order. A farmer, runholder, user or application contractor cannot afford to mess about with damp fertilizer or wet coated seed—but specially an aerial contractor.

Some suppliers may simply not be prepared to work hard and fast to keep your product dry by complaining, e.g., "no one else has ever demanded this"— you may be told. If you are, thank your lucky stars it was in time to dump that supplier and find one who does understand the importance of moisture.

Seed is [inoculated and/or] coated as separate batches of various clover or grass seed cultivars, stored and later combined as an oversowing mix to specific farmer orders then supplied by that seed merchant when ready; while fertilizer is ordered from the manufacturer and collected by a transport contractor when he has capacity available—direct to the farm—often straight to a fertilizer bin sited beside an airstrip for aerial distribution.

Neither seed merchant, fertiliser manufacturer or transport company are normally able or willing to meet exact time frames to secure freshness and beat the weather—to attempt this for all farmers would be impossible for most of them, so the whole ideal to apply freshly processed seed and clean dry easily spread fertilizer—as a prompt delivery [after inoculation] will usually be lost via the constraints of commercial reality. There may be a wait, and a short one can be technically acceptable but there is no place for allowing product to get wet.

Compromise is also achievable. In the past, some seed firm's mechanical mixing equipment has left a lot to be desired. Mechanical mixing of inoculated coated clovers and grass seed were sometimes excessively abrasive causing loss of coating [no doubt inoculant too] creating clouds of dust in the seed store leaving a coating of fine dust over that firm's nicely displayed new merchandise in the showroom adjoining. That type of machinery was designed to mix fluid plain seed and most would do so without mechanical damage to seed itself, but it was not designed for coated seed. An inserted mixer like a giant egg beater will be disastrous for most coated seed. A gently revolving drum with stationary bars—exactly like a concrete mixer, can mix coated seed safely provided it is not overfilled, speed does not exceed 25 r.p.m. and minimum number of revolutions are applied—normally about 5 revolutions after all seed varieties have been tipped into it, will suffice. Seed will further disperse and separate in the air.

Mixing Intelligently

If the mixed coated seed is to be further mixed with fertilizer—specially any unprocessed fertiliser containing variable particle size from fine dust to coarse material with transferable acidity or any other "anti-biotic" component, then the coated seed should definitely only be mixed immediately prior to application. In that event why mix seed varieties [clovers with grasses] earlier? It will minimise damage by taking delivery of coated seed varieties [clovers and grasses] separately from the seed supplier working out weights of each seed variety to be mixed per batch, then bagging them up [all together in each bag—ok unmixed at that point, or] separately in bag weights which match the fertilizer batch weights for each load, be it for spreading by hand, spreader, blower or aircraft. One such seed mixture should be split in half [for aerial work] because when flown on, pilots often fly the first run with a half load to test all potential hazards. These two half mixes can easily be combined to make a full load if need be. Label empty bags clearly prior to filling. Lots of bags?—sure but in a good cause and to be retained, already labeled for next time—or for a neighbour.

Aerial sowing of coated seed separately from fertilizer—or where no fertilizer is to be spread will generally give best spread of seed through the "fantail" distributor fitted to the fuselage of the aircraft, however, where fertilizer or lime is to be applied, it will also be more expensive to overfly the area twice.

Fig. 32 Manual loading of topdressing aircraft may be necessary where a loader vehicle is not available, however, it allows less expensive pre-mixing and bagging a day or two prior to flying on. Bags need to be weighed so the aircraft does not exceed its operating weight limit.

Aerial Oversowing

Even when pressured by more than one aircraft operating from one airstrip for one contract, the farmer needs to make sure that the aerial contractor [and his pilots] understand—that of even greater importance than keeping aircraft moving with a quick turnaround between each load, is the need to mix fertilizer and seed accurately and freshly immediately prior to discharge into the loader's hopper. All participants should work as a team in achieving the most efficient method and routine for which there will be many variations.

The farmer might be wise to set up his own method for loading dry, clean fertilizer into a mixer [Note: we repeat, damp fertiliser can be very dangerous to pilot and aircraft, also harmful to coated seed, and must not be contaminated with sticks, roots, etc.]. When scooping up and discharging a measured quantity [by weight] of fertiliser into a mixer the farmer or loader driver may then discharge a correctly measured weight of plain or processed grass and clover seed into the mixer as well, then discharge each load either back into the skip or bucket then into a holding silo or bin [if more than one skip full is required per aircraft hopper load], or directly back into the scoop of the loader machine where one mix is a full load for the aircraft.

{Photo by courtesy of New Zealand Civil Aviation Authority}

Fig. 33 This waterproofed fertiliser bin with sliding roof over concrete floor is at ground level where this aircraft loader designed for handling fertiliser alone can access it easily. However, the skip here which scoops up fertilizer [top wide mouth] can also discharge it into the aircraft hopper [bottom narrow mouth at other end with flexible chute which flicks into the aircraft hopper] can also feed a mixer in between and collect the complete mixture again for loading. Coated seed needs to be pre-weighed and bagged according to skipful fertilizer lots and discharged onto the fertilizer in the mixing device by the farmer or his staff. The contractor scoops up each measured fertilizer load, discharges it into a mixing device then when seed is added retrieves it for loading aircraft directly, or to a bulking up silo to make up a load.

A front end loader [a hydraulic bucket attached to a tractor which is commonly available in many farming systems] to pick up [a measured quantity of] fertilizer either using a narrow bucket or scoop to discharge it into the mixer [or if a wide bucket then via a loading chute fixed above the mixer] with pre-weighed bags of coated grass and clover seed [to match each fertilizer load] applied on top of the fertilizer while revolving, would be one of many solutions. One level scoop [or bucket] full [of known dry weight to comply with regulations] plus the added weight of coated seed must be calculated for accurately meeting specific aircraft requirements. Shovelling coated seed into fertiliser on the floor will work for small quantities or where no aircraft are involved but is a method unlikely to be able to maintain a reasonable rate of mixing to keep one aircraft busy—let alone two or more.

The mixer would probably need to be sitting on an elevated platform—even on the back of a truck may be suitable. If more than one load of seed/fertilizer mixture is required per aircraft flight it may be necessary to include the above mentioned intermediary silo where more than one mix can be temporarily stored to make up one aircraft load—usually for instance between 900 kg and 1.1 tonne for most [now ageing] Fletcher 400 hp aircraft with a 43^3 ft hopper—on up to 2 tonne for some planes—even twin engined aircraft have been used but usually only operate from a sealed formal airstrip. Huge

{Illustration by courtesy of the New Zealand Civil Aviation Authority}
Fig. 34 Topdressing aircraft being loaded with fertiliser and seed on a rural airstrip.

quantities of fertilizer and seed have been spread on steep hill country simply on foot by fit hardworking pioneer farmers in New Zealand, or from horseback or a blower mounted on a truck or tractor/trailer where much ground can be covered by crawling along ridges spreading to both sides.

This mixing and spreading routine may seem like a lot of work and organization—if it is done well, the results will justify it. Unfortunately, a really good result may not be identified as to inoculation benefit, but a poor one will be–and will also be a waste of money.

If the oversowing is not going to be done correctly—including uniform distribution of seed by skilled pilots or applicators—then don't do it at all until a satisfactory method can be found.

On Farm mixing. Because professionally manufactured inoculated and coated seed can be safely stored a week or two prior to sowing without significant loss of rhizobia [which nevertheless will be reducing as time passes], it may be preferred by some to pre-mix seed varieties (at the farm or airstrip) and bag them into measured batch lots prior to the day of mixing with fertilizer and application.

It will reduce the number of bags required plus the need for calculation during mixing and will be a more relaxed nice clean job on a suitably dry day— perhaps in the barn with a mixer or wide mouthed shovel on any concrete or timber floor, with roof against rain for sure, but also preferably enclosed and promptly used because most fertilizers and lime are unfortunately inherently hygroscopic [they readily absorb moisture from the atmosphere and easily become damp] which adds a further dimension of danger and the need for care.

Wet fertilizer is very difficult to dry out again: if damp fertilizer is refused by the aerial contractor [which is his right] it may have to be returned to the manufacturer's bulk facility for drying. That could be a big problem—an added cost and delay to be carefully avoided. Manufacturers also need to invest some time and money on developing fertilizers which do not dampen easily—without adversely affecting the post sowing solubility of the product.

Granulation of fertilizers such as superphosphate to a particle size of [say] 3 to 5 mm diameter, also preferably both top and bottom screened, reduces flow blocking dust, aids free flow and when lightly mineral oil sprayed resists premature moisturizing—however, not all suppliers can offer granulated product, some granulated products do not solubilize as readily as other fertilizers and not all fertilizer mixes required by farmers or runholders are available granulated [such as [say] dolomite reverted, molybdic superphosphate]. Some farmers may wish to use the spreading opportunity to add [say] selenium granules for livestock health to the fertilizer/coated seed mix—and that too needs to be accounted for in those weight calculations—also kept dry. Any wet product will soon nullify all weight calculations elevating the aerial application danger level enormously.

To mix the seed varieties by weight according to the sowing rate, and then bag off the right quantity of seed for the weight of a single fertilizer load which the proposed mixer can hold to make up a continuously correct seed/fertilizer ratio at the desired application rate for both, will certainly save some time on application day, but weighing each seed variety separately without mixing prior to adding to fertilizer is less damaging to coatings and should not present a problem as long as there are plenty of bags and the person loading the correct number and variety of each coated seed into the mixer on top of fertiliser—is wide awake!. It will help if each mix is separately grouped as to seed varieties, and a clip board used to tick off each variety when mixed into each load—and not loaded until all ticks are recorded.

Mix Accurately

If we have a seed sowing rate of say 18 kg/hectare made up of 8 kg of coated clover and 10 kg of coated grass seed and our fertilizer topdressing rate of application is about 4 cwt per acre [say 450 kg/ha] and the mixer can hold say 1 scoop of fertilizer weighing 150 kg [dry] being the right amount for 1/3rd of a hectare then we need to add 1/3rd of 18 kg of coated seed to each mixer load = 6 kg [2.66 kg clover and 3.33 kg grass seed]. If the aircraft can safely carry [say] 1,000 kg per trip, then we can load 6 mixes of 150 kg of fertilizer containing 6 lots of 6 Kg of coated seed = a total load of 936 kg with a small safety margin and the aircraft will need to distribute that load evenly over 2 hectare [4.8 acres]—usually in a strip pattern [of a measured and known width with allowance for wind drift] then avoid that area in all future drops—a highly skilled and demanding task. In this situation we could save time and bags by pre-weighing coated clovers and grass seed separately into 6 kg bags to be properly mixed later with fertiliser and securely tied or prevented from spill but readily and rapidly dischargeable into the mixer. Aircraft and loader capacities vary as do conditions of application, so each job needs careful planning and good co-operation between all participants.

Aircraft hoppers can be calibrated to discharge fertilizer at a specific weight per minute and that needs to be adjusted so the normal width of the drop by speed of the aircraft will cover (say) 2 hectare from moment of opening the flow to emptying of the aircraft hopper. Where cost is less important than accurate spreading, as mentioned above, fertilizer and seed can be aerially applied separately whereby the much smaller weight of coated seed can be better controlled by the "fantail" spreader. That of course also allows an aircraft to cover a much greater area per load than where fertilizer is involved. It might be good management for instance where time of sowing seed is critical, to do so from a more distant but dry airstrip—to bear the extra cost of distance and sowing seed separately, perhaps because of wet weather and slippery access tracks to an elevated on-site airstrip where the fertiliser cannot be delivered until drying winds allow vehicle access. Or perhaps tonnes of fertiliser are on site but the strip too wet for aircraft and loader, or maybe the access tracks to the airstrip (often on elevated ridges accessed by steep clay tracks) remain too slippery for the loader. Provided the time gap between seed and fertiliser application is not excessive [>4 weeks], then timely sowing of the seed will often be more important than delays until both can be flown on from an on site marginal airstrip.

Aerial Safety

As a safety measure, New Zealand Civil Aviation rules require that any aircraft engaged in agricultural topdressing or spraying must have a jettison system with capability for discharging 80% of its load within 5 seconds in an emergency, if the pilot elects to operate the aeroplane above normal category maximum certificated takeoff weight (MCTOW—likely to be reviewed soon).

Fig. 35 The interior of a topdressing aircraft's holding bin (or hopper) showing the tip of the photographer's shoe in the foreground to provide scale. Coated seed alone is lying in this hopper. On the underside is located a (removable) "fantail" spreader expanding the width of each pass by the aircraft. Pilots need to calibrate speed, height, swath width and hopper opening position to ensure an accurate spread. A feature which can ruin all calculations, is excessive "drift" in windy conditions.

With its unfortunate record of many accidents in New Zealand alone, aerial topdressing and oversowing has been a hazardous occupation. It demands maximum care—no fences, trees, phone or power lines at either end or nearby on grass strips, no livestock wandering across airstrips, any bird hazard dealt with, loading areas drained and hard filled, windsock installed and the strip not in line with low winter sun glare at either end.

Whether sowing by hand, from horseback, by blower or spinner off a truck, by aerial application—or whatever method, the coated clovers and grasses should be accurately, carefully and proportionately mixed and as evenly spread as possible.

5. Optimum Time of Sowing

This is absolutely critical. For instance in temperate climate New Zealand, and in the more difficult winter cold and summer dry and hot hill country areas of the South Island [Northern Southland, Otago, Canterbury and Marlborough] the latter half of August up to mid September [early Spring] has been shown in extensive research to be the optimum time *[a few days earlier at lower altitude but no later even at higher altitude].*

It is specially important in low summer rainfall locations where, coming out of winter when temperatures trigger germination there is still ground moisture plus occasional rainfall to support sensitive early seedling growth, nodulation and establishment.

Sometimes frost lift can beneficially cover as yet ungerminated seed on bare soil patches lifting it plus soil particles on icicles at night then dropping the seed mixed with soil by day. By the time the frosts are finished, that seed is nicely covered in soil. Frost can also be the enemy where, for example, a cultivated seed bed sown in lucerne [alfalfa] near Garston in Southland, New Zealand, a promising crop, nicely germinated and grown to about a 4 cm tap root was, on a clear night after severe frost, lifted clear out of the ground with translucent white roots lying neatly on the surface right across the field.

Fortunately, that is unusual.

Establishment of seedlings should be followed by warmer weather to promote spring growth which permits a short sharp grazing of the sward not only to "inject" essential animal manure nutrients into the soil—particularly nitrogen for support of the grasses until clover nodules begin to release nitrogen into soil, but also to control competing weed growth and consolidate the new pasture.

This 5 point formula has proven to be successful not just in research, but in hundreds of commercial sowings in the Southern High Country and throughout New Zealand—particularly in drought prone tussock country regions as well as in better rainfall areas of New Zealand's North Island's hill and high country.

Warmer Climates

In both sub tropical and tropical locations, environmental conditions and weather patterns will of course be quite different requiring localized field investigation into ideal sowing times related to varieties to be sown, altitude, weather, certainly moisture limitations and excesses in some locations involving monsoonal weather. It is crucial small size experimental field plots are first employed in new localities on which results of local experimentation may be translated to larger more expensive development.

These plots must be kept free of trespass where resultant cross contamination of various inoculation levels [including none], various fertilizers [including none] and artificial effects on establishment could give a false result. Where stray animals [domestic or wild] have access to field sown investigation areas, such areas will need to be securely fence off—more than that where fencing will not exclude large birds, a flock of small birds, reptiles, climbers like monkeys or jumpers like deer. These hazards are sometimes not preventable in farming practice, but in experimental plots where we are trying to get *precise information* about each treatment, removal of just a few seeds, transfer of just a few *Rhizobium* and other minor effects can conspire to give a false answer to the investigation [in miniature]. All intruders must be kept out.

That need not be difficult. A car battery powered electric wire (or fence) [with pulsing energizer] plus netting including overhead netting if need be, can be very effective and is not expensive on a small scale. Field trials however, are too meticulously prepared and laid down—and their results too important to local agriculture and human welfare—to allow interference by "man or animal" which may completely invalidate all that good work.

Developing Nations Must Make a Start

In Himalayan hill country, native grasses grow up to 8,000 ft altitude, and higher, being closer to the Equator than New Zealand thus warmer at higher altitudes. At such altitude, this hill country also has cold and dry winters with wet monsoonal summers. Optimum time of sowing needs to be field tested in such situations and indeed Nepal Department of Agriculture's former Agronomy researcher Netra Bahadur Basnyat {with a M.Agr.Sc. degree from New Zealand} will have collected information about this as he has studied agriculture both in New Zealand and Nepal as well as having received several sacks of top quality New Zealand seed sent to Nepal for field investigation.

However, grassland for grazing animals can be established by surface sowing in any situation where [a] the area can be fenced off [b] existing foliage can be suppressed, partially exposing soil [c] reasonably reliable rainfall or irrigation will support critical post-sowing establishment and seedling growth [d] soil deficiencies can be remedied by topdressing [e] insect predation, fungal disease and bird theft are reasonably able to be controlled and [f] the grassed area can remain fenced off for subsequent grazing and spelling control—and has a reliable source of clean water for grazing animals.

Because drinking water is essential for both livestock and people, alternatives, where available, should be carefully considered. Springs are usually best, then running streams and then rivers, particularly where the intake to a piped system is at high altitude, above contamination level of disease [such as giardia] and where the water can be filtered into alkathene pipe [which does not fracture when frozen] sourced hopefully from such clean sources. Junctions can be taken off and run downhill to livestock troughs which may just be a cut down steel drum, and to holding tanks at villages [then reticulated to individual houses together with adequate drainage]. This may represent a lot of work, but will suffice for many years specially if the alkathene pipe can be buried out of sight and out of harms way. Small intake dams are easily constructed where there is natural ponding and, where freezing is a problem, use of steep fast flowing streams or waterfalls which do not freeze readily if at all.

Fig. 36 On the summit of Mt. Kalang. This altitude at approximately 13,000 ft in the Himalayas is easily accessible as author Bennett found climbing alone and with no special equipment, on a day off from trekking in search of nomadic sheep and goat herds [this photo from a camera stuck in the snow on a timer delay]. Not only alone but also unarmed—[fortunately, no bears or leopards seen that day]. Here lies a wonderful source of water for the dry hills and villages below.

Nepalese, Bhutanese, Indian and Pakistani farmers have huge opportunities to "tap" alpine ponds and creeks which in winter flow too steeply to freeze or rivers too swift to freeze. Above photo was taken mid-winter in Nepal where access to water is freely available. With some planning, local knowledge, alkathene pipe and insulation [where the ground freezes below a metre deep], plus trenching—all aimed at stopping the pipe from freezing, water can be piped, and also stored where overnight freezing is a problem. There are areas all over the world like this—in the mountains of parched Ethiopia and Sudan where huge volumes of water are lost to the Blue and White Nile from surrounding hills and mountains, and in other African Nations, in Russia and Soviet states, China and South America to name just a few.

In New Zealand the cleanest water is pumped up from deep bores supplied by snow, ice and rain falling on mountains which then flows into the ground and into self filtering shingle substructures below the flat land on the plains, cold but very pure.

Fig. 37 Basketweave overnight housing of nomadic sheep and goats in Nepal. It is mainly lambs and kids that are placed in the basket enclosures.

In many emergent and developing agricultural economies there is no fencing or even land ownership rights—nomadic sheep and goat herders, usually younger men from nearby villages, shift their flocks by day to new grassland on public land and then retain the animals from wandering and from predators at night by enclosing them inside long lengths of about 1 metre high bamboo basket weave. While easily transportable to constantly changing overnight camps, this is a burden for the shepherds. Permanent fencing is so much better where improved pastures can follow—but of course land ownership [or stewardship] must precede it. In addition, such fences must keep all predators out at night.

Even if money and materials were available; who is going to develop new grassland which cannot be kept secure and which other herders or other livestock can get the benefit of? For success, people must have the incentive to be rewarded for hard work, protection to invest money and materials to improve the land no matter how modest, for their own and for their families benefit—without which there can be no progress.

Governments may of course elect to allocate land areas for lease where farmers have insufficient money to purchase. Rent payable as a percentage of something grown and sold [above subsistence] may be a slow way to collect tax, but progress has to start somewhere and land proprietorship [owned or securely leased] is an essential starting point. Government must be strong enough and sufficiently well organized that farmers can rely on it to set up fair rules, provide surveyed and pegged registered freehold or leasehold titles legalized in a formal Govt. Register all of which must be free of interference provided occupiers observe rules agreed to at assignment to a title and to that block of land.

Where a start can be made [and it starts with stable Government], successful seed establishment will require certain facilities which a central Government has at least a moral duty to provide.

It is necessary of course to have access to good quality legume seed inoculants as well as good seed plus a method of coating the seed if it is to be surface sown. These facilities and materials may be available from other Nations, but it has to start somewhere and when found to succeed, will, by example grow from that success.

Once again, it may be essential for a developing Nation supported by Aid money and equipment, to set up a "first"—one or two farm development schemes, to experience and remedy any major problems and only then encourage local farmers to do the same.

Simple methods for inoculating and coating smaller quantities of seed are fully described in this book [see Chapter 6].

Tropical Grassland Associations around the world have informative websites on tropical grasses some of which list consultants with experience in this field.

The Tropical Grassland Association of Australia has been a valuable source of information to Graziers in this regard however, sadly, drought, downturn in cattle farming in Australia at time of writing, and departure of many retiring experts from this field of science has led to proposed closure of this Association at the end of 2010 essentially due to declining membership. It would be very sad to lose the knowledge gained by members of this professional organization.

Book sales are made through CSIRO, QPB, at 306 Carmody Road, St. Lucia Qld 4067 AUSTRALIA for access to any past publications which the Association would have formerly supplied.

The Facts about Oversowing Seed

Let us examine the relative merits of Bare seed, Coated seed—and Granulated seed [with fertilizer], at 5 or 10 times seed weight [called "pasture pellets"].

The use of pasture pellets, yet to be fully investigated and developed, has been discussed but is not yet adopted. At this point we have few facts on performance.

Not so with bare and coated seed. After exhaustive field research and wide commercial experience mankind is now beginning to understand which of the two thoroughly investigated methods—bare seed or coated—has proven consistently best on cost and performance for surface sowing specially for legumes where inoculation is usually involved.

The first and most important fact is this: Dissimilar methods of processing the seed as well as variable locations, materials and methods of sowing, will, and have given quite erratic results which have confused many, even some grassland scientists, as we shall show.

To add to the complexity, responses from legumes as opposed to grasses, apart from initial germination, are not comparable. Each have quite different characteristics, qualities and vulnerabilities.

Here, we examine legumes starting with clovers—then in Chapter 4, work on grasses—consolidating the studies of various scientists also the results of many years of actual, commercial experience.

The following summary of widespread research into *Rhizobium* inoculation of legume seed for oversowing is centred [as it must be for subsequent costing] on a specific, typical and fully representative independent, professionally managed examination, in this case, from three typical sites in a classic research study which is vitally important to our understanding on a world wide basis, and indeed is the chief reason for publication of this book.

In this increasingly complex world, we now find some Seed merchants offer variations of the seed/coat ratios exhaustively researched earlier—they take a risk doing so specially if just on a whim [i.e., without full investigation and testing] but provided their new product has indeed been fully tested as to the qualities we now know are essential for good field results and where the bonus has been the simplification of the arithmetic of seed/coat ratios then the industry should remain positive through such changes. Change however needs to be very carefully monitored and tested before farmers buy into it—despite any unsupported sanguine assurances of their local seedsmen. Only facts matter, there is no room in this expensive work for blind emotive or financially captive [i.e., a credit a/c with the seed merchant} commercial loyalties.

The original 4/7ths seed ratio was imposed by circumstance by Wright Stephenson & Co's Christchurch Branch in 1960's New Zealand when it began legume seed inoculation and coating; however the adhesive first suggested when attempting lime coating was hoof and horn hot glue [containing a powerful bactericide against decay]—with inoculation!—on legume seed– which produced some very lumpy unsatisfactory results. Wrightsons™ Seed in Christchurch then had the good sense to approach FCC Ltd (Fruitgrowers' Chemical Co Ltd), an experienced manufacturer [including of chemical prills] to request co-operation in developing a successful process for seed. Following the establishment of CSL (Coated Seed Ltd) being the newly formed joint venture manufacturer for W.S. & Co and FCC Ltd, some large lumps of seed and glue were also produced by Coated Seed Ltd for a short time when it first took over production from W.S. & Co., all expensively dumped of course. A rough start indeed!—which should be avoidable in future with the guidance of this book, specially in developing economies which cannot afford such mistakes, but that is how the industry began in New Zealand, a very different quality of product is processed today.

CLOVERS

In temperate climates where development of new grassland by surface sowing has become an established practice, bare and coated seed with and without inoculation have both been expertly compared in field trials in New Zealand, Australia, USA, Canada and elsewhere by independent scientists using in each study, identical seed and materials for all individual treatments.

One particularly outstanding research effort stood out from the rest in all the literature and test results we have examined. It was substantially motivated by New Zealand's pioneering effort into seed inoculation [and subsequent coating] in the late 1950's to early 1960's by DSIR followed by establishment of CSL, followed by scientific controversy, followed by technical and commercial success which culminated in the sale by CSL of its [then] technology to USA's giant Celanese Corpn. New Zealand was a world leader in production of inoculated and coated seed by the 1970's, confirmed largely by the following study.

This most prominent and important research investigation was undertaken in New Zealand concentrating on one vital and thorough examination, repeated for thoroughness at several typically hostile field situations, where the experiments were independently and professionally designed, sown, monitored and assessed. But [*most importantly*] where miniature seed treatments were prepared by this leading manufacturer and supplied to the scientists fully described and labeled being exact reproductions of existing commercial processes, plus some extra [potentially advanced] processes for comparison. For this research study, all seed, inoculants and coatings were identical for each treatment being supplied by the laboratory of CSL from one source in each case. Storage conditions applied by the scientists were also identical.

The researchers also prepared some of the inoculated and coated treatments themselves [using formulae as earlier promoted by the DSIR].

But before we examine this defining field research, it is important to set the background to it—and the urgent need for it at that time.

Controversy, Confusion and the Danger of Misleading Research

Scientists have frequently prepared treatments for their own research studies which is understandable and occasionally necessary—but some have underestimated the complexity of the science [particularly where it involves microbiology] of formulation by making a definitive "labeling" of such treatments according to the results which THEIR product produced.

A single cake recipe will often produce variable results when baked by different people! There are a multiplicity of methods and materials for processing seed. Resultant products will be as variable as those methods.

In the early days of legume inoculation development, some researchers formally recorded that they had no response at all to inoculation of oversown clover seed and some had no response to coating of oversown inoculated clover seed, others found the coating to have actually depressed seedling establishment in their experiments and a few obtained a depressive effect only where lime was also broadcast. Why waste money, time and effort on coating seed if it produces worse results than plain or just inoculated seed! However, one common factor was associated with all the unresponsive and depressive experimental results—*none of those researchers carried out rhizobia cell counts on the treatments they prepared*, either at inoculation—or subsequently at sowing. Without cell counts, inoculation research was not reliable and its findings controversial, worse, it had potential to be misleading as was subsequently discovered.

We now know successful inoculation is a "numbers game". The number of viable rhizobia retained on seed at sowing is not just "significant" in terms of establishment success on soils devoid of rhizobia, it is pivotal. *Rhizobium* viability is foremost.

But nobody understood that fact during the early 1960's when commercial coating of seed was officially condemned by respected researchers notably in a paper delivered to the formerly[6] conservative but widely esteemed New Zealand Grasslands Association at its location-rotating annual conference in 1966—that annual conference being held at Alexandra, Otago [in the heart of New Zealand tussock country oversowing potential]—which was well attended by local farmers and runholders.

Although just history now, at the time, this 1966 New Zealand Grasslands Conference paper almost spelt the end of commercial seed inoculation and coating in New Zealand.

It is a classic example of the need for meticulous accuracy in research and also an example of the antithesis of meticulous research—being potential for huge damage to society which inferior materials and/or inadequate technique and/or misunderstood results can cause—this particular example, fortunately soon became exposed by further close peer examination.

The 1966 Grasslands Assocn. paper titled **"The Effects of Inoculation, Pelleting [*i.e., more correctly "Coating"*—ed.], Rate of Lime and Time of Sowing on Establishment of White Clover"** by N.A. Cullen, Director of Invermay Agricultural Research Centre and T.E. Ludecke, Scientist, Ruakura Agricultural Research Centre—was delivered to the Conference, we understand, by T.E. Ludecke.

From this paper to the Conference, the following details are quoted [without alteration] in the interests of better understanding and future caution—and with kind permission of the current Executive of the New Zealand Grasslands Assocn:

[Quote:]

In Australia, lime pelleting plus inoculating the seed has been shown to improve establishment, particularly on acid sites [Loneragan et al. 1955], and in New Zealand the use of lime and other pellets is widely recommended [Hastings and Drake 1963; DSIR Report 1966].

And:

"—it has been estimated that 200,000 lb pelleted seed was sown in 1965 [DSIR Report 1966]and there is some evidence [Adams 1964; Lowther 1966] that pelleting has caused a depression in establishment"

Authors note: This is not surprising when we remember that the inoculant quality of much commercially processed seed in the early 1960's, particularly at point of sowing, varied from poor to useless.

And:

"Claims by Australian workers and reports by Hastings and Drake [1963] in New Zealand that rhizobia survival has been aided with pelleting have not been corroborated in trials at the Plant Chemistry Division, DSIR [Triennial Report 1964]".

And:

"In this series of six trials conducted on a wide range of soils throughout Otago, pelleting white clover seed with lime or Gafsa/dolomite in addition to inoculation failed to show any benefit. At two of the sites there was a very marked depression to both types of pellets and some evidence of a depression was noted at the other four sites. Pelleting did not improve either germination or establishment; in most cases percentage nodulation was no better or even poorer than the non-inoculated seed. As there appeared to be a close relationship between nodulation in the pelleted treatments and the untreated seed it is suggested that any nodulation which occurred was largely from native rhizobia present in the soil. The greatly increased percentage nodulation in the inoculated only treatment, however, indicated that this treatment was of considerable value at several sites".

And:

"Samples taken at the time of pelleting were subsequently tested by A. Hastings and shown to be effective although rhizobia survival was low after one month. Commercially prepared pellets sown at Berwick about the same time, however, failed to give better results. No tests were done to determine the relative numbers of rhizobia on the seeds at time of sowing".

And:

"With present knowledge it is recommended that white clover seed oversown on acid soils in Otago should be inoculated shortly before sowing but not pelleted. Present trial work with new pelleting materials, however, could affect this recommendation".

And:

"T.E. Ludecke and J.A. Douglas [pers. comm.) in a trial at the Mt. Burke site found that inoculated white clover seed could be mixed with superphosphate for five minutes without affecting nodulation adversely and that lime pelleting was not superior to inoculation only. When the inoculated seed and lime pelleted plus inoculated seed were mixed with superphosphate for one day clover nodulation in both treatments was greatly reduced. Any contact of rhizobia with an acid fertilizer such as superphosphate nevertheless is best avoided and separate sowing of seed using a spreading device to obtain improved seed distribution is recommended for aerial sowing".

These observations were in contrast to other experience and the recommendations of New Zealand's DSIR (Department of Scientific and Industrial Research). However, the technical people at Coated Seed Ltd also had a different view of the value of coating based on their own research. Technical Director G.G. Taylor was absent overseas during much of this crisis which threatened the existence of CSL, so your author and contributor {Bennett and Lloyd} were required to deal with it.

When suspicion first arose as to the effect of variable rhizobia numbers on seed at sowing, author [Gerald Bennett] asked the other [John Lloyd] if, using his considerable processing expertise, he could prepare for Dr. Lowther some special treatments investigating ONLY the importance of variable numbers of rhizobia per seed at sowing. Not a person of half measures and with the standard 40 lb {inoculant} packet in mind, Bennett optimistically asked Lloyd if he could apply peat inoculant at 40 times the single rate, to white clover!—and still satisfactorily coat the seed. Not a person easily daunted by processing complexity the answer was " yes".

The smallest pack of commercial peat inoculum culture a researcher could buy for clover in those days was sufficient to inoculate 40 lb of clover seed at the single standard recommended rate. Bennett was suspicious of the way in which inoculants intended for 40 lb of seed were handled in the field to inoculate just a few grams of seed for sowing on research plots. The matter became more acute when Terry Ludecke [subsequently Lincoln University soils science lecturer], who had not only established to his own satisfaction that coating of inoculated seed depressed establishment of white clover in comparison with plain seed just inoculated—but also reported those findings directly to a joint official enquiry into seed inoculation held subsequently in the North Island New Zealand.

Having himself carried out inoculation of seed in the field, Department of Agriculture farm advisory officer Bennett knew it would have been difficult for Athol Hastings [DSIR] (when preparing inoculated coated seed for Mr.

Ludecke) to have used more than a "single dose"—or small amount of peat culture on a small experimental quantity of seed, and still successfully coat it with lime or Gafsa/dolomite or any other finely milled material. On the other hand, because it was widely believed that "you cannot overdo it" with inoculating bare seed—which is fundamentally true BUT NOT where treatment comparisons were being made! [possibly also motivated by some "Scottish heritage" well established in the Otago region of New Zealand] where Messrs Ludecke, Cullen, Douglas or technicians, not wishing to waste good culture [1]—very likely applied "plenty" of peat culture onto the bare seed which they could comfortably achieve by hand in a basin—using much—if not most of a commercial inoculant pack {for 40 lb of seed}–onto the few ounces required in their experimental plots—then oversow it by placement of 50 individual seeds per treatment—as they described in the conference paper.

Significantly and as stated in the paper, they did no rhizobia per seed counts on any of the treatments.

Stung into action by these "findings", Coated Seed Ltd development manager [Bennett] then requested of Dr. Lowther a special study in both field and laboratory, supplying him with essentially five carefully prepared treatments.

They were: [all treatments using identical seed and inoculant]

1) Plain seed
2) Plain seed inoculated at the single commercial rate of rhizobia.
3) Plain seed inoculated at 40 times the single rate of rhizobia.
4) Plain seed   "   " the single rate of rhizobia, and coated.
5) Plain seed   "   " 40 times the single rate of rhizobia, and coated.

These treatments were expertly prepared by Lloyd [in the laboratories of FCC Ltd, Nelson, under contract to Coated Seed Ltd, Christchurch], examined independently on arrival at the Department of Agriculture's Invermay Research Centre and were then oversown in a professionally configured randomly laid out field trial managed by Dr. Lowther.

And the result ?

Dr. Lowther's report on the results of this crucial experiment was momentously simple—and enormously decisive, he advised that:

The "40 times rate" of inoculation, whether coated or not, gave overwhelmingly superior results to the single rate of inoculation—whether coated or not.

It dramatically confirmed all the manufacturer's suspicions! More importantly, it further led to laying down of the classic Lowther and McDonald [laboratory and field] comprehensive oversowing investigation [discussed

---

[1] A sealed pack of inoculant once opened must be used promptly—or dumped.

above] at 3 typically hostile field trial sites—being hugely decisive research, extremely important for New Zealand Agriculture and a monitor of the widely used commercial products of Coated Seed Ltd—in protection of New Zealand farmers. That research actually being the focus of this Book in which results are recorded in the summary [Table] below.

Significantly, when you apply something like 40 times the single recommended [and affordable] rate of culture onto bare seed, you are actually "coating" it with peat culture, so that needs to be kept in mind—also that it is unaffordable.

Meantime, we had our answer as to importance of rhizobia numbers on seed!

The number of bacteria per seed at sowing is absolutely fundamental —indeed the deciding factor, to the extent it can override all other treatment effects.

The prime driver of success therefore is:

Rhizobium **survival on seed**—not the coating of it, or anything else.

However, we still had to show that coating [i.e., in our case, Prillcote™ coating] aided that vital survival of rhizobia on the hostile poly-phenolic surface of oily legume seed (particularly white clover) when applied at the affordable, practical, single dose rate. We also needed to establish whether the coating assisted germination or not, and produced more healthy vigorous clover plants—or not.

Fortunately above results of Cullen and Ludecke have now been understood and their findings erased from the record by people such as the late Dr. Jim White, Professor of the Plant Science Faculty of Lincoln University, New Zealand who advised Messrs G.G. Taylor, and G.M. Bennett in the mid 70's that [personal comment] "I have now got to eliminate all that work of Terry Ludecke's from my mind". Sadly, both former Lincoln University staffers White and Ludecke are now deceased.

Ludecke was essentially a soils scientist, not a Microbiologist or Rhizobiologist and then became a University lecturer, no longer a full time scientist. While a technical officer with the Department of Agriculture he had also carried out a series of successful field experiments through the 1960's into identifying widespread sulphur deficiency throughout lower rainfall Otago and Canterbury tussock grasslands. Those experiments were highly informative and revolutionised topdressing programs in those areas winning him much esteem.

In hindsight too, despite some acrimony during an address he gave to that technical meeting recorded above, which Taylor, Lloyd and Bennett also attended in the North Island, Ludecke involuntarily compelled these industry people to get busy to finally and emphatically prove his findings to be wrong!.

His results were not so much "wrong" however as they were misunderstood and almost certainly reflected:

1. A much lower loading of rhizobia on the coated seed he used than the plain seed he mixed freshly with an unmeasured quantity of inoculants, probably on site.
2. An inferior coating process to Prillcote™.
3. Much greater delay between inoculation and sowing for the coated product than the inoculated seed.

At least this investigation contributed to controversy—it also forced further and closer examination.

Subsequent key research has been carried out by World class Agricultural Scientists who's work is fully published in professional science Journals after peer evaluation. Even more importantly, their findings have been indisputably supported by large scale commercial experience, not just in New Zealand but in many other agricultural economies around the world.

The "Bottom line"

Before we look at the technical facts, a highly significant associated fact which has not been examined in conjunction with those results in the past is the actual relative cost, as bare seed, as inoculated seed and as inoculated and coated seed, of a typical oversowing seed mixture, then examine the percentage of: *Actual Seed Successfully Established in Each Treatment*, in relation to that cost—revealing *The True Actual Cost of Each* to the farmer.

In addition, there are other factors to add to or detract from that initial cost of establishment, e.g.,—vigour, health, dry matter production of established clover in the first year and most importantly, the level of nodulation by introduced *Rhizobium* as opposed to nodulation by indigenous strains in the soil—if any.

Level of [elite] nodulation success is a most difficult one to assess.

Scientists look at establishment levels, farmers look at practicality and cost while seed merchants look at profitability—but in the final analysis, we all need to know: what is the real cost of each treatment related to its performance?

After all, we are not going to get any benefit from seed which has failed to establish, indeed when sown, such seed (historically, being most of what was sown) is an absolute financial loss.

We are only interested in the seed which has succeeded to establish.

It follows logically that the higher the % of successful establishment of any particular treatment, the lower the actual cost [or "success cost" if you like] of that treatment [and we should not forget that a very high % of seed sown around the World traditionally fails to establish, not to mention the additional waste of seed incurred by thinning of overdensely drilled crops.

Seed Quantities and varieties for Oversowing

There are more recently bred varieties of both grasses and clovers produced by plant breeders in many parts of the world now—than those species dominant during the height of field research into oversowing during the 1960's to 1980's. Some of these new varieties may be higher producing or withstand more intensive grazing or be more resistant or hardier, particularly in the regions where they were bred—than earlier more traditional varieties, and most of these new cultivars are protected by plant breeders rights.

However, historical key research during the period of establishment responses to comparative seed treatments did not include such varieties, so they are not included here. This intensive investigation must eliminate ALL items which could interfere with absolutely true comparison between seed treatments.

It will be wise for farmers to study the results of independently and professionally assessed field trials with these new cultivars before committing to them. Some contain endophyte protection from above ground insect infestation while others have been bred for better vigour and disease resistance by special chromosome arrangement such as in tetraploid enhanced varieties. Other coatings include chemicals, but are they *Rhizobium* compatible?

Farmers engaged in sophisticated (computerized) recording systems will benefit from retaining the exact detail of relative cost of seed, inoculation and coatings plus actual field performance of new varieties to maintain accurate comparability and a true record related to observations of today. Plant breeders need to prove the superiority of their new cultivars before they can expect farmers to buy them—today, this must include their performance as inoculated and/or coated seed—oversown into hostile environmental conditions too—not just drilled into a worked seedbed.

Best Research Approach

In New Zealand the most important research was enhanced by the wisdom of a small number of grasslands scientists who recognised the importance of including commercially manufactured inoculated and coated seed treatments in their laboratory and field investigations—among the most thoughtful being Dr. Bill Lowther of Invermay Research Centre, Department of Agriculture, Mosgiel, Otago, New Zealand. Manufacturers have skills and experience which are outside the expertise of core scientists—particularly in large scale production. It is a very different matter inoculating and coating a few grams of seed for field research compared to processing many thousands of kilogram of seed per week, sometimes on a 3 shift continuous 24 hr production line, as CSL had later to pursue to keep pace with demand.

Bill Lowther was the ideal person to administer this research too, because those who knew him were in no doubt that his professional integrity left no chance whatsoever for allowing any sort of bias creep into his work.

Manufacturers Need Laboratory Quality Control

The unique availability of one manufacturer's professionally equipped and staffed laboratory, i.e., FCC/CSL, chiefly for factory support, formulation development and quality control, but also able to prepare and supply meticulously formulated small scale treatments in exact replication of commercial product was very important. In addition, it was also able to prepare special research treatments for independent scientists investigating hypotheses and assumptions—and some exploring the boundaries of the science all of which was undoubtedly instrumental in some of the discoveries made by the New Zealand Department of Agriculture.

It is the firm belief of the authors that no Company serious about supplying commercial quantities of inoculated, coated or precision pelleted seed, should attempt to do so without the support of a well equipped laboratory which includes qualified [formally or by experience including microbiology] technical staff capable of unbiased quality control and the ability to grasp opportunity when it arises, often briefly. In a relatively new industry such as this new possibilities arise [for example, new inoculant carriers, new adhesives and coating materials, new, faster and cheaper coating formulations and drying techniques, ability to maximize production for short term large scale land developments—and other special demands].

A further ingredient of this success and an amalgamating factor which brought the scientists and the manufacturer together, the individual who devised the treatments [particularly the crucial "40 times single rate of inoculation" treatment] and organized their supply, was CSL's development manager who, having been a Farm Advisory Officer of the Department of Agriculture was familiar with the Department's scientific centers, its staff and capability, its rules of operation as well as the commercial laboratory's capability. He was also well aware of the hostility of seed to *Rhizobium*, difficulty of survival, of on farm delays and field conditions. Armed with this experience he became aware of the probable reasons that some independent researchers were obtaining depressed results from using seed they inoculated and pelleted themselves. Some of these results nearly wiped out CSL and the industry in New Zealand. That would have had repercussions elsewhere in the world as CSL eventually exported its successful treatments which have achieved widespread International adoption of the industry today.

There are lessons to be learned here—to succeed, firms need good facilities and adequately qualified staff. Researchers need to be aware of commercially prepared product which includes processing expertise outside the experience of those researchers. Universities need to ensure research they undertake is wholly professional. Superficial evaluation "just to get published" can be seriously harmful.

The work achieved by Lowther and CSL has and will save farmers around the world multi millions of dollars just in saved seed cost alone—year after year. In the past, huge quantities of seed had to be wasted to get any sort of result at all—whereas today, much of it does not even have to be sown to get a superior result, it is not required now, therefore its cost is saved. So, because we are getting these better results, how much seed do we really need to buy and sow?

A typical oversowing mix in New Zealand during the 1960's and 70's would be:

Clover seed

4 kg per hectare [approx. 4lbs/acre] of good quality white clover [*trifolium repens*]
2 kg per hectare of Red clover [*trifolium pratense*] or
[2 kg per hectare of Alsike [*trifolium hybridum*].

And for lower fertility sites where complete fertilizer input is unattainable, a wise New Zealand runholder would add:

1 kg per hectare of suckling clover [*Trifolium dubium*] or
1 kg of Lotus major [*Lotus pedunculatus*]

Plus of course grass seed (see Chapter 4)—but we will study just clover performance here.

For the purposes of this study, the basic [and minimal] oversowing seed mixture [above] provides a total of (say) 6 kg per hectare of clover and with [say] 5 kg per hectare of grass seed = a total of 11 kg per hectare. That is just over half the weight of a traditional arable pasture mix, but is intended, at minimal cost, to create the nucleus foundation of a future more vigorous, palatable and nutritious grassland food source for herbivores which then return animal manure to the sward gradually building up the level of fertility, creating an improving financial return with better quality animals leading [specially if assisted by fertilizer and/or lime] eventually to a change in that grassland from largely unproductive native species to a much improved, even good quality pasture. This is further promoted by managed natural reseeding of the existing species, by closing the area from grazing at appropriate times for natural seed production, gradually building up quality and fertility with both animals, and [ideally] topdressing, until, over time, a productive pasture exists.

Adequacy of sowing rate should of course be much more determined by % of successful establishment than by some old tradition set in historical concrete—take white clover for instance—as there are about 1,320,000 individual seeds per kilogram, if we sowed 4 kg/hectare that will apply about 5,280,000 seeds/ha [e.g., per 100 million square centimeter] that provides one seed for each 20 cm² [about 3 inches square] which is roughly the area a good white clover plant and its root system should occupy at establishment.

However, even in our smaller oversowing mix shown here, there is another 2 kg of red clover plus 5 kg of grass seed as well—if they all established, they would be much too densely populated with intense inter-plant competition. However, as we now know [from this research] 100% establishment from an oversowing does not happen. Past sowing rates have more likely been assessed on success rates of <u>less than 10% establishment</u>, but as shown in this Chapter, we can do a lot better than that these days, which means less seed is required than in the past.

Thus, oversowing is sensibly viewed as an initial small but key investment in the longer term future of any ex native pastoral system. To aim for a first class pasture out of the rough by surface sowing into native species in one hit—say within a year, would be an expensive exercise indeed requiring more seed and fertilizer per hectare, closer subdivision with more intensive grazing therefore more intensive water supply, certainly more risk and requiring much heavier initial financing. The whole effort might also fail as we certainly have seen in the past in New Zealand—financially embarrassing the farmer. However, a temporary setback is not as drastic if we progress in small steps until we are confident of success. It starts with small field plots and moves up in size and scale from there. First class pastures can be developed within a year from the rough under arable conditions which usually means easier country on flat or undulating land—and at much greater cost in machinery, fuel and time. That is not the goal for steeper, rougher and tougher country where improvement over time, well managed by patient farmers will eventually become successful but at much more modest cost.

The above seed weight is as bare seed. But as this seed is to be coated, what will the total sowing weights ideally be?

That will of course depend on the % of successful establishment achieved. That is, if I sow 100 bare seeds and 100 coated seeds, the latter [say] doubled in weight by coating, then I will be sowing 2 kg of the coated variety to every 1 kg of bare seed. The coating will be an added cost on top of the same seed weight—so I will need to be convinced that I should pay more for the coated seed, yet only get the same weight and number of seeds.

If hypothetically the coated seed gave 60% establishment and the bare seed 40%, I have paid for 60% of the bare seed which did not grow but only 40% of the coated seed which did not grow. Put another way, of every 100 bare seeds sown 40 grew whereas of the coated seeds 60 grew—then because the seeding rate is likely to be based on earlier experiences with plain seed, where 40% establishment or less has historically been acceptable, I only need to buy and sow 67 coated seeds, e.g., [67 seeds x 60% = 40.2%] to get 40 seeds established—being the same number established as bare seed. Then I have to find out [quite apart from subsequent nodulation success, plant vigour and health] which one of these costs less—100 bare seeds or 67 coated ones?

As coated weight cost is usually less than same seed weight cost, by buying coated I have:

A/   saved the cost of buying 33 seeds [which would not grow] plus their coating cost.

B/   not wasted the cost of coating the 67 coated seeds which did grow.

C/   to balance the cost, I should calculate the cost of 67 seeds {i.e., 40.2%} plus their coating [which did grow] and compare that with the cost of the 100 plain seeds required to achieve the same 40% establishment. And, I need to consider plant health and dry matter production as well.

This is relatively easily worked out—but of course is hypothetical as we have selected percentages which may not reflect actual—therefore these figures don't mean anything in practical terms—but illustrate a point.

There is a better way of comparing treatment success related to true results and actual cost. We need to compare [using identical seed and inoculant]:

1/.   Plain seed weight and cost plus cost of inoculation, with ⌀

2/.   Identical seed weight to 1/. plus cost of inoculation and coating:

- both related to levels of successful establishment identified in carefully executed field research which will then allow us to identify the actual cost of successfully established seed of each treatment—and indeed the cost of the seed which failed. For bare seed inoculated, we must also remember, it needs to be carted up to the site, oversown separately from fertilizer [which can quickly destroy *Rhizobium*] and for which a separate flying [or other] coverage of the development block would be required to spread fertilizer—for that inoculated bare seed treatment only. That may be an added cost which, if so, should be included in final results.

To compare these treatments accurately—as to % establishment and actual total cost, we need to find a classic example of oversowing with every check carried out by independent experts to make sure that the treatments, their application, the location and the results can be relied on as being genuine and reasonably repetitive after which we can then apply costings in relation to results to measure the true cost of establishment of each treatment.

Where do we start?

There has been some excellent work along these lines done in Australia as well as in the US and Canada—however we cannot use the result of every field experiment in multiple locations or indeed use multiple processing methods around the world into inoculation treatments to determine true results and actual cost.

In addition, we need to be confident that if we select one really solid official establishment and growth result from sound field research upon which to measure relative performance and subsequent cost of successful inoculation, establishment and growth, that such information will indisputably represent on farm results, upon which not only farmers may rely, but also future manufacturers [because the principles remain, even where the processes are altered]. Such critical investigation must be conducted absolutely free of any commercial or other bias having been planned and executed by one or more senior, independent, highly skilled professional grasslands experts with results subjected to visual and mathematical analysis—and *Rhizobium* counts carried out on all treatments independently, after receiving them from the manufacturer (to ensure consistent rhizobia loadings).

That narrows down the range of opportunities considerably.

It points quite clearly to that classic investigation conducted by Dr. W.L. Lowther and Technician I.R. McDonald both of Invermay Agricultural Research Centre, Mosgiel New Zealand, the results of which were published in The NZ Journal of Experimental Agriculture Vol. 1. 175-9. Titled [deceptively humbly and simply as]:

"Inoculation and pelleting of clover for oversowing"

(NB. The term "coating" of clover is now replacing pelleting which is a different product).

- whereby an exhaustive examination of all relevant qualities of treatment were undertaken by them in 1972, not just the merits of inoculation and or coating, but of 6 individual treatments including the popular ones at that time and of course plain seed not inoculated as a control measurement.

Every aspect of all treatments was carefully analysed in this exemplary research effort—not matched by any comparable study we have been able to identify. We have elected to base our analysis of true cost on these results—but hasten to add that variations in quality of materials used or methods of assembly will almost certainly give different results.

The results in this fundamental research apply to the materials and their qualities, processes and costs at that time [1970's to 1980's] with some evidence since, that qualities of some inoculants and various coatings in New Zealand and around the world are not the same as these.

At last however, with considerable authority and finality the following examination clearly demonstrates what works well—what does not—and why.

Importantly it demonstrates what is technically and commercially achievable. Furthermore, if farmers cannot now access the same quality of product as used in this study, this book provides the formulae for achieving at least the same quality coatings themselves.

As for inoculant quality, "New Rhizocote" commercial cultures manufactured by Axis Associates Ltd of Nelson, New Zealand attained even higher quality than the standard "Rhizocote" culture used in this field study [which used modified peat as the carrier then but has used more protective diatomaceous earth more recently].

It included counting *Rhizobium* cell numbers per seed on each treatment at day 1—counting them again at various storage times after stress delays to simulate commercial conditions. All treatments were surface hand sown onto existing vegetation in replicated, mathematically approved separately fenced field plots at not just one, but three typical hill country tussock grassland or fern covered sites notoriously hostile to establishing seedlings and more so to rhizobia, the three typical sites with variable altitudes from 420 to 1,070 metre a.s.l. {above sea level}. The results of these sowings meticulously measured by independent expert manual counting of actual numbers of successfully established healthy clover plants eventuating from each treatment—counted both under shade [bracken fern] and out in the hostile open by professional scientists of the New Zealand Department of Agriculture.

To measure vigour and production as a separate item to % establishment, each treatment was also [multiply] cut for analysis of dry matter production.

To ensure identical quality of seed and inoculant, each were supplied from one seed lot by the manufacturer's laboratory together with a sample of standard commercial coated seed called Prillcote™. The other treatments, inoculation of bare seed, lime coating, gafsa phosphate/dolomite coating and inoculation at 5 times the single rate of rhizobia—were all processed by the researchers.

This investigation had become essential due to that previous controversial evidence, some of which was published. The result—critically important.

This field and laboratory experiment was directed at identifying answers to these questions:

1. Is clover seed inoculation important in these native grassland localities typically colonized by fern or tussock?
2. If inoculation is important, does coating the inoculated seed assist inoculation by maintaining rhizobia viability? and if so does that assist establishment and does it also promote more healthy vigorous clover plants?
3. If inoculation is important and coating aids nodulation and/or establishment and/or growth, which of the currently popular coatings has the most beneficial effect?

To eliminate all chance of variable results due to miscellaneous factors such as inconsistent seed, inoculant or coating material qualities, all three were supplied from one source of each by the manufacturer of Prillcote™, being the product of Coated Seed Ltd at that time, to be included [and compared] with other coatings and with simple inoculation of bare seed—and indeed plain seed too with no treatment whatsoever [as emphasised, the seed was taken from one line of "Grasslands Huia" white clover with a good quality P&G {purity and germination} Certificate from New Zealand's independent Seed Testing Station].

Because there was strong evidence at this stage that massive inoculation [high number of *Rhizobium* applied per seed at sowing] could override other treatments such as coating, all inoculation treatments carefully applied the same or similar numbers of cells per seed when processed—or at sowing. Further, to ensure there were no anomalies in relation to numbers of cells per seed for each treatment, the researchers subjected each treatment to a *Rhizobium* count per seed by the plant infection technique [of Brockwell 1963] at day 1 [for the two coatings—lime and gafsa phosphate/dolomite processed by the researchers] and repeated after a time stress at day 17. The Prillcote™ coating was identically tested at days 3 and 20. The plain seed inoculated only was also tested for cells per seed at three time spells, at inoculation, after 1 day and after 17 days.

There was a further treatment being plain bare seed inoculated at five times the single rate of cell loading per seed, which treatment was designed to re-check the effect of higher numbers at sowing—however, that treatment was simply sown at the three sites as a check, not counted as to cells per seed—but sown with the rest and inoculated carefully by the researchers.

The Results are abbreviated here for wider understanding of readers.

Table 7. Comparative study of—levels of inoculation v plain seed, two coating types and a standard commercial inoculated and coated product—all of which were applied to an identical line of white clover seed.—Author.

The treatments were :-   [from - Lowther and McDonald]
1. Plain clover seed
2.   "   "   plus normal inoculation
3.   "   "   "   "   plus lime coat
4.   "   "   "   "   plus gafsa/dolomite coat
5.   "   "   "   standard commercial Prillcote™ coating.
6.   "   "   "   inoculated at 5 times single rate.
⬇

**Treatments numbers
carried down to "Results".**
⬇
⬇

RESULTS

	Rhizobium/seed.		% healthy seedlings*.	Production (avge 2 sites)
⬇	[At inoculation]	[at day 17]	[Average of 3 sites]	[kg/ha dry matter]
	⬇ [x 1000]	⬇	⬇	⬇
1.	none	none	>3	37.5
2.	3.32	1.92	28	150
3.	3.32	1.58	45	298
4.	3.32	2.12	34	301 [1 site]
5.	3.32 [at 3 days]	2.92 [at 20 days]	82	538
6.	[not counted]	[not counted]	45	286 [1 site]

* As a % of emergent seedlings pegged. This does not identify % of seeds which failed altogether which would possibly further condemn bare seed – however. that issue is covered by dry matter production because identical areas of total growth were cut from each treatment then herbage dissected to extract the clover – then weighed. The absolute thoroughness of this research is the reason it has been used as a classic example of actual field performance and for costings.

This was a hugely important result which has had widespread ramifications.

While the result was gratifying for the manufacturer of Prillcote™ [Treatment 5]—by far the most significant feature of this intensive research effort was the fact that the Prillcote™ coated seed carried no more *Rhizobium* cells soon after inoculation than did the other four inoculated treatments at inoculation, meaning the superior number of healthy seedlings [almost double the next best treatment] and nearly *double the dry matter production* of the next best treatment could only be due to the fact that the Prillcote™ coating kept more *Rhizobium* alive in a healthy environment for longer than any other treatment.

At last, we can now introduce actual cost—seldom if ever included in research results. Rather amazing really—because the results don't mean much—at least to farmers, until they are cost/benefit analysed.

THE ABSOLUTE REALITY OF CLOVER INOCULATION, COATINGS, OVERSOWING BENEFITS, SEEDING RATES AND COSTS. *THIS IS TRULY THE "BOTTOM LINE".*

These carefully researched results included using the "do it yourself" coating method promoted some years earlier by those pioneering and innovative DSIR Technicians [Technical Officer Athol Hastings and Dr. Doug Dye] which for this experiment were prepared by Lowther and McDonald using the DSIR formula and those same simple materials, methyl cellulose as adhesive and finely milled lime on one treatment and finely milled gafsa phosphate and dolomite [50/50] on the other, as the popular coatings of the day—which indeed were [as seen in the following Table], both superior to simple inoculation.

Plain bare seed, often used in the past, gave disastrous results at all sites. Maybe that is where earlier traditionally high sowing rates came from!

Prillcote™ commercial coat while easily the best treatment, was "as at" the early to latter 1970's—we should be able to improve on that quality by now, we know we can—but commercially, do we? There is still clover, grass and brassica product being sold as Prillcote™, but no longer by former Coated Seed Ltd [which is now terminated]. The successive manufacturer's processing is clearly not the same and for instance no longer uses Rhizocote inoculants. It should now be a better product, as a result of a technical "breakthrough" introduced by the authors while employed by FCC Ltd and CSL using a chemical offered to author Bennett by Mr. Art Gross, scientist with USA's "Celpril" which purchased original Prillcote™ formulae from CSL. Its successful use now offering a much simpler product to process with potential for even better rhizobia survival. We explain this superior process [called "New inoculation Technique"] in Chapter 6.

Rhizocote inoculant itself has certainly improved since then, however production ceased in 2010 when the FCC and CSL relationship had ended. Also coating methods such as the "new inoculation technique" as indicated above, devised by the authors [and quite different from standard Prillcote™ used in the experiments above] whereby seed is no longer inoculated first—but coated first, dried by warm air instead of CSL's expensive slower vacuum refrigeration driers [raised temperature being acceptable because there are no rhizobia to damage at that stage] then stored in relaxed short or long term storage conditions as uninoculated coated dry product which is then held ready for a rapid machine overspray of culture applied freshly with minimal moisture just prior to sale and sowing giving enhanced results in terms of rhizobia survival, which certainly revolutionized the coating industry in the late 1970's.

However, it is unlikely anyone has repeated the huge and complex workload of Lowther and McDonald simply to test these new products. In our search for the absolute truth about this technology—we are not interested in "secrets and confidentialities", we need to work with authentic results of specific products and methods used, combining this with actual comparative costs at the time this work was carried out. Only then will we obtain a total picture of the benefits of inoculation, of various coatings, of sensible seeding rates and of the genuine final costs associated with each method. All ingredients need to be fixed in time and cost to get a true comparison.

We elected to fix "the date of" these findings as mid 1970's at which time we have kept a record of actual costs of inoculants, seed and coating for accurate comparison. It is the accurate comparison which matters—the specific date and cost does not because the principles learned transcend dates and costs. It is also important to note that the author and contributor have no vested interests in the result of these investigations—we are reporting only the facts.

If inoculated and coated products are better than this today, then the following results will only be more positive in favour of both—but meantime, we are holding fast only to known facts.

This of course is just in reference to clover and to oversowing. We will also look at the same scenario with grass seed and oversowing.

It is entirely reasonable to take this classic field trial as a fair indication of what happens out there in the field, equate the results with actual costs of each process or product and analyse from those field results what the ACTUAL cost of each treatment was at that time [and probably still is in relative terms on undeveloped virgin land].

All contributing costs have increased since then of course [most of it being inflation], Dr. Lowther is now enjoying a well earned retirement, and we do not know what quality the commercial coaters are producing today—including in New Zealand where some seed processors historically under performed in both inoculation and coating. Farmers in New Zealand and overseas should insist on copies of independent tests of any product they are tempted to purchase today [particularly *Rhizobium* counts per coated seed after a delay stress], and if that is not available, our advice is very clear and commonsense—don't buy the product, it will be far safer to refer to Chapter 6 of this book and prepare your own—after all, this is a crucial tool of farm management, you do not need to take risks—you do need to take care.

Further Evidence

These Invermay [Department of Agriculture] researchers also produced an excellent and very positive graph of one of the three sites, e.g., the Wanaka result [alone, but being one of the sites oversown and recorded in the Table below is quite typical of the whole]. The following reproduction of their graph is demonstrated here with kind permission of the Editor, New Zealand Journal of Experimental Agriculture:

<u>Table 8a.</u>   WANAKA SITE   [from Lowther]
{490m a.s.l. stunted bracken fern and fescue tussock.}

<u>Percentage of healthy white clover plants *under cover*.</u>
                                        ☞[ie shaded.].

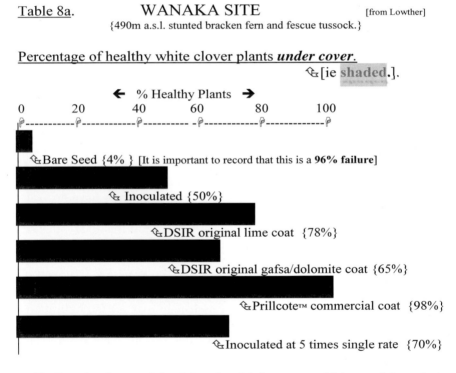

Dr. Lowther has explained that the chief purpose of this graph is to show the spectacularly superior establishment and growth when clover plants are protected from the elements during early establishment—as opposed to their growing out in the open exposed to wind, frost, direct sunlight and snow blanketing [not hoof tramping as no animals had access to these plots].

The research paper showed this result in one graph with black bars for "under cover" results and open white bars for plants growing "out in the open". We consider this graph so important and the black bars so much more visually helpful, that we have produced them as two graphs with the "sheltered" result shown [above] and the more severe "exposed" result as follows:

<u>Table 8b.</u>   [from Lowther]
<u>Percentage of healthy white clover plants growing *out in the open*</u>
                                                  🐾[**no shade**]

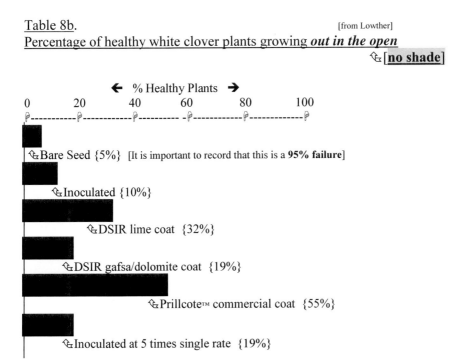

        ← % Healthy Plants →
0   20   40   60   80   100

🐾Bare Seed {5%} [It is important to record that this is a **95% failure**]

🐾Inoculated {10%}

🐾DSIR lime coat {32%}

🐾DSIR gafsa/dolomite coat {19%}

🐾Prillcote™ commercial coat {55%}

🐾Inoculated at 5 times single rate {19%}

If anyone had any doubt previously as to the importance of retaining some natural cover for newly establishing clover plants, the above two graphs will remove all such doubt. It is most important not to burn, slash, spray or in any way remove good clean ground cover which is not a threat to later pasture development. Cover such as open bracken fern and particularly tussock is important—the latter a soil stabilizer as well as providing some wind shelter to sheep. The higher altitude and greater exposure, the more important this is.

It should also be noted, some suppliers of inoculated coated seed still recommend sowing the traditional rate of seed per hectare PLUS weight of the coating [and inoculation]. They are usually seed merchants of course and their job is to sell seed, but maybe some are not aware of the facts—or else their product is not up to the standard employed in the research recorded here. If a new treatment gives consistently higher % establishment than a former standard rate of sowing, then of course you do NOT need to sow as much of it to get an equally good [and usually more healthy] result.

That is commonsense.

If however, a farmer does not use quality inoculated and coated seed of high interim germination or has not adequately prepared the seed bed to be oversown or has failed to topdress it appropriately or sows at the wrong time of year—or does anything to diminish the potential result—then of course he will get a lower % of seedling establishment and in such cases of course he will need to sow a lot more seed to allow for a higher level of failure—probably also at a much reduced level of vigour and growth.

Manufacturing Variations

Seed economies are increasingly important for farmer purchasers because as we breed higher quality seed, costs of that superior seed will continue to increase re-enforcing that we should aim to maximize % establishment and only sow the minimum necessary to achieve a satisfactory result with that more expensive seed. To repeat, if we get an improvement in establishment from various seed processes, then of course we can afford to sow less seed provided it has been processed to the earlier proven standard or better, for a good result—BUT IT MUST BE AT LEAST TO THAT STANDARD OF INOCULATION AND COATING. Furthermore, the process which achieves better % establishment [and hopefully growth], needs to cost significantly less per kg than the value of the extra but untreated seed weight historically used.

In the past, seed sowing rates have often been grossly excessive in relation to the surface area they are capable of covering. Farmers and growers need to be aware of these facts when ordering from seed merchants, some of which in New Zealand still supply coated grass seed as 50/50 seed and coat by weight, but some also supply grass seed simply liquid sprayed against predators [insects, fungi and bacteria]. These are useful treatments too but gain no appreciable weight, are therefore not coated and are not included in these recommendations. Clovers are now supplied with different seed/coat ratios than we are examining here—some with a 75% coat [including inoculation] and a 25% seed ratio, recommending this be sown at the same rate as former bare seed which is supported by the results we see in the Table below with the 4/7ths seed [57%] ad 3/7ths coat ratios—provided the quality of the product is up to the standard of our researched product—and that the conditions of oversowing are similar.

Also, today, lucerne is sold as 25% lime and molybdenum coat, and 75% seed to be sown at the same rate as plain inoculated seed used to be—and that too, may be a satisfactory seed rate reduction, even in cultivated seed beds where establishment is much more certain, provided the quality, freshness and cell loading of rhizobia are equal to the product studied below—the 25% seed sowing weight reduction [as lime coat] being compensated not only by precision placed fine lime around each seed, but also a trace of molybdenum on acid soils.

Brassicas were not investigated during the height of coated seed research in the 1970's, are not legumes, do not associate with *Rhizobium*, are not usually oversown into existing native grassland and do not appear to have been independently researched in terms of coating benefits—but are today offered as bare seed for drilling or broadcasting in arable conditions having been chemically protected from a wide range of predators. Or with a narrower range of protection but as 90% seed and 10% coat, being a product probably better suited to surface sowing onto arable seed beds.

These variations will continue and change with time, but the PRINCIPLES established in the following research will remain forever and need to be carefully followed by the manufacturers of today and the future.

Farmers simply want the *best possible field result for the least possible cost*— and that is what we are about to study.

The Actual Cost [based on above actual results]

We repeat, by pursuing the classic field research of Lowther and McDonald we have meticulously used consistent materials and methods for comparing these treatments—so now, everything is related to a fixed level of quality at a fixed time allowing us to make absolutely accurate comparisons of the actual cost of each treatment, and most importantly, the *true cost in relation to each result*.

Using the exemplary research results [above] and applying those to the sowing of [say] 100 hectare of hill country, the following are the actual costs of each white clover seed treatment: [later we shall include grasses, coated and bare for a full oversowing result].

In the mid 70's a pack of culture to inoculate 40 lb of white clover seed sold for $1.75 including surface postage which on 40 lb [18.14 kg] of seed divided by $1.75 = 103.00 g per 1 cent [or approx. 9 cents/kg].

A kilogram of "Huia" certified p.p. [permanent pasture] white clover seed cost $1.81. A kilogram of Prillcote™ inoculated and coated white clover seed cost $1.60 comprising 4/7ths seed by weight and 3/7ths coating material.

NB: "Actual establishment %" in the Table below is the avge. of "shaded" and "open exposure" (in Tables 8a and 8b above). Costs of transport, fertilizer, aerial or other application, mixing, etc.—are reasonably constant except that inoculated plain seed [not coated] must be oversown separately from acid fertilizers.

Table 9.

Therefore:
WHICH COSTS LESS?

Treatment [White clover]	*Cost @4 kg/ha [seed $1.81 kg]	Cost for 100 ha	Actual establishment %	Cost to get a full 4 kg established on 100 ha
Plain seed	$7.24	$724	4.5	$16,088
Inoculated	$7.24 + $0.36	$760	30	$2,533
Inoculated x 5	$7.24 + $1.80	$904	44.5	$2,031
Prillcote™ coated	$6.40	$640	76.5	$837

*Actual costs charged by CSL to Seed Trade in mid 1970's. Inoculant cost was 36c/kg.

This analysed outcome is of National importance—it is a stunning reward for the designers of Prillcote™ and a vital warning to developing farmers about *wasting money on plain or inoculated seed.*

Because the original DSIR lime and gafsa coats [included in the field work] were purely investigative and ultimately not recommended by Dr. Lowther, they are not included in these results.

While no two sowings will produce exactly the same result, this is as close as we are going to get to an exacting result for white clover under actual field conditions of the various treatments above—AND each actual cost.

This *total analysis*—fixed product qualities, exact performance and actual cost has never before been published to the knowledge of the author.

Now we get a full understanding of the potentially serious financial damage to farmers and runholders who may have followed the 1966 Grasslands Conference advice of Cullen and Ludecke. However, they did qualify their results with the caution that future treatments "may give different results"—they certainly did!

Where we are getting less than 10% establishment from plain seed then that is probably what we have become used to accepting in years gone by [prior to WW2]—and if we can now get 70 to 80% from effective seed treatment, then quite clearly we need a lot less seed.

Nature itself covers risks of new life failure by producing a massive oversupply of reproductive agents—not just seed, but pollen, larvae, sperm, small organism eggs, microorganism division—and other mass reproductive mechanisms, but this "volume" philosophy does not work where the seed is becoming of higher quality and more expensive and where today, "% survival" beats "outright volume" providing large economies with a superior all round cost/benefit ratio.

Further, the % establishment results do not include the additional growth shown in above [RESULTS] Table 7 [almost 4 times that of inoculated bare seed] or healthier plants produced by the best coating treatment in the "% healthy seedlings" and "production" [of dry matter] both obtained by Dr. Lowther in his field research—he reported that % establishment and health plus growth results were closely correlated in the field, i.e., *the treatment with the best establishment also gave the highest yield with the healthiest plants.*

These coating treatments including processing cost are cheaper than seed [by weight] and also pay a big dividend in reducing the actual weight of seed required. In most oversowing situations, sowing the same weight of coated seed [at the former 4/7ths seed weight ratio] as formerly applied as bare seed will be adequate seed per hectare and even at a 50/50 seed/coat weight, may still be adequate provided the level of inoculation, the protective coating and the quality of seed—are all of a high standard.

Above findings comprise a truly dramatic result [seldom experienced in research] which serve to emphasise the importance of sound investigative study and high quality treatments. Though some have tried, most seed merchant companies did not have the expertise necessary to produce top quality product such as the Prillcote™ treatment recorded here [and developed by the authors plus the late G.G. Taylor M. Agr. Sc.]. They needed to employ expert experienced and qualified staff—a fact which senior staff of Wright Stephenson

& Co recognized in the mid 1960's when they unsuccessfully attempted coating seed themselves, then approached Fruitgrowers Chemical Co Ltd , Port Mapua, manufacturer of a wide range of pesticide formulations including granular 'prills'. Thus Coated Seed Ltd. was established and significantly strengthened by the combined expertise of the Chemicals Processing Company [FCC Ltd.], a Minerals processing Company [Lime & Marble Ltd], a substantial Seed Company [Wrightsons™]—and the appointment of an agriculturist from Department of Agriculture with experience in legume seed inoculation and coating of seed.

Those were the ingredients behind the success of Prillcote™—but the farmers of New Zealand also required one more vital component, namely, an independent expert to identify the factual results of various treatments which by good fortune became the vital role of Dr. W.L. Lowther to whom all of New Zealand, particularly New Zealand farmers and runholders should be very grateful. The New Zealand Department of Agriculture certainly made one of its best ever investments when it employed Ph.D graduate [Western Australia] Bill Lowther onto its scientific staff at Invermay Research Centre, Otago in the 1960's.

LUCERNE [*Medicago sativa*] [ALFALFA]
{strain specificity *Rhizobium meliloti*}.

As for white clover, there has been an intensive study of performance of various lucerne seed inoculation and coating treatments around the World during the same period as for clovers—chiefly from 1960's through the 1980's. But no study we are aware of has shown actual field performance of the most popular seed treatments related to their respective costs which finally identifies actual benefit to the farmer. We now do just that.

Lucerne, a native of Iran, like clover, is a nitrogen fixing leguminous species with a natural deep root profile which can access moisture below the level of accessibility of most other forage legumes and grasses. Its roots have been identified growing out of the bottom of riverside banks more than 15 metre below the [parent] lucerne crop growing in the field above.

Among the grasses however, *Phalaris tuberosa* can also access deep moisture. Both it and lucerne have been *extremely* valuable in summer dry, drought prone areas. Unfortunately, both are sensitive to weed infestation and inter-plant competition. In an attempt to deal with vulnerability to weed infestation, the tetraploid genome of lucerne has been modified in USA to resist the herbicide glyphosate ["Roundup"] so that weed infested lucerne stands could be sprayed to eliminate weed competition without damage to the crop. That has been successful—but.

While technically successful, fears of spontaneous adverse genetic modification which might lead to glyphosate resistance spreading {via cross pollination} to other cultivars including the more serious weed species themselves, has led to legal intervention in USA where the concept of *"Roundup ready lucerne"* is under US Department of Agriculture study to determine its risks.

Lucerne of various types and varieties [from upright habit for hay and silage cutting to stoloniferous ground cover for grazing] together comprise the most cultivated legume in the World. It is particularly important in USA, Australia, New Zealand, Canada, Russia, parts of South America, South Africa, Middle East and Southern Europe [France, Spain, UK, Italy, Greece and Turkey], and elsewhere where topsoils dry out in summer and where the crop can be cut several times per summer season for high quality leafy hay [or silage] baled and stored for winter supplementary feed. Its protein rich leaf is also used in salads for human consumption, its concentrates for health supplements and its flowers for decoration. It is also compressed into dried pellets for export as a livestock supplement.

Lucerne Nodulation

While lucerne has such important features—nitrogen fixation, root depth against dry soils, nutrient rich foliage, top quality hay, etc.—it also has one important limitation in that its strain specific *Rhizobium* [*R. melilotti*] do not persist for long in the absence of the host plant specially in soils from mildly to strongly acidic [pH 5.8 or lower] —unlike clovers which have a wider range of specificity including some strains which persist without access to a host even in moderately acidic soils, but tend to be of poorer nitrogen fixing capability.

Lucerne does not accommodate such wide spectrum strains of *Rhizobium* and the strain it must associate with needs to be introduced with every new sowing of the crop otherwise it may fail to nodulate. If so, leaves will turn yellow from nitrogen deficiency [unless topdressed with expensive nitrogen fertilizer which will further depress nodulation], growth will slow and the crop will suffer from suppression by more rapid growing weed infestation.

Even more so than for clover, fields to be sown in lucerne last Century were "inoculated" by transfer of some soil from an existing crop to the new field about to be sown—but also more importantly than for clover, that practice would be frowned on with today's knowledge because it also spreads lucerne pathogens—fungi, viral, bacterial and even insect larvae from the old crop [which is probably failing because of the ravages of one or more of these predators]—to the new crop giving predators a "jump start" in infesting and damaging the new crop as well.

Thus the introduction of *R. meliloti* on seed about to be sown developed around the mid 1930's when experimental and then commercial cultures became available to farmers. New Zealand's chief strain of *Rhizobium meliloti* originated from Uruguay, a substantial lucerne grower and has remained unchanged for several years.

The practice [of inoculating lucerne seed prior to sowing] is now widespread on a worldwide basis—but still vulnerable to the qualities of culture produced, their persistence often under adverse conditions—heat, excessive storage time, exposure of inoculated seed to sunlight and warm air desiccation, loss of viability of *Rhizobium* when inoculated seed is mixed with acidic fertilizers and loss of nodulating capacity from acid or saline soils when sown.

Fig. 38 Lucerne (alfalfa) well nodulated on the left and close to a nodulation failure on the right. This stand (crop) was inoculated with lucerne inoculant badly contaminated with clover rhizobia. The green well nodulated left side was a failure too until recovered when over-drilled with Lucerne "Rhizocote" inoculant granules (calcite chip coated with lucerne inoculant and dried by vacuum refrigeration). *The drying step probably not now required—Ed.*

These problems have been studied in many countries around the world and a fair summary of that research indicates that for successful lucerne crops, the seed should be inoculated with a good quality culture of *R. meliloti* providing a minimum of 1,000 cells per seed for lucerne [which has about 440,000 to 480,000 seeds per kilogram—lets say 460,000], should be freshly sown into soils which are not excessively acidic [pH 5.4 is marginal and should be limed, 6.0 + is much better for survival of *R. meliloti* which with lime coated seed should ensure establishment]. Inoculated seed must not be mixed with acid fertilizer [not in the drill box or in the drill row] unless the inoculant is protected by a lime or similar coating.

Even this [above] reasonably simple formula has seldom been adequately met in some countries and never in others. The most limiting factor appears to be poor quality of commercial inoculants which means further stresses on rhizobia {like delay or heat} will often become significantly fatal. Lucerne nodulation failure can however be recovered in the field.

The most disastrous effect arising from poor inoculants is that users of it loose interest in inoculation and nodulation of this vital crop and either resort to expensive artificial nitrogen, or accept a poorly performing crop perhaps assisted to some extent by grazing and the animal manure nitrogen which that supplies. The technique [and the massive free world supply of atmospheric nitrogen] is then dismissed as "academic", "mumbo jumbo" or some other graphic local term of condemnation [because it did cost money—{i.e., the inoculant} and produced only disappointment]—thus the procedure is abandoned for future crops—and probably by future generations of that farm or farmer. Worse, customary procedure is then hard to change.

Where inoculants have been successful, more intensive studies have taken the technology to new heights of achievement and it is from these that we extract an accurate reading of true cost and benefit to farmers of various seed treatments related to optimum levels of nodulation of lucerne.

The differences in subsequent lucerne production between good and ordinary seed treatment is not just significant, it is huge—when executed correctly. We have not been able to identify independent research elsewhere than in New Zealand, upon which we can reliably base true cost to the farmer, probably for two important reasons which may avoid bias:

The first being that only in New Zealand do we appear to have had a major commercial manufacturer of inoculated coated seed during the 1970's through 1980's supported by chemicals and minerals processing companies with a production and quality control laboratory attached, staffed by highly skilled technicians who also manufacture commercial inoculants. They were able to supply to independent researchers specifically formulated treatments using identical seed and inoculants, as bare seed, as inoculated [at specific levels], and as coated.

Secondly, it appears that only in New Zealand have independent research scientists been sufficiently aware of the importance of advanced skills in processing, that they have been willing to use product supplied by that manufacturer processed to critical standards suitable for research purposes—rather than attempt to prepare treatments themselves with limited processing expertise.

Some technicians have made the mistake of indelibly "labeling" seed treatments according to experimental results obtained in their work, based on limited if any professional processing experience whatsoever. That is presumptuous—particularly for a scientist.

We recognize greater expertise around the world in terms of pure *Rhizobium* research than in New Zealand, even specialized alfalfa seed establishment research, but for a clear concise result allowing comparable costings, we don't apologise for using the New Zealand research results due to a uniquely effective combination of technical and commercial skills and very good [maybe unique] co-operation by processors and independent scientists and technicians.

Firstly however, let us take a look at a lucerne study carried out in the laboratory on a sandy silt loam sown with identical seed and inoculant as two treatments:

1/. As inoculated seed—and
2/. As inoculated and Prillcote™ lime coated seed.

Effect of Lucerne Seed Treatment on Subsequent Production

Scientist K.R. Middleton of the New Zealand Ministry of Agriculture's Ruakura Agricultural Research Centre, Hamilton, conducted this examination of above two treatments sown in the early 1970's as follows:

{We also have directly comparable New Zealand seed and inoculant costings for these sowings which adds a further important practical and economic dimension to the trial results}.

There were (most thoroughly) 12 replications of each treatment. Bare seed was inoculated by the scientist at point of sowing with the commercially specified rate of culture supplied to the scientist together with standard Prillcote™ formula inoculated and lime coated seed. Resultant growth was cut on three dates and dry matter weighed.

Table 10. Lucerne Yield [as Dry Matter Per Acre] [from Middleton].

Date of cut	Bare seed inoculated	Prillcote™ inoculated coated seed
22nd March	643	1920
27th April	957	2531
28th May	1588	2159
TOTALS	3188 lbs	5439 lbs

This yield is more than a tonne/acre higher for lime coated seed than plain inoculated in the first 6 months growing season alone. Once again, it is a result which can only be expected where quality of materials used is high and where the soil involved is slightly acid giving the lime coated seed an advantage. Lime broadcasting is important for acid soils, but where that requirement is marginal, not available or not feasible, the small quantity of lime in the lime coated seed, is vital.

Another examination of processed lucerne seed performance—also chiefly as simply inoculated compared with inoculated and coated seed under research conditions was undertaken by scientist Dr. Ron Close of the former DSIR, Lincoln, New Zealand in conjunction with Mr. John Whitelaw B.Agr.Sc. Managing Agronomist for a Canterbury export lucerne pelleting Company plus Mr. Geoff Taylor M.Agr.Sc. who was Technical Director of Coated Seed Ltd.

Lucerne Inoculation and Fungicide Seed Treatment Research

In addition to a comparison of the two basic lucerne inoculation treatments, this research study also investigated the effect of incorporating fungicides into the coating of inoculated and coated seed. Reason for this being that lucerne in New Zealand has become subject to infestation of a range of diseases which are particularly virulent where the crop has been grown repeatedly on the same land, or where weed infested older stands are "renovated" prior to renewed seeding. Such renovation carried out with selective weed control, liming, fertilizer and surface tillage [spiked harrows or light grubber to loosen soil—but not grubbed where bacterial wilt is evident as it accesses the plant via damaged crowns]. Young seedlings are very vulnerable to a wide range of fungous disease around the world, but in New Zealand *fusarium* and *verticillium* wilts, *pythium,* downy mildew and *phytopthera* have become significant problems of the lucerne crop. Introduction of resistant varieties has been helpful and seed treatment offers some respite.

In this experiment sown under normal field conditions in Canterbury New Zealand, the fungicides demosan and benlate formulations [in combination providing a wider spectrum of control than either alone] each at 1% [formulation—of seed weight] were incorporated in the seed coat as an additional treatment to standard Prillcote™ coated seed and an additional treatment had thiram included after a pre-soak in water at 30 deg. C for 3 hours—then Prillcote™ processed as for the standard coated seed treatment. The results of this research were published in the Proceedings of the *Fourth Australian Legume Nodulation Conference* held in Canberra titled "Studies on the Nodulation of Lucerne in the Field" Paper No.12, p. 2.

While we are more concerned here with the direct comparison between lucerne just inoculated and Prillcote™ inoculated and coated as a basis for accurate costing of the two methods from actual [professionally managed] field results, we also include the benlate/demosan treatment and the thiram treatment results which are interesting and also significant.

A further reason for importance of fungicides is the use of precision drilling of lucerne where quite a small weight of seed per hectare can be sown to a uniformly spaced pattern of plants [saving seed waste and avoiding inter-plant competition]—which precise sowing needs to be with high quality seed, effectively inoculated and coated. Where fungicides may also maximize % establishment, farmers {or their precision drilling contractors} can achieve a pattern of one lucerne plant established at every 2 , 3 or 4 cm^2 which in theory should provide the best lucerne plants free of intense inter-plant competition—provided they all grow.

In view of the increasing damage to lucerne from soil and seed borne pathogens, from the mid 1970's the lucerne crop has also suffered from the ravages of insect infestation in New Zealand such as blue-green aphid which not only sucks the sap of lucerne foliage, but injects a toxin causing leaf deterioration and drop, and also bacterial damage including bacterial wilt [*Corynebacterium insidiosum*], fortunately, since this fungicide work was undertaken there has been an important advance in selection of resistant lucerne cultivars as well as introduction of lucerne varieties with a wider growth tolerance to extremes of heat and cold than the former traditional New Zealand "Wairau" lucerne cultivar.

However the merit of plain inoculation versus inoculated coated seed has remained largely unresolved as to "microscopic" examination [i.e., very close examination under identical conditions conducted by scientists] and totally unresolved as to the actual cost of each. This question is now addressed.

There were four replications sown in the field. 6 months after sowing all plants were dug and examined for nodules with the following results:

Table 11. Lucerne [alfalfa] from Close, Whitelaw & Taylor.

	Effectively nodulated plants
Inoculated seed	27%
Prillcote™ lime coated	80%
Thiram pre-soak then Prillcote™ coated	79%
Prillcote™ with *1 demosan and *2 benlate in the coat	85%

Both manufactured by Du Pont *1 Demosan is an organochlorine [Chloroneb] effective in controlling soil and seed borne diseases particularly damping off and *pythium*. *2 Benlate [benomyl] is a systemic fungicide which has been most effective over a wide spectrum including downy mildew, *pythium* plus other seedling diseases however, may no longer be in production having been found severely toxic in contaminated drinking water in USA. Replacements such as Topsin M [Thiophanate methyl] may be satisfactory for many uses but would need to be tested in conjunction with *Rhizobium* prior to use as a seed treatment like those in the above table.

Despite this result, lucerne establishment results considered satisfactory have been obtained around the world over many years from simple seed inoculation—dependent of course on good quality inoculants, fresh sowing after treatment and sowing with lime—but not mixed with acidic fertilizers prior to or during drilling. Liquid inoculants [soil injected below seeds] are of increasing importance specially where peat is not available for solid inoculant manufacture.

Nevertheless, as indicated by the above field result [of Close, Whitelaw and Taylor] in Table 11 prepared and laid by experts, we cannot ignore the fact that the inoculated and coated seed was three times more successful than plain inoculated seed, and that this result is likely to reflect the majority of actual commercial sowings on the farm.

In the 1970's, sold ex Coated Seed factory to the seed trade, "Saranac" lucerne cost $6.76 kg and a packet of Rhizocote "New Formula" lucerne inoculant for 100 kg of seed cost $8.80. At the same time, Prillcote™ inoculated and lime coated "Saranac" seed, cost $4.48 kg [being 4/7ths seed and 3/7ths coat by weight].

Therefore, using the % of successful nodulation and establishment from the Close, Whitelaw and Taylor research result, the following costs can be attributed to the basic two treatments—just inoculated—and Prillcote™ inoculated and coated.

Table 12. [column 2 is from Close, Whitelaw and Taylor].
LUCERNE SUCCESS & RELATED COST

1	2	3	4	5
Seed treatment	Nodulated & established.	Cost per Kg	Cost to drill 10 hectare @ 10kg/ha	Cost of achieving a full 10kg of <u>seed</u> successfully established on 10 hectare
Inoculated plain seed	27%	Seed $6.76/kg Inoc.$8.80 /100kg	$684.80	$684.80 = 27% success therefore :- 1% = $684 ÷ 27 = $25.3 x 100% success = **$2530**
Prillcote™ Inoc. & coated	80%	$4.48	$448	$448 = 80% success therefore :- 1% = 448 ÷80 =$5.6 x 100% = $560 [being 4/7ths seed] ÷ 4 =$140 x 7 = **$980**
Prillcote™ Inoc. + *thiram	79%	N.A.	-	-

N.A. Means, not commercially available or costed at that time.

*Thiram is a standard seed treatment against seedling diseases such as damping off, smuts, seedling decays, rot, phoma and ergot. It is approved for use in USA by EPA [the Environmental Protection Agency] for Sugarbeet and Vegetable seed labels.

This confirms that inoculated plain lucerne seed was almost three times more expensive than Prillcote™ inoculated and coated lucerne seed. Put another way, the Prillcote™ was not much more than 1/3rd the cost of inoculated seed to achieve the same numbers of seeds established. Neither of these take into account the better health and greater production of Prillcote™ as shown in K.R. Middleton's research [above].

The [Close, Whitelaw and Taylor] researchers found that fungicides benlate and demosan may be safely incorporated into the lime coat of inoculated lucerne seed which suggests other [replacement] fungicides may also be tolerated by *Rhizobium meliloti* and possibly clover rhizobia as well.

In column 5 we have adjusted the variable of product weight because that column measures cost of establishing a full 10 kg of seed and if there are 460,000 seeds in a kg of lucerne then the coated seed will have only 4/7ths of those numbers [262,857]. Aware of that difference we have increased the cost of the coated seed to equal the same WEIGHT OF SEED in the plain seed treatment—and still, the Prillcote™ is not far off 1/3rd the cost of plain inoculated seed.

In actual practice, because we can expect better % establishment from professionally [and freshly] inoculated and lime coated lucerne seed, a farmer can sow less such processed seed saving considerable cost at each sowing,—that has to be important to any farm budget.

Some manufacturers used to process legumes at the 4/7ths seed and 3/7ths coat ratio whereby a standard recommendation was to multiply the normal plain seed weight traditionally sown in the past [say 10 kg/ha] by a factor of 2 = 20 kg then divide that by 3 = [say] 7 kg in round figures. Thus where 10 kg of plain seed was sown, we would then sow 7 kg of seed plus coat weight. That is then 4/7ths of the full weight we need to sow as coated weight so we divide the 7 kg by 4 [to get 1/7th the total weight] = 1.75 kg then multiply by 7 to get the full weight = 12 kg of coated weight to be sown [where 10 kg of inoculated bare seed was sown in the past]. That has been fully successful in practice.

To reduce the actual seed numbers in the coated treatment by as much as the weight of coating to make 10 kg total weight sown—same as plain seed—would be a severe reduction, possibly a little risky commercially even though the research result above says we can certainly do it. That research is conducted under closely controlled conditions—seldom achieved under farm conditions.

Today, manufacturers wishing to distance themselves from the mathematical challenge of the former Prillcote™ ratio of 4/7ths seed and 3/7ths coat, have turned to more easily workable ratios—retaining the 50/50 seed/coat ratio for grasses but for instance adopting 25% seed and 75% coat for clovers and 75% seed with 25% coat for lucerne—and no doubt variations of those ratios according to manufacturer. It is important to be certain of the ratio when planning seed sowing rates and to be certain that inoculated products do carry the full loading of *Rhizobium* [as discussed in Chapter 2] and have not been allowed to deteriorate in storage—and indeed were up to standard in all respects when processed.

New Zealand had National standards until the 1970's for inoculation [i.e., minimum *Rhizobium* per seed] and it is tragic for farmers that the testing station, its standards and small staff are now gone. If their closing down was intended as an economy measure—it was very much false economy. It is a facility which could conveniently be managed by an existing laboratory such as Landcare Research—and it needs to be re-instated in New Zealand.

Today, lucerne has become less popular in some countries because its ability to cope with dry surface soil and drought has been subjugated by its attraction to the destructive forces of insect, fungi and bacterial predators. Nevertheless, its deep tap root moisture seeking habit plus nitrogen fixing abilities are so important that the ongoing search for predator resistant varieties and organic or chemical methods of control need to continue. It is a world wide vital crop.

Broadcasting Lucerne

In field conditions where full cultivation is not possible or desirable [very stony, broken, temporarily wet, or wind-erosion prone], where cost is prohibitive [in developing countries], or time is critical [i.e., large areas which cannot be cultivated in time to meet ideal sowing dates], simple broadcasting of lucerne seed, mixed with lime [calcium carbonate, CaCo3] on lower pH soils, can achieve satisfactory results—not likely as good as when seed is drilled into worked soil, but a good result in relation to more modest cost.

In that event, it is almost certain that use of lucerne seed both inoculated and lime coated will give superior establishment results to that of seed just inoculated. The seed and its bacteria are already covered with the seed coat and both are protected from sunlight, wind desiccation, premature germination from marginal moisture plus bird and/or insect depletion of this nutrient rich oily seed.

It is too expensive a crop to waste, but in many situations, the broadcasting of 12 to 15 kg/hectare of well inoculated and coated seed possibly followed by chain or spiked harrows to cover as much seed as possible, will produce a satisfactory stand of lucerne which however, is vulnerable to weed infestation, so if cultivation is not possible or desired, then a cleanup of weeds prior to sowing will be a highly desirable pre-treatment either by mechanical or chemical means.

Identifying Nodulation success and failure

All legumes have nodules on their roots [except some tropicals which have them on stems above ground] which root nodules when bright pink in colour indicate not only successful nodulation but a favourably high level of nitrogen being converted to plant use as well. Plain white or grey nodules indicate that while rhizobia are active and have formed nodules, they are usually of inferior quality where the level of nitrogen they obtain from the atmosphere is minimal—if any. So colour intensity of nodules is a useful indicator of the level of nitrogen being "fixed" by that nodule and the strain or strains of *Rhizobium* it contains.

Where there are no nodules at all—such as in a failure of lucerne seed inoculation because the inoculant culture used was "dead"—or for a number of other reasons, it is still possible to recover that disastrous situation by drilling inoculant granules into the crop after the soil has been moistened by rainfall or irrigation [see Chapter 2 "Inoculants"].

It is a relatively simple matter to test success or failure of nodulation by digging up one or more legume plants, carefully cleaning the soil away [washing is best] until nodules are exposed and with a razor or sharp knife cut some nodules in half to expose colour intensity. Not only are more pink nodules an indication of superior strains of *Rhizobium*, but the size of nodule is also an indicator because large nodules can fix more nitrogen.

At present level of knowledge, there is limited opportunity for displacing existing inferior indigenous strains of *Rhizobium* in soil, with elite strains other than by sowing seed inoculated with those superior strains and hoping at least some of the better strains will form nodules [because legume plants can host several different strains of *Rhizobium* concurrently]. If research can find an economic way to displace inferior strains with elite ones—or at least allow the elite to function independently of other strains, that would be a major world wide breakthrough in natural nitrogen supply. Current application of huge quantities of manufactured artificial nitrogen is a serious concern in terms of environmental damage and climate change. Application of biological nitrogen is not.

Tropical Legume Seed and Other Coatings, Potential

Tropical legumes such as soybean, chickpea, serradella, sainfoin, desmodium, siratro and sulla are preferably also inoculated at sowing and for which *Rhizobium* Mother culture is available to manufacturers—no doubt there are others. While the inoculation of these cultivars is now a well established practise, coating them in addition does not appear to have attained the same importance that it has for the more hostile conditions involved in oversowing of temperate forage species.

Research in USA has found that certain fungicide and insecticide treatments of soybean seed have not been harmful to concurrent seed inoculation provided the agrochemicals are dry on seed when the seed is inoculated. Whereas in Brazil where soybean is also a major crop, investigations have been conducted there to avoid negative effects of agrochemicals on inoculants by use of liquid inoculation along the furrows where seed is sown. This has been successful—though is usually more expensive than simple slurry seed inoculation and can only be used in arable applications.

The basic principles which apply to clover and lucerne rhizobia [demonstrated above] will also apply to tropical *Bradyrhizobia* despite that the latter are slow growers more tolerant of acid soil conditions—but sunlight, heat, desiccation, acid fertilizers, delays between application to seed and sowing and seed toxicity itself, will all impact adversely on both genera.

Where soybean is drilled into an arable seed bed, pre-coating the seed would allow subsequent inoculation [oversprayed and not dried] under factory conditions for managing large volumes [of seed] plus the probability of a safe subsequent period for transportation and temporary storage pending sowing, as well as provision of a small but likely valuable starter fertilizer in the coat—plus fungicide or trace element inclusion on the seed and under the coat as well.

In summary, large volume soy bean sowing could be pre-coated over chemicals then dried. At sowing, oversprayed with inoculant—minimal moisture, then drilled.

To our knowledge that work has not yet been done.

It may be because the facilities do not exist for inoculating and coating large volumes of soybean seed quickly and inexpensively—or because the product would be more expensive than alternatives. Due to phenomenal growth in seed coating capability throughout the USA in the latter 20 years, including by former Celpril which purchased New Zealand's Prillcote™ technology in the 1970's, it is unlikely the former reason still applies. It is more likely that the latter reason is a factor—i.e., cheaper methods—but also maybe not yet as fully investigated as our studies [above] of coated clover and lucerne inoculation and establishment success. The TRUE comparable COST has been elusive, yet has now been identified for clover and alfalfa [despite earlier condemnation by "experts"!]—but, has it been sufficiently examined for soybean?—or has it too been condemned by experts. Comparing costs of seed treatment is only half the story.

Superiority of coated/inoculated soybean would probably include economies of seeding rates and, maybe better inoculation levels achieved which, as we have seen with lucerne in an arable situation in New Zealand, are both quite superior to simple inoculation. If they reduce seeding rates [waste]—that could more than cover the cost of coating. Then if superior inoculation meant there is no need for artificial nitrogen as well—there is very likely no contest. Precision agriculture now being pursued will sort this out.

That work will need to be done on soybean in USA or Brazil or Indonesia, the latter with a population of nearly 240 million, producing around 600,000 tons of bean a year, yet is a large scale importer of beans from the US—or maybe in Argentina the third largest grower of beans in the world.

However, results of research cannot be transposed from the lucerne crop in New Zealand to the soybean crop in the US—because there are significant differences between the two, not least being the differences in growth response to symbiotic nitrogen fixation in various soils.

The Future of Legume Coated Seed

Coating other legume seed may well develop in the course of time and for various reasons of which colour coating for varietal identification may become one. New and more effective additives to coatings—another.

Any valuable seed which may suffer bird loss on a large scale will also be a candidate for protective coating—primarily to achieve disguise plus taste distortion, but with the added opportunity of precision nutrient placement, other chemical treatment inclusion, and of course inoculation if it is a legume which like soybean and lucerne has a short host-free soil survival index. A danger however is that a "cocktail" of chemicals can cause inoculation failure.

From above research and many field results viewed by the authors, it is clear that coating seed for various purposes is so economically important that it is now a permanent part at least of New Zealand Agriculture and also is or will be of many other agricultural economies in the future. Ongoing education of new generations of farmers will be essential to maximize the benefits.

New processes and further developments will probably increase the importance of seed processing, but of much greater immediate concern is the need to develop high quality *Rhizobium* inoculants around the world, have them affordable to underprivileged farmers and graziers together with locally effective legume seed after which the benefits of coating can follow.

We have shown they can be a highly productive combination and that their world wide potential offers a powerful ingredient to superior livestock management and increased food productivity in an under nourished world.

References

(1) Hirsch Ann, M. 2009©. "Brief History of the Discovery of Nitrogen-fixing Organisms". Internet article. Quote:
"The microorganisms were first isolated and cultured from nodules of a number of different legume species by Martinus Beijerinck (1888) of Holland. Over time, modifications to the culture media were made to ensure easy isolation and growth of the nodule bacteria, which were placed in the genus *Rhizobium* (*rhiza* = root; *bios* = life). Since that time, *Rhizobium* has been found to consist of a number of distinct genera, including *Bradyrhizobium, Sinorhizobium, Azorhizobium, Mesorhizobium,* and others".

(2) Lowther, W.L. and P.M. Bonish. 1977–78. "Commercial Clover Seed Pelleting In New Zealand". Paper from Australian Legume Nodulation Conference.

(3) Francis Ivor. 1997 (Dissertation). "Companies should be run by executive directors, not boards heavy with independent directors" stated Dr. Francis in a "Christchurch Press" Business article Sept. 3rd 1997. Author of "Future Directions" [published by Pitman for the Australian Institute of Company Directors] Dr. Francis, born, educated and graduated in Christchurch New Zealand followed by a Harvard Ph.D and lectureships at Cornell, Otago and Auckland Universities is now a consultant in Sydney where he observes Australasian boards do not compare favourably with those of the United States nor with Japan where almost all directors are executives.

(4) Bennett, G.M. (author) [2014 Dissertation]. Molybdenum becomes increasingly available to plants at increasing levels of pH which happens when lime is applied. But in some situations where molybdenum is deficient applications of molybdenum alone can remove the need for lime. In other situations where soil acidity is the limiting factor, lime (calcium carbonate)—or dolomite (magnesium carbonate) will be required to lift pH and slowly improve soil texture.

(5) Hay, R.J.M. Lincoln University, New Zealand [M. Agr. Sc. Thesis; Summary circa 1970's] "The effect of coating and pelleting on germination and establishment of some grasses". Published as an appendix to Proceedings of the New Zealand Grasslands Association.

(6) Bennett, G.M. [Dissertation 1970's]. The author formally offered the presentation of a technical paper on this work on several conference occasions—but presumedly because of the controversy and/or his presumedly biased "commercial association", his offers were repeatedly declined [to the detriment of scientist understanding], mainly by a committee reputedly influenced by Senior agriculturist Paul Lynch, a very good man—but also very conservative. In today's more enlightened technical environment, many papers are delivered to this conference by people commercially employed, also often uniquely skilled in their specialist fields which developments must be accessible to all scientists.

4

Grass Seed Coating

Grasses in Hill and High Country Oversowing

Research workers have identified the fact that specific grasses are more productive over a longer period of the year in more hostile, higher altitude or hill country environments than [currently available] legumes which are more sensitive to lower temperatures plus frost damage and cannot be exclusively relied on to provide feed for livestock during cooler winter months.

Some grasses are more tolerant of frost and periodic light snow which, together with a lower soil temperature growth index, supplies the majority of winter forage in these situations. Introduction of plain bare grass seed on lower easy contour or flat country has not been a problem where cultivation and sowing under protection of soil is usual, also by direct drilling into clover established tussock grasslands, but establishment of grass seed into more hostile conditions without cultivation by simply sowing onto the surface of acidic soil—worse, into existing native foliage, often at altitude and under low rainfall conditions, is a much more demanding exercise where success is more likely when grasses are oversown with legumes together on undeveloped flora—e.g., good ground cover such as tussock grassland, bracken fern, cleared scrubland, etc.

This is a tremendously important subject, because, much of the uncultivated and undeveloped land around the world, many millions of hectare, currently considered too hostile in one way or another—i.e., too infertile, too acidic, too dry, too steep, too hot or too cold to grow anything—often in locations where there is no money for substantial development, nevertheless can be inexpensively developed. Such development at least to a preliminary stage, is achieved by simple surface sowing of grass, clover, vetch {a legume} or other seed to grow forage for grazing livestock. On the face of it, this can easily be seen as too difficult, but in each tough location there is likely to be a "recipe" for optimum results which needs to be identified—as it has been in New Zealand (a useful guide).

REVERTED SUPERPHOSPHATE COATED GRASS {1 KG SEED = 2 KG COATED}.

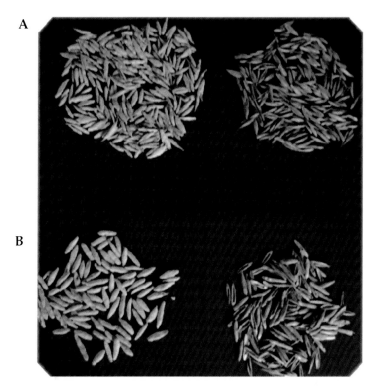

Fig. 39 A. Cocksfoot. Left coated seed. Right bare seed.
 B. Ryegrass. Left coated seed. Right bare seed.

In earlier days grass seed has been sown in New Zealand and elsewhere (with clovers) as plain bare seed on steep and/or rough country by hand or from horseback often into cold ashes of bush and scrubland burnoff using "cheap" seed (it had to be cheap presumably because much of it was not expected to grow!), even dressing plant "offal" was sown, long since recognized as a failed farm management tool and worse, a certain method of introducing noxious or other weed seed into steeper unploughable and inaccessible situations where subsequent major weed control expenditure can become a huge threat to economic survival of that land.

Today, at least in New Zealand, there is a completely different approach. We now use high quality seed but less of it, further enhanced by nutrient coatings providing a small but important finely milled, precision placed, ballistically improved coated seed product which has shown improved surface sown establishment success. Not all coated seed sowings will be automatically more successful than plain seed would have been—conditions dictate results, for instance, where soil fertility is high and large livestock numbers can be deployed after oversowing for sharp hooves to press seed into the ground—the ground being soft, then bare grass seed may perform just as well as coated. That situation is however unusual in hill country.

Coated grass seed provides a fertilizer "drop" wherever each seed lands, also creating more ballistically effective seed units capable of higher terminal velocity when dropped from aircraft (or from any sower) with better foliage penetration and an important degree of moisture control (the seed stays dry until its absorbtive coat captures and retains [around the seed] a more regulated supply of moisture during critical germination and early seedling development). Also, these objects awaiting germination do not look or smell like protein rich seeds as sought by birds.

This would seem a more promising approach—but, also more expensive?

When we study the tables (below) of performance and cost between bare seed (and we mean good quality bare seed—not rubbish), and identical seed coated—when based on painstaking investigation of independent grassland scientists, it will become clear that easily the cheapest way to go, is to:

1. fence into manageable block sizes to allow periodic concentrated animal grazing to "inject" animal manure nitrogen onto the new grass. Ensure availability of good clean water supply within the block. "Rough up" the soil surface as much as possible using concentrated livestock, even for a short period before sowing [and again after if possible] even where there is little feed for them—they are a tool of farm management, but used with care. Fencing off an area may seem "impossible", in fact, it is essential.
2. buy high quality grass seed.
3. coat it to specific formulae (either reliably purchased, or do it yourself as in Chapter 6). Include inoculated coated clovers.
4. sow less seed per hectare than traditionally (because more grows).
5. oversow it only at the optimum time of year, with appropriate fertilisers.
6. oversow on to prepared surfaces modified earlier by concentrated and forced livestock grazing and trampling (as described in {1} above) first applying major weed control where required, possibly burning where dense foliage cover must be removed (as in scrub weed or gorse lying dead from manual cutting or heavy roller crushing after herbicide destruction). *This is of course essentially the same programme as for clovers.*

But at the beginning, while the case for coating grass seed with phosphates and other nutrients looked promising, it had to be proven not only unharmful, but sufficiently beneficial to justify the cost of processing.

Coating Effects

The initial question for CSL technicians was simply this: *Does throwing a coat over a grass seed interfere with its vital germination process?* because if it does, the rest is inconsequential. As there are no sensitive bacteria involved in grass seed coating [as in legume inoculation and coating], more readily plant-food-available and more acidic nutrient coatings are able to be used without damage to grass seeds. Rock phosphates from deposits around the world such as Gafsa from Africa and [now exhausted] Nauru from mid Pacific have been used but are not as readily available to plants as superphosphate which however is at the other end of the spectrum—just too acidic for direct seed contact. It is also originally rock phosphate—but becomes "superphosphate" when treated with free sulphuric and sometimes phosphoric acid to yield about 20% phosphate— or P_2O_5. To permit it to be used for grass seed coating, its acidity is modified by mixing with highly alkaline calcium carbonate (limestone) whereby both ingredients are ground down to fine particle size for faster and more complete chemical reaction becoming reverted superphosphate which can be coated onto grass seed without damage. Like all seed coatings, reverted super or other nutrient or inert coatings such as talc, if preferred for any reason, all need to be milled to fine particle size around 95% passing through a 300 mesh British standard sieve. Details of coating materials, specifications and some sources of supply are detailed in Chapter 5.

Thus Coated Seed Ltd. began its processing research programme by examining the effect of coating (and precision pelleting) firstly on germination— not just grasses and legumes, but of all commercially important varieties, e.g.,—the ryegrasses, cocksfoot, timothy, browntop, danthonia, chewings fescue, prairie grass, *phalaristuberosa* and other grasses, plus clovers, lucerne, lotus, soybean, lupin, and pelleting of onion, brassicas, parsnip, carrot, tomato, lettuce—and many more. Today, these are all either nutrient (or lime) coated, precision pelleted with neutral fillers or film coated with soluble polymers and all have shown in countless tests that, professionally processed, they germinate as well as, and often faster and more completely than [identical source] plain seed—specially under moisture stress.

This Chapter is however, specifically about grass seed: on which we shall concentrate.

Grass Seed Germination. Bare Seed v Coated

To coat grass seed was a reasonably obvious idea, even though it does not include the inoculation complexity of legumes. Because, in addition to the several theoretical advantages stated above, a coating of finely milled mineral nutrient might be a suitable depository and protection (of seed) for inclusion of chemicals such as fungicides or maybe trace elements like boron, molybdenum and copper for plant health, or selenium, cobalt and manganese for animal health. This theoretical notion would have become stalled "at the first gate" if coatings had interfered in any way with essential qualities of sound, timely, robust germination.

Thus *the effect of various coatings on the quality of germination* became the first study undertaken by FCC Ltd. laboratory on contract to New Zealand's Coated Seed Ltd. at commencement of Prillcote™ operations in the early 1960's.

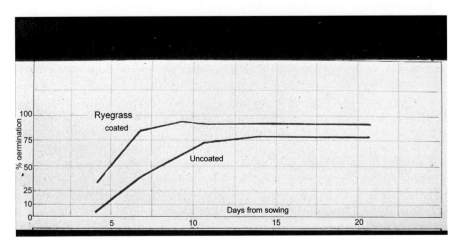

Fig. 40 Above is a typical graphed result comparing speed and extent of germination between bare seed (see "uncoated" lower line) and coated at 50/50 seed/coat ratio (top line) of identical perennial ryegrass seed under identical surface sown conditions. Bare seed reached 76% in 14 days while coated reached 90% in 9 days.

Variable Results Reported from Miscellaneous Sources

While these many results of tests conducted by CSL were both positive and encouraging, others reported some quite different results.

In a field study conducted in Australia by scientist P.M. Dowling of the NSW Department of Agriculture's Research Centre at Orange[1], two types of coatings [inert and nutrient] on *Phalaris aquatica*, *Festuca arundinacea* (both grasses), *Medicago sativa* (lucerne) and *Trifolium* subterranean (sub clover) were sown 29 times in the field over 2½ years. In addition, he conducted an examination of the effect of coatings on seed germination in petri dishes.

Peter Dowling reported that germination of the grass seed species was depressed by the coatings and that the beneficial effect of coatings on field performance was minimal—better in the warmer months of the year (due, he thought, to higher rainfall) however, response varied with the species of seed coated. He referred to comments by another researcher in Australia (Mullet) who likewise found a reduction in both total germination as well as rate of—for coated when compared to identical plain seed. However, yet another

Australian researcher (Johns) had identified the importance of moisture tension with coated seed in laboratory germination tests.

Two hypotheses were discussed in this Dowling research paper as to the difference between these results and those in New Zealand where coated grass treatments both in the laboratory and in the field were reported by independent researchers as being superior to plain seed—firstly that the Australian field site had a reasonably thick (5,000 kg/ha) cover of vegetative litter over the surface which may have acted as a moisture barrier, being more critical for seed already enclosed (coated), whereas in New Zealand, he suggested the field sites were bare ground or depleted pasture.

The second hypothesis advanced by researcher Dowling was that in New Zealand selective bird theft of plain seed may have depleted that treatment leaving more coated seeds to germinate and survive.

The seed used by Peter Dowling was Prillcote™ processed and supplied by Coated Seed Ltd. (from New Zealand)—both bare and coated from exactly the same line of seed, as was the coated seed supplied to New Zealand researchers who however reported the established numbers of plants from coated treatments to be quite superior.

How can this be?

These were identical coating treatments, both oversown (i.e., surface sown), both dry sites, which eventually get some rain, yes, different countries (but the seed doesn't know that!)—leaving only the question of VERY different seed beds—being the most obvious reason for differences in results. In New Zealand, the classic oversowing sites are tussock grasslands at about 500 to 1,000 m [a.s.l.], of low fertility, generally acidic soils (pH range about 5.2 to 5.8), not usually bare soil but with sparse native grasses and weed, no clovers, one or two tussocks per square metre but all reasonably upright growth providing important cover for establishing seedlings. There was certainly not a matrix of dead foliage lying on the surface which Peter Dowling has indicated was the case at his trial sites following herbicide defoliation which is commercially practiced in Australia for such oversowings but is seldom if ever practiced prior to oversowing in New Zealand (large scrub weed such as gorse, broom or sweet briar, excepted).

5000 kg/ha of litter on the ground is only .5 g per cm^2—an area smaller than most adult fingertips, but if that is feather light dry matter which, after herbicide treatment and for days or weeks is lying under the Australian sun—it represents quite a lot of bulk {volume} on this tiny area; (perhaps a reasonable analogy would be a small very dry lock of hair, 1 cm wide by 1 cm length, covering a fingertip area). If continuous across the seedbed, that degree of litter would prevent much of the coated seed from initial and even eventual contact with soil—maybe less so smaller (½ the size) slippery bare seed. That could account for some of the differences, and probably causes a % of the coated seeds to fail in New Zealand too.

An important feature of the coated seed is that it is slightly hygroscopic—i.e., absorbs moisture from the atmosphere which in temperate conditions can assist retention of moisture aiding germination—but in fiercely hot conditions may work in reverse—triggering the germination process, then denying it when all moisture is gone under hot sun and drying wind. In that event, bare seed may fare better by not commencing germination until it positively rains. While conjectural, the grasses in New Zealand behave differently.

That difference in these seed/coat results were most likely due to differences in farm management practices creating dissimilar seed bed conditions. In practice of course, after such sowing (in the Australian situation), had a sheep flock been forced back and forth across this seed bed, the results could well have been quite different—but that is not usually an option with precisely managed research field plots.

As for birds, the New Zealand sites tend to be up to and even above 1,000 m in altitude in barren treeless tussock grassland locations where birds are sparce particularly in early spring when these New Zealand trials are sown, because few plants are in (birdfeed) seed head at that time of year. While certainly a possibility—specially for small scale trials where a couple of adventurous birds discovering the plots could drastically alter the odds and render the research valueless, this is unlikely to be a factor where, as has been the practice, seedlings and sometimes even seeds themselves, coated and bare, are actually pegged (using nails or similar as was also done in the Dowling experiment). In New Zealand, and almost certainly in Australia, the trial plots are always securely fenced off from grazing animals—including netting against rabbits.

Other researchers reported[2] superior growth from nutrient coating of seed compared to plain seed or lime coating.

R.J.M. Hay (Lincoln University, New Zealand), in a M. Agr. Sc. thesis summary recorded in proceedings of the New Zealand Grasslands Assocn. that in a controlled environment cabinet which imposed low moisture conditions, coated seed (processed by Coated Seed Ltd. at its standard 1:1 seed/coat ratio by weight), improved germination three-fold compared with bare seed.

Then there is fine turf, or lawn grass seed, bare versus coated. On a commercial website recently, a "garden doctor" received this query:

"My lawn seed is slow to germinate".

His reply:

"This is a common problem. It may be because you have chosen a coated grass seed and these are slower to germinate but produce a more reliable lawn eventually".

That comment tends to assume that all coated lawn seed is the same from different processors or seed firms. That is most unlikely.

Neither is it likely for any other coated or precision pelleted seed

The quality of seed, its % interim and total germination, the quality of coating, the adhesives used which may include a mixture of soluble and insoluble polymers, method and heat of drying, length of storage, fast or slow sales turnover and shelf life—are all factors which can have a significant impact on quality and speed of germination—they vary with each manufacturer, each retailer and every different line of seed.

"Chewings fescue" in particular has a tricky germination threshold; named after that famous Southland New Zealand farming family from "Five Rivers"where slow growing, low pH tolerant fescue was traditionally farmed—is more scarce today where some of these farms have been limed, pH lifted and are now producing ryegrass and clover pastures. Tricky, because fescue can lose germination viability quite rapidly—from good soon after harvest, to "gone" the following year—which is the case whether bare seed or coated, but with the selling novelty of more colourful coated seed on show at retail outlets, some stored too long, some coated with semi-insoluble adhesives which slows moisture absorption, it is not surprising that above criticism is made.

Historically, Coated Seed Ltd. was the inventor and original supplier of fine turf colour coated seed in the 1960's in New Zealand using only Browntop (*Agrostis tenuis*) and Chewings Fescue (*Festuca rubra* ssp. commutata), but these days processors are using several other recreational grass varieties. The author and contributor of this book worked together (in FCC Ltd. and CSL) to pioneer and establish colour coated fine turf seed for easier identification of ground spread and inclusion of a fungicide and bird repellent—a product which became successful and is now widely marketed by numerous processors. However, to achieve this success co-author Bennett first challenged formulations expert [and contributor] Lloyd with the task [again!] of being able to successfully devise a coating method to process tiny browntop seeds. A principle of successful coating is that the seed must be larger than the coating particles—otherwise the process might finish up coating browntop seeds onto coating particles! Not phased by this degree of difficulty, John Lloyd soon devised a method, and coated lawn seed was "born" (start with a super-fine coating material such as a very finely milled talc—then when particles are large enough, continue with normal filler).

In comparison with more complex inoculated clovers for instance, as this fine turf seed product was relatively simple and more lucrative to produce for the small scale high margin home market, several commercial competitors commenced producing colour coated lawn seed.

Slower germination of it than experienced people had become used to with plain seed did damage the saleability of the product. While not in itself too critical, slower and reduced germination allowed establishment of early weed competition before a first cut could be made on the new cricket pitch, golf green or lawn.

But that is due to an entirely different range of reasons from PM Dowling's Australian experience. Retailers should request fresh germination certificates (as coated) for any coated lawn seed they buy in for sale. Be firm—no certificate, no sale—also, no sale if total germination of the coated product has fallen below about 70% unless it is repacked with instructions for sowing double the usual weight per m². Sale of poor germinating product with excessive contaminants and no warning of inferiority on the packet is likely to breach both the Consumer Guarantees Act and the "Fair Trading Act" in New Zealand of which the latter is enforced by The Commerce Commission.

As can be seen in above illustrations, germination of Prillcote™ coated and precision pelleted seed was usually found to be faster and often better as a percentage of seed sown—than bare seed. There has never been a problem with germination of any commercial product processed by Coated Seed Ltd. due to the effect of coating—every order was sampled for random testing over the many hundreds of tonnes of coated seed produced over many years [up to 1980 when author Bennett resigned from Coated Seed Ltd.] and probably beyond.

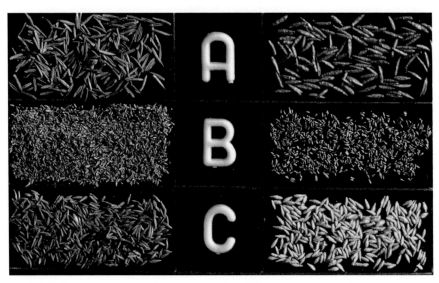

Fig. 41 Illustration shows: Leftside: bare seed, Rightside: coated and coloured.
A) Fescue, B) Browntop, C) Dogstail
All lawn seed.

It must be emphasized that not all coated seed is the same, nor has other product been so exhaustively tested as was Prillcote™ manufactured between 1970–1980. The success of Prillcote™ and its publicity in this book is not intended to provide a commercial "untested free ride" to other or subsequent manufacturers or to changed formulae of Prillcote™. New processors [and any changes in past processes] need to be independently re-tested all over again—and results made available to farmers, landscapers and sports ground developers. Establishment of an independent testing authority— even an extension of the National Seed Laboratory (now called AsureQuality Laboratory) for all inoculated and coated products is a service positively required—specially to protect New Zealand farmers. Arrival of superior rhizobia with superior infectiveness will require the whole of New Zealand to be "re-inoculated". Not if—but when.

Grass Seed Oversowing [e.g., dropped on surface of the ground]

We next examine the results of grass seed oversowing with good quality forage seed—both bare and coated.

As discussed, grass seed processing does not involve inoculation or microbiological technology—rhizobia are not involved. Well, not yet—research in the US has claimed actual nitrogen fixation in non legumes [grasses, brassicas, wheat, barley, rice, etc.] by *Rhizobium* transconjugants with patent rights applied for in 1993. However, no further development appears to have been achieved commercially. This subject clearly needs further investigation because of its enormous potential for the future. Subsequent research[3] has examined identification of the signaling device in cell structures which can initiate nitrogen fixation.

While in its infancy, probably even before its technical time, this is a field of investigation which may well become colossally important.

If ultimately, biological nitrogen fixation outside legumes did become a practical reality, it would become a stunning and huge leap forward.

Field Research

In the oversowing seed mixture for the mid '70's (detailed in Chapter 2), in addition to the 4 kg/ha of white clover and 1 kg/ha of Turoa red clover [bare seed and coated for comparison], we would certainly have added grass seed, e.g., non legumes—typically:

Grass seed

4 kg per hectare of a good perennial ryegrass [Lolium perenne]
1 kg per hectare of a good quality Cocksfoot [Dactylis glomerata]
Both with high quality P & G Certificates (e.g., purity and germination)

And for lower fertility sites where complete fertilizer input is unavailable, unaffordable, unknown or impractical, a wise farmer in the 1970's to 1980's would add:

1 kg per hectare of [say] Crested Dogstail [*Cynosurus cristatatus*]

Our next job was to find a similar, independent and expert judgement of authentic field performance as was achieved for legumes, by searching for and studying results obtained by independent scientists around the World who have examined grass seed establishment by surface sowing of both plain and coated seed. There were very few and none involving Prillcote™.

Fortunately however [as for legumes in Chapter 3], the most exacting, dependable and relevant research we discovered being another truly classic, independent, professionally managed and closely controlled field investigation was laid down at three typically challenging tussock grasslands sites by two respected research scientists in New Zealand.

Not only are we able to demonstrate their research findings here, but we are uniquely able to take their results an important step further by subjecting them to a table of actual cost—the cost of actually getting 4 kg/ha of ryegrass or 1 kg of cocksfoot established under typically hostile field conditions. Direct comparison was possible because all items were locally supplied.

One might say it was our good luck that this research study also took place so accessibly for us in New Zealand, or more cynically, that this is suspiciously so!

The author and contributor, who have long since had absolutely nothing to gain from Prillcote™ or Coated Seed Ltd. products being promoted by such independent research, might answer instead, that availability from within the New Zealand Agricultural Science fraternity of scientists capable of sorting out the facts was hugely important not only to CSL and its FCC Ltd. formulations support, but to the New Zealand farmer as well—and now via this book—the agricultural world. The authors gain nothing from the facts publicized here.

Close liaison by CSL's development manager (formerly with Department of Agriculture) among Grassland and *Rhizobium* Scientists (within New Zealand and also many overseas) plus his offers to provide them with professionally prepared seed treatments able to meet mathematically designed exacting field research layouts, led to critically important combining of skills, better understanding of obstacles and methodology for dealing with field difficulties—plus ultimate success of those scientists in sorting out the technicalities without having (also) to become expert seed processing manufacturers themselves. This success certainly reflected back on them leading to some excellent research publications. CSL learned from it too of course, but the greatest benefactors of all were New Zealand farmers and runholders, then ultimately the "Taxman".

The scientists in this instance who planned, executed and closely examined field sowing of both ryegrass and cocksfoot, both plain seed and coated, were Dr. E.W. Vartha and his Associate Peter Clifford both of the [former] New Zealand Department of Scientific and Industrial Research, Lincoln [sadly both deceased in recent years] who reported in the *Tussock Grasslands and Mountainlands Institute "Review"* No. 16, 1969 published by Lincoln University of Canterbury, New Zealand, the following establishment results at one of the three typically challenging environments, being at "Dusky Station" in

McKenzie Basin, South Canterbury New Zealand where "Occurrence of grass plants two years from sowing" was reported, to-gether with the following summarizing observation by the researchers:

Quote: *"The superiority of coated grass seed has been shown not only at Dusky Station, but also at [dry] Grampians Station and [wet] *Glentanner Station sites in the McKenzie Basin"*

Table 13. Occurrence of grass plants two years from surface sowing [from Vartha and Clifford].

Grass Variety	Plain Seed	Prillcote™ coated
Cocksfoot	8%	24%
Ryegrass	12%	45%

Many runholders and farmers were shocked by these results—specially for plain seed, and quite disastrously—cocksfoot, remembering how much of it they had sown [and wasted] in the past! Even the best treatments provided a 76% failure of established seed in cocksfoot and a 55% failure in ryegrass.

Nevertheless, these runholders and farmers had traditionally [before coated seed availability] sown plain seed [if any] in the past which would have given them more like the 92% failure of establishment (with cocksfoot shown above), and 88% failure with ryegrass—probably worse if they used dressing plant "offal". New Zealand Runholders were unaware of the extent of such losses. Their only recourse being to buy and sow more seed. Some grass seed varieties were however relatively inexpensive in the early days [1800's to 1940's) being stripped from cocksfoot and fescue growing along public roadsides when the target species had gone to seed. It would indeed need to have been cheap, as it would also contain a wide range of weed seed and all sorts of foreign matter as well as being of unknown germination viability. Fortunately, New Zealand has progressed a long way from those early primitive, but historically inevitable steps in land development.

Dr. Vartha cautioned that in his opinion not all sowings of coated grass seed would show above superiority even in oversowings into existing foliage on wet or dry native grasslands. Much depended on extraneous factors such as the level of fertility and importance of the nutrient coatings, the weather at or soon after sowing, bird theft, nitrogen availability from legumes because if the

---

*Glentanner Station, situated on the Ben Ohau Range along the South side of the mighty Pukaki River, is a High Country Run on the way to Mt. Cook Alpine Village with its Internationally acclaimed "Hermitage Hotel" where Mt. Everest conqueror Sir Edmund Hillary practiced mountaineering. While generally summer dry East of the alpine divide, it is close to the Alps with Westerly driven alpine rainfall "overflow". It can periodically get very wet as testified by the author who worked on Mt. Cook Station directly opposite Glentanner but on the North side of the Pukaki River which was often crossed by men on horseback to retrieve wandering cattle—the hardy horses and cattle swimming in near freezing water.

legumes fail, the associated introduced grasses will usually fail too—however, some of those effects apply equally to bare as well as coated grass seed.

On the positive side for coated grass seed, it has several advantages such as double the weight [at 50/50 seed coat ratio] but not double the mass thus increasing terminal velocity when dropped onto ground which is helpful in securing penetration of existing foliage onto the soil surface. If the seed does not reach soil surface it is probably not going to survive. Opposed to that, an astute farmer with adequately subdivided paddocks or blocks can aid seed/soil contact by an intensive short term trampling of the area just sown, which of course would aid coated seed as well.

Other advantages are the moisture repellent effect of coated seed where marginal moisture [night dews, brief drizzle or light frost] may trigger germination of bare seed followed by a dry period—off and on, whereas coated seed containing weakly plasticized polymers in the adhesive formula mix which though intrinsically water resistant nevertheless, purposefully diluted with other more soluble adhesives were easily moisturised under light rainfall—a feature probably unique to the chemically astute Prillcote™ formulae at that time.

Nutrient coatings on Prillcote™ grass seed were also "finely divided", e.g., ground down in particle size [as for clover coatings] by Raymond Roller Mill to at least 95% passing a 300# British Standard sieve [about the same mesh size as a nylon stocking] which greatly increases the surface area of such materials therefore increases their solubility, plant food availability and rate of adsorption after breakdown.

Wherever each grass seed landed from oversowing, a small but vital few grams of fertilizer fed its roots right there where it lay. On hillsides, the moisture which removed the coating might tend to run the coating material downhill away from the seedling—that has not to our knowledge been studied, but the coatings are reasonably stable, slough off the seed husk slowly and even where lost downhill, may benefit another seedling—after all take cocksfoot at about 500,000 seeds per lb which is about 1.1 million per kilogram, even sown at 1 kg per hectare [1 ha = 100,000,000 square centimeter] provides about 1 seed per 100 square centimeter [10 cm x 10 cm—or about 4 x 4 inches] which surface area a good mature cocksfoot plant would need [to grow strongly] accompanied by a ryegrass or two and hopefully a clover plant—but otherwise relatively free of inter-plant competition.

Such theory must be modified by the realities of % germination, establishment failure, hard seed, and uneven spread. Also the fact that on that same 100 cm^2 there will also be clover and ryegrass seed becoming established in addition to the cocksfoot—indeed clearly too densely populated even at a minimal total (say) 5 kg/ha of clover and 5 kg/ha of grass seed oversowing mix—*if it all grew and were uniformly placed*—neither of which happen in reality. Tussocks indicate healthy land and are soil stabilisers as well as good shelter both for newborn lambs—and newly establishing seedlings, but any seed

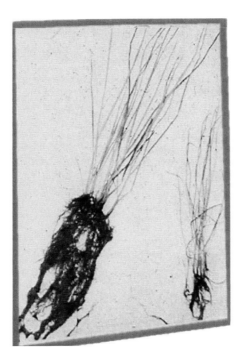

Fig. 42 Cocksfoot [Dactylis Glomerata] reverted superphoshate coated seed (left) and plain seed (right) grown in low fertility soil, same seed and no other topdressing.

which lands in a tussock is probably lost as indeed those that land on rock or in a stream or on a weed being held up off the ground.

Actual Cost

As shown earlier for white clover, the following table shows actual cost of seed and coating at a fixed point in time—namely when the research was carried out in the '70's plus the actual cost of each treatment based on these results. Today, costs have altered—mainly in a uniform way due to inflation so the relationship of costs of seed and their coatings plus factory processing have remained relatively stable. Therefore the following results are still likely to be a good indication of the extent of cost comparison, but if not, the result remains a constant to which variable costs may be re-calculated.

We have calculated what it actually costs to get a full 4 kg of white clover established per hectare on 100 hectare—and now the same can be done for 1 kg of cocksfoot. Once again we need to qualify these results with the observation that they are only a guide, however this is the best guide available to illustrate actual cost to the runholder and farmer being based on carefully prepared, professionally analysed actual field responses from the totally independent Government's Department of Scientific and Industrial Research. It is equally possible that at other sowings and in more recent years, the actual results may be worse—or better, but we have here a genuine "fix" on a typical result.

We must emphasise again, this is the result using a particular formulation—we explain that formulation and others in Chapter 6.

Based on Prillcote™ and the Vartha/Clifford results: WHICH COSTS LESS ? - { or produces more per $ }

Table 14. Ryegrass seed sown @ 4 kg hectare   [from Vartha and Clifford].

Ryegrass	$/kg*[1]	$/100ha [at 4kg/ha]	% *[2] established	Cost for each 1% established	Cost to get 4kg establ. on 100ha
Plain seed	$1.81	$724	12%	$60.33	$6,033
Prillcote™ coated	$1.60	$640	45%	$14.22	$1,422

*[1]This was the actual cost around mid '70's ex factory of CSL [Wholesale to Seed Trade].

*[2]These % are seeds (see Vartha/Clifford results above) which were wire pinned at sowing or germination—then checked for survival at 2 yrs. These percentages cannot be related to numbers of seed sown for a specific weight of product because coated seed has only half of the seed weight that bare seed has. They are percentages of bare or coated seed pinned at sowing. It is our opinion that pinning seeds which have found ground contact and are visual, may have tended to disadvantage coated seed because more plain seed can be expected to fail to reach ground visually prior to pinning, than heavier, more penetrating, nutrient supplied coated seed. In other words, the percentages of survival at 2 years in the table—are of seeds which had a chance to survive, so it doesn't measure outright survival of all seed sown.

Table 15. Cocksfoot seed sown @ 1 kg hectare   [from Vartha and Clifford].

Cocksfoot	$/kg*[1]	$/100ha [at 1kg/ha]	% established*[2]	Cost per 1% established	Cost to get 1kg establ. on 100ha
Plain seed	$3.06	$306	8%	$38.25	$3,825
Prillcote™ coated	$1.99	$199	24%	$ 8.29	$ 829

*[1] Actual cost around mid '70's ex-factory of CSL. [Wholesale to Seed Trade]
*[2] Pinned seed after sowing—measured at 2 yrs against the initial count (see research results above).

It should be noted that in all these field investigations, CSL supplied the seed—it drew all seed from <u>one line of p. & g. tested seed</u>, coated part and supplied that together with identical bare seed to the Researchers.

Summary

The full picture is of course provided when all above results are combined to determine actual total cost of establishing:

5 kg of white clover
4 kg of ryegrass
1 kg of cocksfoot—on [say] 100 hectare.

We would have added Turoa red clover or alsike in the 1960/70's where topdressing was adequate and accurate, or for lower fertility conditions subterranean clover [particularly where dry] or Lotus major [where wet] as additional legumes, plus crested dogstail as a low fertility demanding grass, but these species have not been examined under the same precise and exacting conditions as the top three, or as is necessary to be confident of results. Red clover and subterranean clover results will probably be similar to white clover and dogstail similar to cocksfoot—but because the work has not been done to our knowledge, they cannot be included here where we are dealing only with facts.

The following Table shows the ratio of actual cost between bare seed and coated, from actual results, in achieving establishment of a full 5 kg of white clover, a full 4 kg of ryegrass and full 1 kg of cocksfoot on 100 hectare of hill country in an annual rainfall range [as stated above] from 450 mm in the traditionally driest localities [which is poorly spread] up to about 700 mm average [winter wet and summer dry].

Table 16. Ratio of Actual Cost of Seed Treatment.

Seed treatment grass and clover	Combined cost of 100% establishment	Cost Ratio
Plain seed, grass and clover	$25,946	8.4
Coated Seed [clover, also inoculated]	$3,088	1

So there we have it—bare seed wasover 8 times more expensive to use, in terms of an equal result, than processed seed. Even this is not a complete comparison because we have already [above] identified **that inoculated and coated white clover plants were clearly healthier and in the first 8 months from sowing produced a whopping 14 times increase in the weight of dry matter, to that produced by bare seed.** Rarely in Agriculture, do we get such overwhelming differences in performance of a fixed item—due to its treatment.

Providing the quality of inoculation and coating are of the highest standard, anyone oversowing undeveloped land, at least in New Zealand these days, should be using professionally coated seed—or DIY as in Chapter 6.

Arable Sowing

In arable situations where we have cultivated soil on flat or undulating country—usually of higher fertility where seed is drilled into a competition free seedbed [not dropped on top], quite a different set of criteria apply.

Firstly, many of the benefits of coating for oversowing—become superfluous. Better ballistics and seed penetration from coating, polymer protection from marginal moisture absorption and premature germination, close proximity of a small but vital quantity of fertilizer as in coated seed, protection of inoculant for longer periods exposed to sun, wind, etc.—are all unnecessary when seed is protected by placement into soil.

Both grass seed and inoculated legumes to be drilled can be held a few days after good quality coating whether it is carried out by the farmer in the shed on a rainy day [following the coating formulae and instructions in Chapter 6]—or ex-factory by a seed processor with a reputation for quality product. It can be drilled from a seed box alone with fertilizer drilled down a separated tube—or it can be mixed in a single box seeder provided it is mixed at a calculated seed/fertilizer rate which sows the seed at a constant rate per hectare. If mixed, it is always wise where inoculation is concerned to mix only coated seed—or use alkaline fertilizer—both mixed freshly just prior to drilling, and as soon after inoculation as possible.

There are however benefits from coating grass seed [to be drilled] for other reasons—to include fungicides, insecticides and/or trace elements; for legumes, more positive inoculation and nutrient or Ph application around the seed in lower fertility, or into acid soils with a lime coating—as is now common practice when drilling lucerne [alfalfa].

Additives

Fungicides such as Captan and Thiram have for many years been included in fine turf and other grass seed coatings. They are effective in combination against "damping off" disease, botrytis, scab and various rots—in addition to which Captan is claimed to have some bird repellent qualities although we cannot find definitive research in support of that claim. Such research has been conducted however into the value of Mesurol as a seed (film) repellant against birds and has been found to be statistically effective, non-phytotoxic and at 1% a.i. economically viable in saving almost 50% more seed [than "control"] in pine tree seeding research. But Mesurol has not been used as an additive to coating in the research we have seen.

There are many other repellants against seed loss to birds, but they include a range of sticky materials and toxic gels—a formulation obviously not suitable as coating additives. The active ingredient however may be.

Other fungicides included in seed coats such as Demosan and Benlate (the latter now withheld or terminated due to toxicity in water in the US) have been proven beneficial to establishment of lucerne yet not harmful to seed or inoculant. No doubt many more chemicals have been and will be tested around the world in attempts to show good control as well as environmental compatability. Where they demonstrate valuable success, they will usually be patented in an attempt to recover at least research cost and at best show a profit, but patent protection expires at maximum around 20 years after which the patent then becomes an advice document to others for quite legal copying of the process—unless of course it has already been superceded by superior products. Patents also have to be protected which is not a simple matter. Some manufacturers prefer secrecy at least until research cost is recovered.

Inclusion of chemical additives can be inexpensive in terms of processing cost of coated seed because the primary processing cost rests with the coating itself. Taking this opportunity to add fungicides, insecticides or bird repellant in small quantities [but precisely placed] adds little more to the finished product cost than the cost of the chemical itself. However, there is no point including any of these chemicals in seed coatings unless independent research can show them to be—safe for transport and to handle, environmentally acceptable to humanity and of course economically worthwhile for the farmer.

Insecticides are more challenging. They usually have to remain active beyond germination to the point where insects have become attracted to emergent succulent seedlings—difficult to achieve physically and timewise, but also many are toxic to *Rhizobium* and indeed seed itself. However, seed is usually not as sensitive as rhizobia cells when concentrated sufficiently for either rhizospheric or systemic action. Some chemicals are severely toxic to seedlings even diluted by seed coatings.

Fig. 43 Photograph—Department of Agriculture Ruakura Research Centre 1976.
Two replicates each of:
Left: chemically coated grass seed and
Right: Bare grass seed
The seed, time of sowing, post-emergence care—were all absolutely identical except for the chemical treatment for protection against *Soldier fly* in the coated seed treatments on the left (in above Illustration), showing severe damage to the unprotected bare seed treatment on the right.

Many of the organo-phosphorus chemicals plus chlorinated hydrocarbon insecticides such as DDT which might have been useful as a non-leguminous seed treatment, are so toxic and soil persistent they have quite sensibly been banned in most agricultural economies. Though not as persistent, other chemicals such as dieldrin which has been used in seed treatment was also widely banned many years ago (including in the EU) or severely restricted (in the US, Australia and New Zealand plus many other Nations).

Nevertheless, seed coating companies in the US and elsewhere now promote seed coat (or film coat) insecticide protection of a wide variety of cultivars including grass, also maize, sugar beet, cereals, oilseed rape, sunflowers, peas and beans, brassica and coated clover seed (with inoculation). These treatments are claimed by processors to control a wide range of field pests including aphids, springtail, stem weevil, thrips, whitefly, ground beetles, leaf miners and more, which indicates a degree of acceptability of insecticides for this purpose and that they are environmentally acceptable, user safe, toxicologically effective and worthwhile in relation to added cost. Most importantly due to precise placement with seed, far less chemical is required to achieve a similar result to less desirable soil saturation.

Chemicals such as *thiamethoxam* introduced into New Zealand in 1997 (also approved for use by the Environment Protection Authority in the US) has systemic qualities and is used successfully on seed but with a limited "protection life" reportedly of about 6 weeks. There are products derived from active ingredient *imidacloprid*, also a systemic which Bayer marketed as "Gaucho" (since 1991) but has been more recently superceded by second generation "Poncho", both of which have been used widely as in a "film coat" (with little appreciable seed weight gain from application), rather than included in built up coated seed.

A relatively new but rather obvious process now available at least in USA is a three layer coated seed product with a benign coating overlaid with a toxic chemical additive and then covered again by a harmless coating to provide safety to handlers—processors, truckers and farmers. The chemical is thus encapsulated until washed off *in situ*.

Unfortunately, we cannot relate any empirical results of intensive studies on control of specific pests related to specific chemicals, specific seed, field and sowing conditions, and ultimately cost/benefit analyses of such treatments (if any) because we are not aware that work has been done (as it certainly has been done in seed inoculation and coating treatments).

It is important that this work should progress—at least on the most serious predators around the world so that waste in food production is minimized and that in developing economies future serious losses which cannot be afforded, are substantially eliminated. These independently and professionally managed studies are essential for farmers to be convinced that they should pay for specific seed treatments—not just fungicidal additives but also insecticidal chemicals, as well as coating itself, plus inoculation which is thoroughly proven.

Insect control is a big subject, a huge challenge for entomological scientists, and while no doubt some research is in progress, chemical manufacturers need to be encouraging both scientists and university researchers by supplying them with a range of appropriate chemicals, funding individual investigations from tax deductible profits and publicizing any key development as promptly and as widely as possible—particularly on the world wide web.

This publicity does work (most professional farmers in New Zealand and elsewhere now own computers and are connected to "The Net"). A classic example of this publicity at work happened when "Poncho" replaced "Gaucho" virtually in just one season in the US. Change in farming practice historically took many years to complete. To-day, even rainforest Brazilian villages have satellite internet access.

Chemical manufacturers have an important opportunity here—to part fund and supply test product to research organisations with qualified staff who are (and must remain) absolutely and fiercely independent of commercial outcomes. Researching staff then have an important opportunity to (a) research (b) patent then (c) publish their findings, selling off manufacturing rights to all who are interested thereby facilitating the commercial aims of the supplying chemical manufacturer, assisting its product to be used commercially by those willing to pay a modest licence fee to use it. This will provide for cost recovery and funds accumulation by the researcher or University or laboratory to underwrite other projects or studies for which limited or no funding could be expected.

Pasture Pellets

As discussed in Chapter 3, large pellets made of finely divided nutrient materials in which seed are attached or enclosed, is a field of potential development which deserves much more study.

As for legumes, grasses may also be included in pasture pellets for spreading both seed and its essential fertiliser onto the surface of uncultivated and often low fertility soils, at ratios of (say) 5 parts pellet to 2 seeds or 10:2 or 15:3 or 20:4. The same principles apply with grass p/pellets as they do for legumes, as fully explained in Chapter 3—so will not be repeated here.

As grass seed does not (yet) rely on inoculation with rhizobia bacteria and unless included with inoculated legume seed, grass pasture pellets present the additional potential for including a small but possibly vital quantity of a seed tolerant "starter nitrogen" fertilizer which may ensure better grass seedling survival—possibly critical until legumes are able to supply biological nitrogen to them. Biological nitrogen supply in winter cold, summer hot and dry climates (common around the world) could take more than a year in some situations, by which time grasses may well have failed.

Master of Agricultural Science Graduate R.J.M. Hay studied pasture pellets and coated grass seed many years ago as part of his Lincoln University (New Zealand) thesis requirement.

In a summary presented to an Annual Conference of the New Zealand Grasslands Assocn. and reported in its "Proceedings" titled *"The Effect of Coating and Pelleting on Germination and Establishment of some Grasses"* the following facts emerged:

In a controlled environment cabinet under low moisture conditions, a coating to seed ratio of 1:1 by weight improved germination three-fold compared with bare seed.

But he also reported: *"Large pasture pellets containing ryegrass seed supplied phosphorus at a rate which gave the same dry matter yield as bare seed with 43 kg/hectare of surface applied phosphorus (about 700 kg/hectare of reverted superphosphate").* The p/pellets weighing much less than 43 kg/ha. There is our aim again!—we aim to fertilise the plants—not the hectare which would both permit waste and grow more or better weed.

These pasture pellets were supplied by Coated Seed Ltd.[4], proposed originally by author Bennett but made by their designer—technical manager John Lloyd. The concept however, did not progress into commercial production—or even significant scientific research to our knowledge because factors such as cost, large scale manufacturing equipment required, large handling equipment (turning seed merchants into fertilizer suppliers) demanded a whole new—and expensive approach—it was not popular. In addition, in CSL's own research, p/pellets did not perform as well as had been hoped—but that was a typical challenge which we would expect when refining such an important product. It would require a co-operative effort of both the seed and fertilizer industries.

To our knowledge the concept has still not progressed, indeed became forgotten in the production success of Prillcote™ but also by the administrative events which largely and eventually led to the termination of the FCC/CSL formulation and research facility but chiefly its path finding developers.

That was not however, a failure of the concept.

While a fairly obvious idea and clearly in need of much further refinement it should be investigated more searchingly because it has even greater potential for transforming low fertility inaccessible or steepland soils into forage grassland economically (with the right grass species)than more conventional coated seed plus fertilizer.

The product could quite possibly be produced in a suitable granulation plant which would manufacture a very dry but hygroscopic granule (with actual moisture retention properties built in)—also extruded and specially shaped with weight distribution designed for the ballistics of penetrating native foliage/litter (not round which might roll or bounce downhill). Such pellets would contain a suitable non-phytotoxic nutrient loading eliminating topdressing and supplying *Rhizobium* safely—or containing a vital slow release nitrogen value and oversown separately where hugely successful inoculated coated clovers are to be used concurrently. Application with drone aircraft may be a vital part of its success.

Great precision is required for this concept to work—precision not only in concept, but in materials, formulations, costings, laboratory production to supply those engaged in proving research, and when successfully developed and substantiated, precision in commercial production. This product may be too big for New Zealand—or, New Zealand's potential use may be too small to justify the cost, but those are questions which could only be decided when the concept and its magnitude have been fully developed.

Grass species

Some grasses will be more suited to both coating and oversowing conditions—in his research R.J.M. Hay found that ryegrass had three times the viability of prairie grass and almost twice that of cocksfoot under identical conditions—that was in a controlled environment cabinet where low moisture stress was imposed (as we know often occurs in the field). So, once again precision is required to ensure that the most suitable grass seed varieties are used—lest we (condemn the product) when it was the method that failed [i.e., the human factor].

New cultivar research might do well to breed and select from within this framework to achieve a "Groundbreaker" ryegrass or cocksfoot with (eventual) lower fertility tolerance, hardiness and when germinated, grown and matured, also possessing seedhead multiplication excellence both in hostile field conditions—as well as in higher fertility seed bulking up regimes—importantly still retaining a good measure of livestock palatability.

Axis Associates (see index for contact details) have the technology and experience to advance the pasture pellet and granulation concept and may even be willing to manufacture small research quantities by arrangement for scientists or plant breeders keen to investigate this (theoretically) huge potential.

References

(1) Dowling, P.M. 1978. The New Zealand Journal of Experimental Agriculture 6: 161–6.
(2) Scott, J.M., C.J. Mitchell and G.J. Blair. 1985. The Australian Journal of Agricultural Research 36(2): 221–231, CSIRO.
(3) Drozańska, D. and Z. Lorkiewicz. 1978. Genetic transformation in *Rhizobium trifolii*. Acta Microbiol Pol. **27**(2): 81–88. Lorkiewicz Z, is a Scientist, Department of General Microbiology, M. Curie-Skłodowska University, Lublin, Poland. Article published in "PubMed" of the US National Library of Medicine.
(4) Lloyd, J.M. Circa. 1970's. Provision of many research treatments published in ref.(1) above—and in various other publications in acknowledgement of J.M. Lloyd's formulation expertise. There was a verbal understanding between Coated Seed Ltd. and Researchers it supplied with professionally prepared seed treatments {at no cost} that, results of their investigations with these treatments would eventually be made available to CSL {at no cost}—an arrangement which worked without a problem. The results of that research however were strictly the property of the researchers from whom CSL obtained permission to relate their findings generally, when published. The researchers retained publication rights on results of their work using treatments devised by development mngr Bennett and expertly prepared by John Lloyd at FCC Ltd. for CSL.

5

Precision Pelleted Seed

Introduction

Horticulture, Market gardening, Vegetable production—call it what you wish, feeds billions of people around the world as of course do cereals, wheat, rice, soybean, fish, fruit and meat. These are all "staple foods" in the regions where they are predominantly grown or harvested, all vital to survival of mankind.

In horticulture's vegetable production development, there has been a significant move away from labour intensive scatter and thin line hand sowing of seed even in China which grows[1] [*amazingly*] nearly half the commercially grown vegetables of the world—also in India, another large producer where Germains (Seed) Technology Group are now facilitating the commercial distribution of seed coating polymers.

A world trend has been toward machine transplanting of nursery grown seedlings either precision sown by various ingenious methods[2] into trays or minipots for transplanting or into larger pots where minimal disturbance of roots later at transplanting will not damage or wilt the seedlings. Also, precision seeding of less transplantable crops (i.e., parsnip, carrot, beetroot) directly into the field enabling economies of expensive seed as the quality and cost of newly bred seed varieties increases.

Precision drilling reduces the cost of labour by permitting optimum spacing of plants (variable row width as well as inter-plant spacings) for access by implements for faster weeding, fertilising, spraying, irrigation and harvest. Precise placement also reduces wasteful gaps in seed rows and minimises manual infilling after establishment.

The forestry industry has gained important advantages from precision drilling.

See definitions at the end of the text.

Fig. 44 Precision pelleting of *Pinus radiata* forestry seed has given superior results in seed placement to plain seed via belt drive precision drills as in this forestry nursery where the four beds on the right each of four rows of *Pinus radiata* were drilled with Prillcote™ precision pelleted seed. At a distance these 4 x 4 beds appear similar to the beds on the left, however, on closer inspection the pelleted rows have quite superior placement using less seed with faster growth due to freedom from inter plant competition. More recently, pneumatic drills may place some varieties of plain forestry seed with acceptable precision, however film coating with additives may still be preferable. A higher % of uniformly sized seedling trees for sale is the final deciding factor.

When evenly spaced, plants get equal opportunity to access soil moisture, nutrients and light. This freedom from excessive inter-plant competition—a detrimental consequence of scatter or random sowings, does grow healthier more vigorous plants with better resistance to disease—than plants growing on top of each other with ultimate stunting of a percentage of the crop—not to mention those wasteful periodic gaps along the row which allow ingress of weeds. Under good management, the gaps can and should be filled at early seedling development by manual transplanting—using thinned plants from dense clumps which however usually causes wilting and slows maturity—is also a major and laborious task in large scale scatter sowings but a much faster and easier task with fewer gaps or "doubles" {the latter requiring thinning} where seed has been precision placed by accurate calibration of a precision drill.

Multiple seed pellets are called "doubles" in fact some may have more than two seeds per pellet {a fault, but a small number may be unavoidable and is thus acceptable], also precision drills do pick up two, but rarely more, smaller or variably shaped pellets in one belt or suction hole—instead of a single. With good quality pelleting oversized "doubles" and [very rarely] two or more bare seeds attached and pelleted like that [for which parsnip is notorious] are first minimized by careful screening over a three level vibrating mesh screener.

The top mesh retains joined or multiple seeded pellets (doubles) which pass off the end of the top mesh into a container where, if significant, these oversize pellets can be separated gently to individual seeds again {carefully crushed—or washed and dried} then re-pelleted. The main correctly sized product however passes through that top screen onto a lower screen which retains the true pellets to pass off into a bag weighing and bag sewing facility. Any dust or minor particles pass through the lower mesh and are discarded to waste.

Fig. 45 A small pellet screener with a sturdy frame to withstand continuous vibration, to which is fitted a receiving bin (top right) under which is the sloping screener beds with replaceable wooden frames for various mesh sizes (see some frames at the right not yet fitted out with woven wire mesh screens). The small electric motor (bottom right) drives a belt fitted to the main drive wheel which in turn drives an eccentric drive shaft (off centre to create vibration). Another axle or shaft drives the vibrator for the receiving bin to ensure free flow of product which runs down the central chute over the mesh with correctly sized product passing through the top mesh to bagging off on scales, lumps or oversize retained on top (discharged to rubbish or re-cycling) and any chips or dust falls through screens being discarded to a dust bin.

Precision Drilling Demands Precision Care

Of equal significance however is the importance of recognizing that where we sow only those seeds required as plants, it is essential that every potential cause of seedling failure be prevented or minimised, otherwise even a 20%

failure as a precision crop will usually represent a poor result indeed. Thus the seed must be top quality with a high interim germination percentage, time of sowing and field conditions for sowing need to be favourable (not flooded, wet or drought prone), predators—insects, virus or fungi must be controlled before or promptly after sowing and during establishment, and water—if not by rainfall, then by irrigation must be applied when dry. If these factors cannot be adequately provided for, then it will be safer to revert to the old methods and sow a massive excess of seed to aim for an acceptable percentage survival

Fig. 46 A. These carrots have been drilled the old fashioned way by scatter sowing of bare seed. There is a wide variation in product size, shape and maturity with a high % of market reject and many stunted plants.
B. These carrots were sown with precision pelleted seed which was drilled with a Stanhay precision drill producing a largely uniform sized crop reaching maturity simultaneously and with a high marketable %.

in the same way nature produces massive volumes of "seed" within the human, plant or animal genus to ensure survival of the species—but most of which fails. In commercial cropping the result can be untidy with overcrowding, inferior product with *high reject market quality levels*, large gaps (which can also be manually filled at extra cost but seldom are in such situations), damaged and stunted product where insects or disease are not controlled or irrigation not applied.

In other words, there are now essentially two principal types of cropping—both in agriculture and horticulture. They are:

1/. **Precision crops** where exact minimum seed is sown, all factors are carefully considered and all potential risks understood, foreseen, provided for and competently managed and;

2/. **Conventional crops** where the basics are provided, ample seed of average quality traditionally sown, many details left to chance and a final result reliant on historical percentage survival.

The latter, still by far the most common will often be cheaper to apply, but the former, with the greatest potential by far, will usually be more profitable—after all factors are considered.

There is a worldwide need for a series of careful studies into the cost/benefit relationship of "scatter" or traditional seeding compared to precision drilling—also comparing bare seed and pelleted seed precision drilled over a range of the most widely grown vegetable and horticultural crops—in key locations.

Effect of Pelleting on Germination and Establishment

Extensive experimental work conducted in New Zealand by FCC Ltd. [Fruitgrowers Chemical Co Ltd.] on behalf of CSL [Coated Seed Ltd.] during the 1960's showed no detrimental effect of pelleting on either speed of—or final germination % compared to identically grown, identical bare seed. On the contrary in these extensive greenhouse comparisons the evidence favoured pelleted seed for speed of germination with several seed varieties, suggesting moisture retention by the pellet may have been more consistent than for seed alone as well as providing more uniform temperature control. In potentially dry soils, pelleting protects seed from premature germination which, if followed by dry conditions during critical seedling radicle emergence will lead to large scale seedling failure—the crop may need to be drilled again. On the other hand, saturated soils not well drained may allow bare seed establishment—but will often give poor results from pellets even where fungicides are included. Where the pelleting material stays wet and remains on the seed without washing off, such conditions will tend to block out essential oxygen in original heavier pellets—but may perform more satisfactorily in (for instance) perlite pelleted seed where oxygen bubbles occur and the pellet splits open when damp. However, we have no information that such comparative tests with various products have been carried out professionally and independently.

Where the pellet provides a host matrix for addition of fungicides of which Benlate has been valuable but may no longer be manufactured, widely used Captan and Thiram have proven over many years to be an effective combination in covering the maximum spectrum of seedling fungous diseases. Pellets carrying these protectants have a distinct advantage over bare seed in the face of "damping off", pythium, seed rots and seedling blights (mainly suppressed by Thiram) plus fusarium, botrytis and downy mildew plus a further range of seedling rots (mainly by Captan). In combination, they have some insecticidal, rodent and bird repellant qualities as well. Prompt breakdown in soil preventing toxic run-off to waterways plus tests indicating that these chemicals are not carcinogenic to humans at agricultural levels of use, may leave them in the "hazardous substance" but not "dangerous" range of seed treatments however, pelleting has the additional advantage of being able (provided the processor adopts it) to seal in these fungicides within the pellet usually eliminating harmful dust and only releasing the chemical when such pelleted seed is safely in the soil and dampened by rainfall or irrigation.

Precision Drills

Essentially two quite different types of drill currently compete for popular use. Each have advantages and disadvantages. They are:

(A) punched hole belt drive seeding units such as the Stanhay Mark 11 followed by the Jumbo machine which have been in use for over 40 years; due to limitations for some users [such as contractors who drill at higher speed] belt drives have the disadvantage of requiring slow drilling speed to permit optimum bare seed selection capability and accurate field operation (i.e., to avoid belt "slippage" and belt vibration "popping" seeds out), and despite that they cost 10 to 20% less, have been substantially (though not entirely) superceded in many countries by:

(B) pneumatic drills—also now made by Stanhay Webb Ltd. [UK] (in fact comprising the majority of their sales for about the latter 15 years) plus many other manufacturing brands [most with current websites] throughout Europe and the USA including Kverneland, Nodet, John Deere, Kuhn, Agricola, Monosem, Kongskilde, Agrolead, Amazone, Becker, Alibaba group, Prosem and of course several Asian products. Others including Gaspardo also manufacture direct seeding drills—these latter are not conventional pneumatic precision drills, but employ air to mobilise continuous seed and fertilizer flow with a new degree of accuracy and reliability directly into minimum tillage, even into litter strewn seed bed conditions with substantial fuel and time economies, compared to conventional seed bed preparation.

The latest of the precision drills such as Kverneland have optional electronic monitoring whereby drill operation can be controlled from the tractor cab employing a colour monitor. This sophistication can include actual seed monitoring via a visual sensor, being extremely valuable because with enclosed seed units, one could be tractor driving a long time before discovering a malfunction with no seed having been sown at all from one or more seeding units. Worse, the driver may have no idea when or where it [or they] stopped sowing (for this reason, brightly coloured pellets—specially white, are invaluable for checking pellet placement in the soil whereas brown or grey pellets are often almost "invisible"). This sort of electronic sophistication is appropriate for larger growers and contractors who are able to cover large areas with this equipment in intensively cropped localities because such service can run day and night (with relief drivers) in air conditioned cabs using earphone entertainment to overcome boredom as well as with electronic surveillance of continuous operation all made possible by reliability of modern machinery. This is particularly important where work is highly seasonal and growers all optimistically want their fields drilled simultaneously and promptly by contractors when conditions are optimum.

These more recently developed pneumatic drills have the additional important advantage that they can drill a wider range of natural seed variability (in size and shape) with acceptable sowing accuracy which no longer require pelleting for singulation, and they can also work faster in the field. Not all seed varieties however are included in this new pneumatic drill bare seed capability: *Parsnip, carrot, onion, brassicas, celery, pepper, tomato, tobacco, begonia, impatiens, marigolds and petunias* are some of the crops more accurately sown when precision pelleted—furthermore, when so pelleted professionally, can be drilled equally accurately with a belt drive as a pneumatic drill except that, as mentioned above belt drives are much more sensitive to excessive ground speed and vibration thus operate with greatest accuracy for most crops at tractor speeds of 1.5 to 5 km/hr—very slow for a tractor but ideal for single and double unit manual operation. The belt passes through a seed hopper filling successive belt holes with one seed or pellet of a size designed to fit into that hole size. Seed or pellets pass through a flow controlling choke into the belt chamber agitated by a wheel which also presses seed units through the belt holes after singulation on a flat or grooved base plate and then dropped to a small furrow on the soil usually covered by trailing chain harrows.

The manual Stanhay drill can be a single seeding unit or two units attached (as shown below) hand pushed, for smaller scale growers at quite minor cost.

For different crops, this requires use of multiple belts of different punched hole sizes to accommodate each seed variety and chokes which control seed or pellet flow.

Drill vibration and speed can reduce accuracy, whereas pneumatic drill suction tends to hold seed or pellets more firmly until the vacuum is released.

Fig. 47 Two Stanhay 820 belt drive precision drill units connected up for 4 person push pull operation] at minimal cost.

This concept is ideal for small to medium sized operations where more people are available than machinery and where costs must be kept to a minimum (These photographs by kind permission of Stanhay Webb Ltd. Grantham, UK [See more detail at: www.stanhay.com/products].

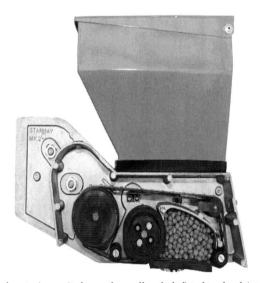

Fig. 48 Stanhay seed metering unit shows the endless belt fitted to the drive wheel (left) passing clockwise [round to right] and down around two tensioning wheels then travels left across the flat or grooved base plate under the pellets which drop singly into punched holes designed to let pellets drop through.

Fig. 49 A Stanhay 870 precision drill belt drive unit able to be coupled up with others, from 1 to 18 units tractor drawn.

Nevertheless, while pneumatic drills are generally superceding belt drives (specially for contractors and large scale growers) they can cost about 1½ times more than that of belt drives which are able to be fitted with gangs of seeding units on multiple tool bars covering large areas quickly.

The large units fully loaded, require increasingly powerful tractors to pull them and drive vacuum and hydraulic gear as well. They also require the support logistics of accompanying truckloads of seed or pellets plus fertilizer—out of danger of getting damp—certainly not wet. This large scale equipment must shift continuously with the tractor/drill units. Many of these drill/tractor combinations can be hydraulically compacted and mobilized for safe road transport essential for busy larger scale contractors.

For smaller scale contractors the Stanhay drill and tractor unit [see below] mounted on a truck is also a highly mobile and versatile combination which was "state of the art" in the 1970's and while superceded technically today by larger electronically monitored and largely pneumatic drills, is still a very viable option for medium scale farming groups or contractors who have a limited and seasonal demand for precision drilling.

In developing economies, belt drives will be viewed favourably and if managed carefully will provide good results. Furthermore, single unit manually-pushed belt hole drills are available for the smallest scale operations.

While some of the larger seed varieties can be drilled with an acceptable degree of spaced singularity as plain seed (not pelleted) nevertheless, even light pelleting, encrusting or film coating of most of these varieties does render

Fig. 50 The illustration (above) shows a "Kverneland Accord" pneumatic precision drill which has an important rapid change wheel for various sized seeds and pellets (Illustration by courtesy of www.kvernelandgroup.com/accord and their New Zealand Agency "Power Farming New Zealand Ltd.").

Fig. 51 Contractor's mobile Stanhay Precision drill.

them more flowable as well as more uniform in shape and size. Larger bare seeds can be individually selected satisfactorily by precision drills without pelleting providing an acceptable degree of precision. However, when taking the cost of expensive seed into consideration plus labour saving cost (less infill and thinning from fewer "doubles" and misses) plus superior crop quality, the relatively modest cost of encrusting, film coating and even the more expensive precision pelleting of certain seed may nevertheless be found economically valuable with some larger seed crops and field situations.

The New Zealand Experience

In New Zealand, the commercial scene has changed in recent years. The sole professional precision pelleting firm from the beginning in early 1960's was Coated Seed Ltd. [CSL]—its product called Prillcote™ as were its forage seed coatings except film coating and encrusting which were not produced by CSL in its formative years.

Fig. 52 A simple pelleting drum capable of producing about 2 Tonne per day of precision pellets when operated by a skilled technician. Costing $1,000 in 1979 it is constructed of 1/8" mild steel plate, the bowl diameter being .7 metre wide and .8 metre depth with a clear interior and the bottom lip of the bowl being 1 metre from the ground. It has an exterior chain drive from a 1 h.p. electric motor geared to rotate the bowl at 25 r.p.m. The frame comprises 1½ inch mild steel legs which are adjustable in height (threaded) and the base plate holding the drive cog is reinforced. This simple machine has given good service and is all that an operator needs to commence a simple pelleting operation but requires skills which can be acquired using some reject seed about to be dumped, to develop and practice the skills essential for continuous commercially acceptable product. This bowl is also suitable for processing inoculated coated legumes and coated grass seed.

Fig. 53 This bowl is specially suitable for processing smaller runs of precision pelleted seed but also for preparation of coated seed and research treatments over the whole range—as used by Axis Associates, Nelson, New Zealand. A similar bowl to this can be purchased from Disha Engineering Works, India called a lab coating bowl [http://www.dishaengineering.co.in].

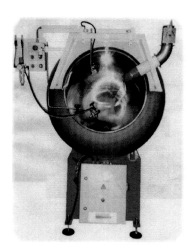

Fig. 54 A fully equipped precision pelleting drum fitted with electrically controlled and adjustable liquid spraying device (left) plus pelleting solids by aerated tube (right) and is well illuminated by a fixed light (centre top).

This machine [type 9100 20 00] is marketed by Seed Support [which is a trade name of Synthesis B.V. in the Netherlands], of Hengstdalseweg 44, The Netherlands.

This Company trains their new clients in the art and science of pelleting and pelleting quality control. This technology is now available worldwide. Contact is (Willem van Lith. email: willem@seedpelletingequipment.com) photo: "Seed Support".

The method of precision pellet production during those formative years was by manual revolving drum processing—one highly skilled person operating a revolving open ended cylinder similar to a concrete mixer (without internal agitators) applying consecutive coatings of a finely atomized adhesive spray directly onto seed—being careful not to spray the drum making it sticky which would not only "grab" coating material, but also seed, congeal both and ruin the procedure.

Much like pan coating of confectionery (of which there are many *expired patent* descriptions published. Confidentiality of this process terminated long before CSL commenced production], the operator in seed pelleting by this method must be really skilled in judging the correct alternate applications of adhesive then coatings and eventually determining the point at which to stop building them up when pellets have reached optimum size. Small rectangular punched hole test sieve [deep trays] are used to check sizing while pellets are still damp and still in the processing drum. Some seeds within certain varieties vary so much in size that two and even three finished pellet sizings are provided to the user.

Fruitgrowers Chemical Co Ltd., developed chemical "prill" processing expertise by producing DDT coated prills to combat grass grub damage to pastures—the most severe damage to already stressed dry stony "light land" pastures in New Zealand. Better quality pastures also suffered their ravages but faster recovery on good soils minimized the visual effect. Unfortunately, while the product succeeded initially, soil tests revealed hundreds of thousands of acres (now hectare) had become persistently contaminated with (now banned) DDT[*1] which disqualified such land from dairy production for many years— also condemning meat if it tested adversely. This contamination has thankfully all but disappeared—it was a lesson learned well in New Zealand. On the plus side, the same "prill" manufacturing principles were adapted to produce successful inoculated and/or coated seed—hence the trade name Prillcote™.

FCC Ltd. went through transitions in ownership after the decease of its founder T.J. McKee then later became involved in environmental issues concerning its chemical manufacturing and toxic waste disposal activities at Port Mapua near Nelson, New Zealand. These and other events proved terminal for the Company and some associates such as Farm Chemicals Co. Later, Fruitgrowers Chemical Company, taken over by the Newmans Group disposed of its half share interest in Coated Seed Ltd. to the other half owner— Wrightson™, a well established Stock & Station Agency with substantial Grain and Seed Merchant capability and with International interests- all of which the Coated Seed Ltd. activity was only a small part. At that point however, there was no longer an "impartial" processor to the New Zealand Seed Industry; Wrightson™ became owner of many of the former resources of Coated Seed Ltd. in a "Division of technologies" agreement. It was the combined personal and professional staff expertise of [principally] G.G. Taylor, J.M. Lloyd and G.M. Bennett that had earlier carried Wrightson™ into a premium position as partner in New Zealand's top seed coating facility—indeed acting as a "commercially neutral" processor to the New Zealand seed industry via its vehicle CSL which accepted and processed seed from all seed merchants who

---

1* DDT is the chemical: dichlorodiphenyltrichloroethane

chose to use the expertise. Some did not, sadly depriving their clients of the most successful product available.

Coated Seed Ltd. had in fact become not just a world leader—but almost certainly *the* world leader in seed inoculation and coating technology proven by unrivalled field results [as recorded in this book] identified by totally independent research during the 1970's through 1980's. Nowhere else was this technology as far advanced—including Australia (with its world class *Rhizobium* research personnel) and USA—which, via a large US Firm eventually purchased CSL's processing expertise [excluding inoculant manufacture and all other published technology]. Logistics of commercial use, specially of inoculants, suited temperate New Zealand more than in the vast expanse of frequently hot, dry conditions experienced for instance in most of Australia and some parts of the US.

Wrightson™ continued processing but began to move away from the "Coated Seed Ltd." concept. Continuing the seed coating activities of former CSL within its own genre in the same mode as for seed dressing and trading of which it was a major independent operator. Wrightson™ eventually discarded the "Coated Seed Ltd." facility which is no longer a registered Company.

Precision pelleting at CSL had been a manually skilled activity operated exclusively by senior processing technician Mr. Norman Hill who was approaching retirement by the mid 1980's and, not only that, but because it concerned mainly vegetable, horticultural and forestry seed which did not fall into the historical spectrum of the New Zealand Grain and Seed Merchant's traditional core expertise, there was an apparent lessening of interest in precision pelleting. Consequently, when in the early 2000's Wrightson™ was merged with another large New Zealand Grain & Seed Merchant- Pyne Gould Guiness Ltd. to become in combination *PGG Wrightson* becoming easily the largest Grain and Seed merchant organization in New Zealand—and of formidable size and marketing strength on an International basis, it disposed of the precision pelleting activity of former Coated Seed Ltd. in February 2009.

While the new firm *PGG Wrightson* has now fully discarded "Coated Seed Ltd.", the registered trade name Prillcote™ remains commercially viable—its International success obviously too important to discard lightly—however, current product clearly differs in several ways from extremely well researched Prillcote™ of the 1970's to 1990's—particularly its inoculated legume seed coatings.

Author (Bennett) who edited the Prillcote™ *"Technical Manual"* published in the 1970's included in it a table of seed suitable for sand pelleting plus the normal weight gain and size of finished pellets—[a similar table is shown below] which was widely distributed at that time—almost 50 years ago, so it is not new.

Pellet Properties

That Table [below] providing weight increases and pellet diameter sizes relates to drum processed pelleting using graded sand, clay and talc providing weight gain targets and finished diameter sizes which applied to the Prillcote™ process at that time. These were "heavy" pellets which are still preferred internationally by some growers using machines with a long drop to the furrow where faster drop and less bounce are important. Today, there are a multitude of processors around the world who have developed a range of seed weight increases, varietal target diameter sizes and pelleting materials to suit local conditions and new methods of pelleting—including much faster rotary spinning disk technology now rapidly replacing drum pelleting in larger production. The spinning disk method is more automated, more mechanically regulated, faster and does not require the same level of operator skill.

Modern Pellet Contents

Improvements are ongoing as one would expect including new drilling machine sophistication and automated nursery methods of seeding into transplant containers. Some pellets are now quite lightweight using materials such as diatomaceous earth and perlite as the main filler plus surface stabilizing binders of which hydroxypropyl cellulose, a water soluble hardner and lubricator, or polyvinylpyrrolidone are likely to be suitable, but more expensive than some simple starches and gums which may be found suitable. An inexpensive binding agent called pullulan (currently, at time of writing, $37.50 to $125 kg but used at less than 1.5% of the pellet weight) like other binders is used in pharmaceutical tablet manufacture—very likely suitable also for finishing pelleted seed. The aim of these binders is to provide a smooth, polished, dust free surface [particularly important for user safety where chemicals are included with or in the coat]. A smooth abrasion-free (almost lubricated) movement in the mechanically agitated seed box of drills which vibrate as they travel across paddocks at contractor speeds, even those with inflated rubber tyres, is very important for smooth drilling operation. More about formulations however in Chapter 6 herewith.

In determining optimum pellet sizes, the following seed weight increases and finished diameter sizes were found to be those most suitable for maximum drilling precision in tests and trials conducted with Prillcote™ pellets and mainly with Stanhay continuous belt precision drills in the 1960's to 1990's.

This guide may still remain valuable for developing economies to adopt initially because the cost of drum pelleting production for punched belt drills, using local screened sand as the main filler is definitely less overall (than rotary pelleting with lightweight fillers, modern polymers and using pneumatic drills)—but while cheaper, requires more skill both in pelleting—and drilling.

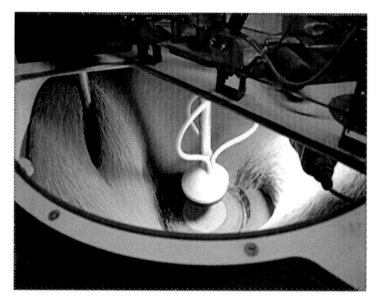

Fig. 55 The Rotary coater in action (Photo by courtesy of "Seed Support" of The Netherlands).

An important aid to grower acceptance and understanding in New Zealand has been the availability of the test bed calibration unit of the *New Zealand Institute of Agricultural Engineering* [NZIAE] at *Lincoln University.* Calibration is an essential aid to accurate belt drive use. Precision drilling results will be poor unless correct drill settings are used for belt hole size, choke [size and type] plus ground speed. Once a grower has found the ideal combination for a pellet size and shape, provided the seed variety is the same, the actual seed size similar and the pellet finished to a specific and constant sieve sizing, then a quick check by the grower to confirm that pellets are "the same as last year"—may be all that is required to retain the same settings. Number of belt holes per metre and single or double rows of holes per belt should remain constant for any given crop. A contractor will of course have a range of accessories to match various crop seeds and pellets.

To assist development and for a specific time, NZIAE offered this calibration service directly to growers who sent in a matchbox full sample [of recently pelleted—or bare seed about to be sown] together with a nominal fee of around 50 cents. The NZIAE then ran that sample through their test set up (a continuously rotating conveyor belt about 10 metre length and 150 mm wide run by electric motor governed to simulate the variable speed of a drill passing over the soil, except in this test [indoors], the drill unit remains stationery, the seed box and punched hole belt is set in motion, while the large conveyor belt (i.e., "the ground") coupled to an electric motor passes at various speeds beneath the seeding unit. With the Stanhay seed metering unit [see illustration below] mounted at one end, variable belt hole sizes and hole spacing per belt plus choke and ground speed configurations were compared to find the ideal drilling combination for those pellets and a test sheet was sent back to the grower or contractor with the Institute's recommended settings.

This facility was set up by NZIAE Senior principal Research Officer Mr. J.S. Dunn who dedicated a great deal of time and attention to the new technology plus encouragement to growers as well as to Coated Seed Ltd., the sole New Zealand precision pellet manufacturer through the 1960's until the turn of the century. His inspiration led to inclusion of forestry seed precision drilled nursery crops at Ford's Nursery, Oamaru while an innovative Lincoln University Senior lecturer, Bob Crowder of the Horticulture Faculty developed and demonstrated highly successful precision drilled paddock scale dwarf tomato crops using pelleted seed, also other precision drilled crops.

These two [Lincoln University associated] pure research and research/ teaching scientists were exemplary in their fields—not content with written or lecture room demonstration they both got out there right into the industry on a commercial size scale and made sure it all worked in the field. It is often said—"there is no better way to teach than leading by example".

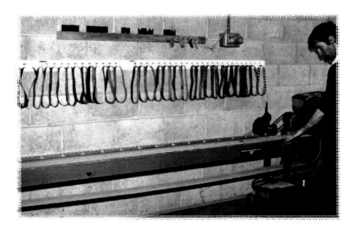

Fig. 56 The earlier precision drill calibration machine of the New Zealand Institute of Agricultural Engineering at Lincoln University, Christchurch, New Zealand. Note the wide range of punched belts hanging on the wall for testing best sizes as well as bases and chokes above. Engineering Technical Officer David Jamieson controls the "ground speed" by use of a variable speed electric motor driving the conveyor belt under the seed metering unit [instead of the metering unit passing over soil] to visually count successful individual pellets or seeds plus misses, doubles or triples per hole. By trying different combinations (including speed) the most suitable setting for that crop can be determined for the Grower. The travelling conveyor belt is made slightly sticky with vaseline or grease to hold pellets in place where they drop, for counting.

When a satisfactory precision placement pattern is obtained, such result would likely be about 96% correctly placed singles, 2% doubles [twin pellets at one location], 1% triples or multiples and 1% gaps or misses. That would be a good result. At speeds exceeding about 4.5 mph, drilling accuracy with belt drive machines deteriorates significantly.

The precision pelleting interests of former CSL were sold in 2009 as stated above, to *Seed Enhancements Ltd.* [SEL] [*www.seedenhancements.co.nz*] P.O. Box 333, Pukekohe, New Zealand to which company any grower may now send their bare seed for pelleting.

In the 1970's minimum orders were set by Coated Seed Ltd. at 500 g (1/2 kg). The cost was $2 per kg of finished weight for orders of up to 20 kg of raw seed supplied. The cost for orders of over 20 kg of raw seed per order reduced to $1.65 per kg of finished weight as a "large volume" discount. A further reduction to the wholesale rate applied to orders over 220 kg of raw seed supplied by seed companies for processing. This method of costing is

Fig. 57 Typical of bare seed ideally pelleted for best precision in belt or pneumatic drills is Anise seed *of* 3 to 5 mm, genus *Umbelliferae* (Carrot family), not to be confused with "Star anise" though both have aniseed flavoured fruit.

Fig. 58 Procoat™ parsnip pellets (processed by SEL. Pukekohe, New Zealand.

fair and equitable because with sand coated pellets, weight increase relates directly to the time, materials and skill required for each variety.

For instance 1 kg of lettuce seed was traditionally increased in weight by about 35 times by precision [sand] pelleting to make approx. 35 kg of pellets. At $2/kg of weight increase will cost $70. There is a lot of work and screening to bring all seeds to size, and the smaller seeds need to be additionally processed to reach optimum size. This is a substantial quantity of lettuce pellet, i.e., 1 x 20 kg pack and 1 x 15 kg from just 1 kg of seed. Whereas 1 kg of pine tree seed will increase only about 5 times original weight to make 5 kg which costed just $ 10.

Prillcote™ precision pellets certainly set an industry standard through the 1960's to the early 2000's which was widely adopted and similar to the detail shown in the Table (below)—however, as a result of technology development, this sand based product, no longer produced in New Zealand to our knowledge has been replaced by a range of new pellets using advanced technology produced by "SEL" [Seed Enhancements Ltd.] of Pukekohe. Sand pellets are nevertheless still produced for instance in USA.

The parameters on seed pelleting at SEL have changed from those of former Coated Seed Ltd. Whereas CSL's minimum weight of bare seed per order was 500 g, SEL will accept as little as 200 g for most seed varieties and as their pellets are much lighter than CSL's, the cost is not always assessed on weight gain but also on bare seed weight to be pelleted—as well as on seed count per gram or kilogram.

SEL has its own range of finished pellet diameters and consequent weight gains per variety which detail is more fully discussed in its website (shown above).

Their range of pellet sizes and seed-to-pellet weight ratios are, they advise, identical[3] to Propell™ being Germains' UK product (ppl in USA).

Fig. 59 These squash seeds are film coated—not precision pelleted (processed by SEL Pukekohe). [SEL photo].

SEL Managing Director Mr. Andrew Culley advises sales and performance of these products are doing well in New Zealand but that the market is now small [pneumatic seed drills have removed the need for pelleting some seed varieties]. Some varieties of seed do need full pelleting including parsnip—the latter, processed by SEL are called Procoat™ parsnip pellets [illustrated above].

Nevertheless, sand is commonly available worldwide and may still be preferable in developing economies for precision pelleting—inexpensive, but demanding skill in application.

Seed Varieties Pelleted

Standard Sizes of Finished Pellet and approximate Weight Gain optimum for (sand) pellets used in belt drive drills.

Seed Variety	Approx. weight Increase (gms)	Approx. diameter of finished pellets (mm)
Anise	12–16	3.00–4.25
Asparagus	3–6	4.75–5.50
Cabbage	10–15	3.25–4.00
Carrot	14–20	2.50–3.25
Cauliflower	10–15	3.25–4.00
Celery	15–20	1.75–2.50
Chicory	12–16	2.75–3.50
Cucumber	5–10	4.75–6.25
Fodder Beet	3–4	3.5–4.75 & 4.75–6.25
Gherkin	5–10	4.75–6.25
Leek	7–8	3.25–4.00
Lettuce	30–40	3.25–4.00
Lucerne	1–1.5	2.00–2.75
Mangold	3–4	3.5–4.75 & 4.75–6.25
Onion	7–8	3.25–4.00
Parsnip	12–20	4.00–4.75 & 4.75–6.25
Parsley	12–16	2.5–3.25
Petunia	60–70	1.75–2.50
Pinus Radiata	3–6	3.5-4.75 & 4.75–6.25+
Poppy (Iceland)	25–35	1.75–2.50
Pumpkin	3–4	4.75–6.25
Radish	10–15	3.25–4.00
Rape	8–16	3.25–4.00
Red Beet	5–10	(not standardized)
Silver Beet	5–10	(" ")
Spinach	5–10	(" ")
Sugar Beet	3–4 Monogerm	3.5–4.75
Swede	8–16	3.25–4.00
Tomato	10–20	3.50–4.25
Turnip	8–16	3.25–4.00

The list above shows most of the seed varieties which are or have been commonly pelleted or film coated for better precision in belt drive drills because their natural shape is either awkwardly elongate, discoid, some flossy/bearded or simply uneven and difficult for singulation. Also because of variable bare seed size.

These sizes and weight gains do not consistently apply to pneumatic drills where some bare seed varieties perform sufficiently well in pneumatic [vacuum suction] drills either as bare seed or film coated.

Any processor using graded sand for pelleting (with a firm binder) by a revolving drum method [easily the cheapest way to get started] for use in a belt precision drill—from a single push drill to a multiple seeder, tractor powered, will find the above Table invaluable.

This list is however now substantially modified where two essential changes apply today—firstly where pneumatic drills do not now require some of the above seed varieties to be pelleted, and secondly, because more recent pelleting materials are lighter thus even where those products may be pelleted to similar sizings as in this Table, weight gains will usually be quite a lot less at the same pellet size.

Pellet Variations

Light weight pellets have advantages such as lower freight costs and friendlier sack weights. They also employ "splitting" or "slumping" qualities—depending on materials used, both vital for oxygenation of the germinating seedling. Polymers employed offer refinements such as water sensitivity; for instance a water sensitive polyelectrolyte is formed by combining a basic polymer with an acidic polymer which then does not "dissolve" the pellet [when in contact with moisture], instead it swells and disintegrates falling away from the seed [slumping] thus allowing it access to vital air/oxygen as well as releasing pellet contents to use of the seedling—i.e., fertilizer coating, fungicides and/or insecticides which are safely contained in pellet form under the dust free exterior coat.

Other enhancements include Landec-Ag's "intelligent" polymer applied as an exterior plastic coat which excludes water {from the seed} at lower temperatures, e.g., >55°C being too cold for many seed varieties to germinate anyway, thus avoiding a false start to germination with danger of fungal growth, but at temp's exceeding 55°C the plasticizing polymer admits water when soil temperatures are also satisfactory for safer germination.

Exfoliated vermiculite (similar to perlite) is a micaceous mineral, essentially a magnesium/aluminium/iron silicate with a platelet-like structure which entraps tiny bubbles of air at exfoliation heating (e.g., >1,200°C) expanding the platelets (creating a better binding and smoother coating material) providing an ideal near neutral pH and lightweight oxygenated material which when milled and screened to a consistent particle size is valued for many industrial uses as a filler including seed pelleting[4].

Perlite currently used in coating and pelleting when in conjunction with a moisture responsive polymer adhesive, splits open in soil when wet after drilling, exposing seed to oxygen and moisture thus facilitating safe germination.

Yet other pellets provide a carrier for chemicals aimed at suppression of fungi and/or insect predators, which protection at the vital early seedling stage can be the difference between a good crop—and a disaster—particularly for precision drilling where the minimally required number of plants have been sown. It should be emphasized here that a precision drilled crop with only a fraction of the seed employed which would be used for a conventionally scatter drilled crop, has MUCH greater potential for failure where threats such as disease or insects are not taken care of. To achieve a good precision drilled result, ALL factors must be managed "precisely". Precision is required in all aspects of the technique.

Pellets containing protective chemicals such as Captan and Thiram fungicides which have proven a most valuable combination for many years, do not alone or in combination harm any seed we have examined. While insecticide treatments [more likely than fungicides to be harmful to seed, people and animals] are able to be encapsulated out of contact with the seed itself over a benign protective first coating but under a benign subsequent outer coat, not only to provide safety for handlers from toxic dust, but also to offer better protection from phytotoxicity for the seed itself. Nevertheless, chemicals to which seed is at all sensitive such as certain organo-phosphorous and chlorinated hydrocarbon insecticides (many of which are banned today in agricultural use anyway), are likely to cause damage to germinating seed despite a protective layer of inert or nutrient coating applied between seed and chemical—as a pellet. In general terms, any chemical which shows a degree of toxicity to seed varieties when directly exposed to that chemical in moist conditions, should not be incorporated in a seed film, coat or pellet for obvious reasons. It is unlikely seed or seedling phytotoxicity will be validated by insecticidal advantage.

While in the process of pelleting anyway, it is a relatively simple matter to add colour to the pellet either for seed variety identification (specially where the seed is covered in opaque coating materials and is unseen)—or for identification of any additive in the coat or pellet. Particularly where additives are in any way toxic to people, specially children or animals an identifier is required, as even supervising adults may not be aware of such danger unless it is signaled in such a way.

In due course, a standard for colour coding will almost certainly be established, possibly included into existing International regulation governing the requirement for seed acceptance or certification prior to export—or import. Included in such certification, may be the requirement for correct and adequate identification by colour. Where seed is unseen [covered by coating], identification by colour at least into broad categories—i.e., coated clover and lucerne (inoculated or not), coated grass seed, vegetable and forestry seed pellets as a basic level identifier, then inclusion of fungicides or insecticides, etc. as a second tier identifier, would seem not only possible as safety and practical indicators—but also a sensible, easily applied characteristic for maximum efficiency and safety at relatively minor cost.

Sowing Rates at Specific Spacings

With expensive seed, growers need to work out carefully just how much seed they need to buy and if it is to be pelleted, then that added cost emphasises the importance of not exceeding requirements. However, if found to be in excess, professionally manufactured pellets will usually keep until next year. Most seed loses its viability gradually over time but some varieties are notorious for doing so quickly (like parsnip) thus where we are aiming for maximum precision with as close to 100% successful viability as possible, it is not going to help if the seed itself has dropped from a high interim germination count of (say 98%) to 85% whereby we are then stuck with 15% vacant placings before we even start to drill.

How it Works: See Table 17 below. Select the in-row seed spacing (across, in red)→select the row width (down the left, in mauve)→find number of seeds required per hectare (follow from the red column down to black cell (at row width crossover) →divide that number by no. seeds per kilogram = weight of bare seed required [therefore weight of seed to be pelleted].

Calculation Guide {The following is a demonstration only, not intended as practical advice}.

Example: Let us take the top row [across] as our example. Let us say the overall row width is to be 20 cm (centimetre) then alongside that 20 cm we see there are 50,000 metres per hectare (at 20 cm row spacings). The crop we are to sow [say onions where our most marketable bulb size is (say) 7 cm diam. will require 1 seed per (say) 8 centimetre (see top row in red) which means there will be 12.5 seeds per metre (see the blue row under the red 8 cm). As there are 50,000 metre per hectare x 12.5 seeds per metre we will need 625,000 seeds (see .625 million in the table under 12.5 seeds per metre) to precision drill each hectare. If these onions have a seed count of (say) 300 per gram = 300 x 1000 (grams per kilogram) = 300,000 seeds per kg. Thus, as each hectare requires 625,000 seeds ÷ 300,000 = 2.084 kg of seed needs to be pelleted per hectare.

In the table above (weight gain for Prillcote™ pellets) onions are increased in weight by up to 8 times bare seed weight therefore we need to drill 2.084 kg x 8 times weight increase = 16.672 kg of pelleted onion seed per hectare. However, those are the heavier sand pellets finished in clay or talc as made to Prillcote™ formulae—whereas *Seed Enhancements Ltd.* in New Zealand [who do not produce Prillcote™] now use much lighter weight pelleting materials. In addition, vacuum drills do not need the full 3.5 mm to 4 mm diameter pellet size which gives a near spherical onion pellet essential for best performance

Table 17. Metric Placement Table
NUMBER of SEEDS PER HECTARE [*mil = million.*]

No.Seeds/metre	Spacing in Centi/m →	4	5	6	7	8	9	10	11	12	13	14	15
	row width Row/m →	25	20	16.66	14.3	12.5	11.1	10	9.1	8.33	7.7	7.14	6.66
50,000	20	1.25mil	1mil	.833mil	.715mil	.625mil	.555mil	.5mil	.455mil	.4165mil	.385mil	.357mil	.333mil
40,000	25	1.0mil	.8mil	666,400	.572mil	.5mil	.444mil	.4mil	.364mil	333,200	.308mil	285,600	266,400
33,333	30	833,325	666,660	555,327	476,661	416,662	369,996	333,330	303,330	277,663	256,664	237,997	221,997
28,571	35	714,275	571,420	475,992	408,565	357,137	317,138	285,710	259,996	237,996	219,996	203,996	190,282
25,000	40	625,000	500,000	416500	357,500	312500	277,500	250,000	227,500	208,250	192,500	178,500	166,500
22,222	45	555,550	444,440	370,218	317,774	277,750	246,664	222,220	202,220	185,109	171,109	158,665	147,998
20,000	50	500,000	400,000	333,200	286,000	250,000	222,000	200,000	182,000	166,600	154,000	142,000	133,200
18,181	55	454,525	363,620	302,895	259,988	227,262	201,809	181,810	165,447	151,447	139,993	129,812	121,085
16,666	60	416,650	333,320	277,655	238,323	208,325	184,992	166,660	151,660	138,827	128,328	118,995	110,995
15,384	65	384,600	307,680	256,297	219,991	192,300	170,762	153,840	139,994	128,148	118,456	109,841	102,457
14,285	70	357,125	285,700	237,988	204,275	178,562	158,563	142,850	129,993	118,994	109,994	101,994	95,138
13,333	75	333,325	266,660	222,127	190,661	166,662	147,886	133,330	121,330	111,063	102,664	95,197	88,797
12,500	80	312,500	250,000	208,250	178,750	156,250	138,750	125,000	113,750	104,125	96,250	89,250	83,250

Table Edited by author

Row width Row Metres

[cms]   [per hectare]

in a punched hole belt drill, because pneumatic drill suction can select pellets which are not quite spherical nor of a truly uniform size quite consistently. The drill is fitted with an adjustable toothed scraper alongside the seed disc which removes the occasional "double". This is not so with [punched hole] belt drive drills where two or more pellets can jump into one hole quite easily if pellets are elongate rather than spherical—and more so where pellets are of variable size.

Above calculations leave one question unanswered however. That is, how do we reach a decision that the row width is consistently 20 cm—when we normally need access for tractors and other equipment—maybe for laying out irrigation but certainly for harvest and probably for spray application against insects, and/or disease and/or weeds? It is an additional factor which needs to be included in the calculation.

Row Width

Spacings. Assuming we are still dealing with the onion crop mentioned above. The selected spacings are: 8 cm between seeds (e.g., each onion can grow >4 cm each way along the row before it touches and interferes with the adjoining onion. That provides for >8 cm diameter onions. As we are aiming at (say) an ideal 7 cm diam. table onion for our local market, we also don't want to "stunt" the crop—so 8 cm allows a small margin for "error" placement and is about right. Then we need space between rows which in theory could also be 8 cm because if 8 cm is OK along the row, it should be OK between rows too—however, for many practical reasons [access to sunlight and moisture, foot access for thinners/space filling planters, weeders, room for root development and because row width might be dictated by the type of drill being used] it is commonplace to sow slightly wider spacings between rows. Therefore we shall select (say) 12 cm between rows [usually requiring a double or even triple tool bar drill for such close spacings on precision drills. Seeding units are mounted on fore or aft toolbars offset to our relatively close spacing—not mechanically possible on a single bar) and as our tractor (and any wheeled and towed gear) has a closest wheel spacing (front or rear) of (say) 1.4 m centers, we can drill the following pattern [1.2 m (between inside of wheels) = 120 cm tractor wheel spacing ÷ 12 cm row width spacing = 10 rows per pass then leave a space of (say) 20 cm for machinery and other access. This means we have 10 rows x 12 cm = 120 cm plus an access lane of 20 cm = 140 cm total for 10 rows—or *14 cm per row*. Because our Table (above) commences at 20 cm between rows—being one of the closest spacings [i.e., maize, beet and many of the vegetable crops such as lettuce and tomato require much greater spacing between rows] we can get a reasonably accurate estimate of the quantity of seed we need to send away for pelleting by calculating the increase in density from 20 cm rows down to 12 cm rows being a percentage of the 2.084 kg of seed we need.

e.g., $\dfrac{20 \times 100}{12 \times 1} = \dfrac{2000}{12} = 166.6\%.$

Therefore if we sent 166.6% of the 2.084 kg of seed needed for 20 cm row spacings = 3.472 kg (rounded) of seed away for pelleting (irrespective of the weight of resultant pellets), we should be able to place one pellet per 8 cm along the rows with 12 cm between rows allowing a 20 cm access lane between each ten row block—to cover 1 hectare with 3.472 kg of seed—pelleted or not [3.5 kg rounded].

A more simple reading of the Placement Table (above) would be a crop where (say) seed needs to be dropped at 12 cm spacings along the row with 40 cm spacings between rows—in which case we do not need access lanes as that row spacing allows full access and at 25,000 metre per hectare across the Table to the column under 12 cm seed spacings within the row (in red) and under that 8.33 seeds per row metre (in blue) at the crossover cell of the Table [40 cm rows—left, and 8.33 seeds per row/m, down] reveals that 208,250 seeds (and pellets) will be required to drill 1 hectare at those spacings. We must then find the seed count per kg and divide that into 208,250 to find the weight of seed we need to drill—or to send to the processor for pelleting.

Seed Bed Conditions

As required in every other aspect of precision drilling, good preparation and firm control of all associated factors is essential to fully support the significant investment in precision drilling to ensure best results.

Ideally, the seed bed will be a well worked loose and friable deep soil which is organically rich and naturally well drained, weed and pest free, level surfaced, detritis free, nutrient adequate, of neutral pH, with a sunny aspect and adequately sheltered from damaging winds.

Realistically only a relatively small area of the earth's surface measures up to that ideal.

Some of these targets (above) are of nature and others are able to be developed. Best practice farming will be aimed at developing as many of these features as possible—often developed over a long period of time (i.e., soil structure and fertility, shelter, etc.) while meantime growing crops to the best advantage possible. This often means father/mother, to—son/daughter farming over generations, requiring full control of that land, but many Nations do not have a secure land tenure system—e.g., indisputable ownership (or at least long term lease with credit for improvements), without which personal incentive to spend time effort and money toward such long term goals, is understandably simply not there.

Satisfactory crop results can however be achieved under more marginal conditions, with care.

While it is not a technique suited to stony soils where stones alter the placement of seed, very small stones—or small patches of larger stone may not be sufficiently detrimental to abandon precision drilling.

Sandy soils can blow away taking seed or pellets with it; where this is a threat, either optimum time of sowing to achieve establishment during calm periods, or drilling only inside good shelter—either trees or more closely spaced shrubs and even better—rows of fruit or nut producers alternated with rows of precision drilled vegetable crops is a maximum use technique and if irrigated, can further counteract wind movement. In warm climates deep rooted mango and guava which can stand dry conditions may be suitable while tamarind, citrus and many other fruit and nut trees can provide a second crop, also providing shelter in single, double or triple rows when effectively spaced.

In cool climates, hedge shelter is common as are tree rows but a disadvantage of trees on very good land is that their roots will physically prevent cultivation several metres from the trees (specially macrocapa, pinus and eucalypt species), as well as depleting moisture and nutrient values from those adjoining soils. They also require maintenance such as regular topping and side trimming which can be expensive without the right machinery—and detrimental by blocking sun thus sustaining frost, ice and snow some distance from the shelter on the shady side.

Wet soils of course must be drained to be suitable, and acid soils neutralized with broadcast lime [calcium carbonate) or dolomite [magnesium carbonate] and nutrients also applied after soil or plant tests have identified deficiencies.

Soils low in organic matter—also nitrogen, may need to be cropped to a nutrient and structural boosting interim facilitator such as bitter blue or sweet lupins, which, being legumes and nitrogen fixers can be grown quickly to early maturity then grazed by stock (cautiously to avoid bloat) and/or ploughed under making valuable compost for the ensuing crop—specially valuable for sandy soils. Such lupin seed or other legume facilitator seed should be inoculated with high quality *Rhizobium* culture at sowing.

Conclusion

Essentially, the course of progress has been that centuries old methods of scatter sowing of horticulture crop, vegetable and forestry seed by mankind (as in Nature) is now being steadily improved by precise spacing of much higher quality [often hybrid] seed with close to 100% germination capacity, being more expensive per kg than earlier seed but less expensive per hectare when only the seed we need to grow is planted in the first place; and because it is then essential to avoid failures, is given every support possible to ensure growth to maturity. To achieve that degree of precision planting together with acceptable speed and cost, machinery has been developed which can place several rows of seed at precise spacings both between and along the row to provide from minimum to maximum inter-plant competition as a tool to achieve optimum size and quality of crop for each particular market. In pursuit of this ideal, difficulties in variable size and shape of natural seed which machines could not select individually with acceptable accuracy—have been overcome by processing the difficult seed varieties into spherical pellets of uniform size by use of (first) the punched hole continuous belt driven precision drill whereby hole size is optimum for one seed or pellet to occupy as the belt passes through the seed chamber, then on through the funnel to the soil below, and more recently the use of pneumatic drills at higher cost but which are faster and less demanding of seed size and shape.

The punched hole belt machines which required the majority of horticultural crop (and forestry seed for nursery culture) to be pelleted first was slow (often 2 m.p.h. for good results) and even then allowed a percentage of misses and/or "doubles" which performance was soon challenged by introduction of constantly improving pneumatic drills. These can: a) select a wider range of seed without the need for pelleting by use of a suction wheel fitted with [variable sized] suction holes under vacuum, more positively able to grab one seed or pellet per hole holding it firmly for transport—until the vacuum is released, the seed dropping to its soil bed at faster speed despite vibration associated with drill movement over uneven ground.

Parallel improvement came with mechanization for rapid pelleting of those seed varieties which still needed to be pelleted for a satisfactory degree of precision placement. Instead of heavy reliance on personal skills of the operator applying alternate applications of adhesive and pellet filler being graded sand finished with clay or talc, more recently new rotary pelleting machinery now pellets the seed faster by electronic calibration with much less reliance on human skills. These new methods use lightweight coatings such as diatomaceous earth or perlite finished with a binding agent to give a pharmaceutical-like smooth, almost lubricant, firm and dust free pellet. Alternatively on larger more easily selected seed a simple plasticized film coat is applied—able to include plant protection such as fungicides and/or a limited range of insecticides for optimum seedling establishment as well as providing a smooth, "lubricant" finish for optimum seed chamber performance.

While the latest technology is leading the way, that will continue driven by introduction of important *economies* of increasingly expensive seed plus higher expectations in the marketplace for quality produce. Faster, more mechanised techniques are more expensive than the earlier, more manual sand/talc pellets which will no doubt be more attractive and readily usable by those horticultural economies newly starting out on the precision planting system. The faster pelleting, faster more accurate drilling of larger areas demanding powerful tractors and high performance drills will be adopted in the more financially established horticulturally intensive localities—specially by contractors who need to cover vast areas seasonally, however, there will always be a place for small scale less expensive but equally sophisticated crop culture by the still developing agricultural economies and by small scale farmers.

Even better machinery may further reduce the need for pelleting prior to precision drilling, however, while that would seem desirable, because pelleting does cost money, it is also true that protection of seed from fertilisers and/or plant protection additives in the pellet can in many situations improve establishment and increase quality to such extent that final returns easily outweigh the extra cost of processing the seed first.

Need for more Research

These features need to be determined by independent researchers studying comparisons in the field with great care to find the true cost/benefit relationship of bare seed v pelleted [on germination, establishment, rate of growth, freedom from insects or disease, time to maturity, quality and market value of final produce, etc.].

Further, pellet types v other pellet types [i.e., pelleting materials v others], plus or minus various additives, pellet v film coats/encrusted seed, fertilisers drilled with bare seed v pelleted seed—and other studies await close attention. Growers need this information regionally where conditions vary—there is a big job of research awaiting agricultural and horticultural scientists worldwide. A consortium of seed, polymer, fillers and chemicals suppliers enabling major research efforts would be ideal.

Not only do the Growers need this research—the whole World needs it—urgently in some poverty stricken localities where child mortality is shockingly high due to food shortages and food related disease.

Definitions

There are several terms used to describe specific seed processing methods. Their casual use has caused some confusion around the world. The following brief definitions followed by further explanation may help to simplify and hopefully standardize the issue:-

Seed cleaning	removal of weed seed, inert matter, dirt and dust from raw harvested seed.
Seed pigmentation	surface colouring of seed for identification and/or repellant effect and/or for sales promotion. Pellet pigmentation is likely to become much more important than just on seed.
Seed treatment	one or more applications of a chemical or biological seed protectant.
Film coating	a thin usually glossy and transparent polymer coating on large and/or rough seed to aid fluidity.
Encrusting	thicker than a film coat but also a polymer for product fluidity plus achieving some uniformity in shape and size of seed and may include an enhancement additive.

Seed coating (e.g., grass seed) a nutrient and/or protective and/or ballistic coating on any seed but chiefly refers to grass seed or brassicas.

Seed inoculation and coating inoculation of legume seed with rhizobia nitrogen fixing bacteria plus protection of both seed and bacteria using some or all of—a sealing polymer over seed, a peat or diatomaceous earth based inoculant, a peat or diatomaceous earth coat, a nutrient or lime coat—plus creating a ballistic value.

Precision pelleting continuous loading of (usually inert) coatings on variably sized and shaped natural seed until uniformity of shape and size is achieved by test sieve to meet criteria required for reliable singulation in precision drills.

The following is a more complete description of above definitions:

Precision Pellet: *A single seed which has been substantially modified by application of overlaid inert or nutrient particles, from being either a naturally elongate, ovoid or discoid shaped seed and of widely variable size, into more spherically shaped and more uniformly sized units—called pelleted seed. The chief purpose is to standardise shape and size [approximately], thus facilitate accurate individual selection and pick up by precision drills for optimum placing in the field. These shaped pellets can be uniformly and ideally spaced to variable settings. This permits for instance utilizing natural interplant competition such as close spacing to achieve small pickling onions and succulent baby carrots, or for uniform growth, uniform ripening and "once over" machine harvesting of field sown dwarf tomatoes), or by utilizing wider spacing for larger product size, e.g., optimum market appeal of "jumbo" silver beet or lettuce. Precision placement can also minimize the need for machine or manual thinning, manual infilling of gaps, create machinery access lanes to fit any wheeled configuration, hugely economise on expensive hybrid seed (which alone will often cover the cost of pelleting) and provide a carrier for chemical protectors. They are precision pelleted seeds for precision placement in the field or nursery.*

Example. Flat or discoid parsnip seed, often attached in flat stack multiples is first singled as seed (by agitator, without mechanical damage), then built up, rounded and smoothed by pelleting for punched hole belt, wheel or pneumatic selection.

For full size onions a wide spacing is appropriate from which the correct quantity of seed required to be pelleted can be worked out (see Table placement

Fig. 60 A typical uniformly spaced onion crop achieved with precision pellets and a precision drill.

above). For pickling onions the spacing would be much tighter using inter-plant competition to maintain small size.

- **Inoculated and coated Seed:** *Plain seed, usually clovers or lucerne (alfalfa) to which an application of finely milled nutrient [or inert] coating has been made usually up to 1:1 seed/coat ratio chiefly for protection of root nodule bacteria. In addition to the nutrient coat sometimes other additives [trace elements or protectants] are included to support successful establishment (a) of clover onto native growth surfaces, usually uncultivated hill country—or via direct drilling—both into harsh seed bed environments and (b) of lucerne and other legume crops drilled into good soils but where the rhizobia are absent. There is no attempt to achieve uniformity of shape or size. The rhizobia bacteria inoculant can be applied initially after a sealing polymer film coat—or added onto the dried coated seed as a final minimal moisture spray without further drying.*

  Example. Clover seed to which *Rhizobium* inoculum culture has been applied either after a film coat followed by the main coat, or at completion of coating plus finely milled lime or other nutrient coating which is rhizobia tolerant such as gafsa phosphate and to which molybdenum or other trace element can be added.

- **Coated Seed:** *Applies to any seed which is simply coated [also up to a 1/1 seed/coat ratio] for support of establishment and growth. In addition to nutrient coating (i.e., phosphate, calcium or magnesium) to assist growth and ballistic superiority for aerial application (in which an inert coat will also suffice), this coating can be used to carry (and seal in for safety of handlers) chemical or biological protectants which are not phytotoxic and which cannot destroy rhizobia because this product is not inoculated.*

- **Encrusted Seed:** *A term not formerly used in Australasia, but is used by large Firms like Germains of California (establ. 1871 now represented worldwide) which markets machinery and polymer technology for seed processing.*

Encrustment refers to two qualities the subject plain seed does not have—partial uniformity of shape and size for high quality spaced drilling—or smooth seed surface for "fluid" movement in the drill—usually both. So encrusting provides a building-up film coat [from double to 5 times seed weight] for very small or difficult shaped seed where more than the ordinary coated seed 1:1 weight gain is required—but not right up to pellet size (which starts from about 5 times seed weight), thus saving full pelleting cost. It may also simply be a smoothing and polishing coat for larger sized seed which makes it more free flowing in precision and other drills.

Example. Small seed built up to more manageable size—or awkward seed (like sugar beet) made more uniform, smooth and flowable. The seed retains its shape but is smoother and a little more uniform in size and shape.

- **Film Coat :** *Plain seed to which a thin film of cellulose or a polymer has been applied "dry"(pure—not an aqueous solution) or for greater accuracy sprayed on as an aqueous solution, often becoming transparent and when dry creates very little weight gain (about 3% of seed weight and not more than 20%). Chemicals used may be included under or with the polymer in the fungicidal, insecticidal or repellent (birds, mice, etc.) range. The film also"encloses" any chemical toxicity for safety of workers and handlers which may be the sole reason for using such film. This was the most simple and first used of the seed additive processes.*

Example. Pumpkin seed film coated with a fungicide and coloured for identification.

[Also, see an illustration of film coated squash seed] (above).

- **Seed Treatment:** *The simple application of a chemical or other additive to protect seed in store or in the soil. Fungicides, insecticides and repellents are applied uniformly to seed, manually on the floor, or plastic sheet or often in large volume treaters. These additives may include a volatilizing polymer for surfactant coverage and evaporative drying or sufficient adhesive to cover and enclose the seed without attaching to each other and may include a colour for safety identification. Chemical dusts applied after minimal adhesive do not require drying but other treated product may be warm air dried when the additive is applied [tipped or sprayed] as an aqueous solution.*

- **Seed Pigmentation:** *Application of a colour dye usually for identification and usually applied with a seed treatment.* There are a good range of dyes available and their use is likely to grow with safety consciousness and the need to identify seed under an opaque seed coat or pellet.

- **Seed Cleaning:** *The process whereby raw harvested seed often containing other seed, weed seed, chaff, straw, inerts, dirt, stones and dust is run through a seed dressing mill which by various means—vibrating steel mesh riddles with punched or slotted vents of specific sizes, air blowing and suction and even* *viability separation by photometric sorters at high speed for individual (mainly large) seeds like peas and beans—is all designed to present a finished product of close to 100% pure harvested seed—or grain.*

There is some unavoidable overlapping of these descriptions, but to call a coated seed a "pellet" may be technically confusing within the Industry—and within the scientific community as well as is describing a "film coat" as coated seed or even as "a pill".

Today, the added weight depends much more on the pelleting material used (i.e., heavy sand—or featherlight perlite) than on the added volume, or mass.

Take as an example lettuce seed—if processed with (heavy) graded sand, will require between 30 and 40 times the seed weight {increase} to attain an acceptably uniform and spherical shape of 3.25 to 4.0 mm diameter required for belt drive precision drills and as ideal {if not essential} for pneumatic drills as well. But if the lettuce seed is processed with (featherlight) perlite for instance, it may only require 5 to 10 times the seed weight gain *to reach what we are mainly interested in*—a uniformly shaped and sized pellet of between 3·25 and 4.00 mm diameter. As weight gain is now much less relevant than size and shape, the author suggests that where diameter and uniformity are paramount for precision drilling, that these are all "pellets". But that where application of additives is for seed protection or enhancement (without regard for size and shape) that these are all called a "coat". "Filming" is self explanatory. As "encrustment" can contain a bit of each—"film", coat and pellet, it may be best left as an encrustment!

The term "pill" has been used in the industry but seems imprecise for describing seed processing and is fully described by Wikipedia Free Encyclopedia as follows:

Pill

Pill or The Pill may refer to:

- Pill (pharmacy), referring to anything small and round for a specific dose of medicine. The term is used colloquially in several ways:
  - A tablet or capsule which replaced dosing via a pill.
  - American slang for recreationally used prescription drugs, particularly when used in plural as in "popping pills".
  - "The Pill", a general nickname for the combined oral contraceptive pill
  - Pills is a nickname for the recreational drug MDMA, also known as ecstasy.
- Pill (rapper), American rapper
- Pill (textile), a small ball of fuzz on cloth formed by rubbing or wearing, or the creation of such fuzz balls

- Pill, Tyrol, a municipality in Austria
- Pill, Somerset, a community in the United Kingdom
- Pill, a colloqial abbreviation of Pillgwenlly, electoral ward in the city of Newport, South Wales
- "The Pill" (song), a 1975 song by Loretta Lynn
- "The Pill", a 1999 episode of *That '70s Show* TV series
- "Pills for Breakfast", a song on Faith No More's 1985 album *We Care a Lot*
- "Pills", a song by Bo Diddley
- Pill, a slang term for someone with a bad attitude.

But none of these definitions include seed or the seed industry. It may be wise to disassociate the seed industry from prescription drugs, contraceptives, pop songs and ecstasy drugs and stay with coats, films and pellets!

References

(1) FreshPlaza Internet publication. Circa 2013.
Quote: "China is not only the most populated country in the world but when it comes to production of fresh vegetables and fruit it is the world's number 1". "Of the estimated (commercial) production of fresh vegetables in the world, China produces half". FreshPlaza is an International fresh produce marketing organisation which sends out over 100,000 newsletters every day internationally and has some 30,000 subscribers (as at 2014).

(2) Bennett, G.M. 2014. [author]. Dissertation. A common method is the use of a flat tray with holes punched or drilled through it at optimum seedling spaces with hole sizes to accommodate (but not allow through) a specific variety of seed. The back of the tray is enclosed and airtight fitted with a connection for an air suction device—old domestic vacuum cleaners are often used. Seed is sprinkled and brushed across the face of the tray, vacuum is applied, surplus seed tipped back into the seed container and the tray placed face down on a nursery soil seed bed—often drainable plastic trays. The vacuum is released and all seeds are perfectly placed to grow. A thin cover (sand, fine soil or clean sawdust) is applied and the bed lightly warm water sprinkled and kept warm under glass or covered at night.

(3) Culley Andrew. 2012. Managing Director of SEL Advises that both Trade names Propell™ and Procoat™ are protected by Registration in New Zealand.

(4) Perlite Institute Inc. Harrisburg, PA 17110, USA. As published in its brochure "Perlite Volcanic Glass as a Glass Flake Filler".www.perlite.org

6

Practical Application

Coating and Pelleting Materials

Since the 1960's when coating and pelleting of seed began in a significant way internationally, many quite varied materials have been used over the years as the actual coatings. They all needed to be capable of responding satisfactorily to grinding (or milling) down to fine particle size of which the industry standard was then based on the old *British Standard Sieve* sizes. Most commonly used particle size was that of (approximately) 95% passing through a 300# BSS (*300 mesh British Standard Sieve*), a woven metal wire mesh containing 300 holes per linear inch, measured from the centre of any wire exactly 1 inch in a parallel direction. The actual hole size (and number of holes per inch) therefore being somewhat variable by the thickness of wire used, allowing minor variation.

This original basic standard normally provided for a hole opening size of about 53 µm (53 microns) however today, several more exacting measurements have been formalized in terms of wire diameter (i.e., the *SWG* criterion = *standard wire gauge or Imperial wire gauge*) which has more closely provided for uniform mesh hole openings and therefore more consistent particle sizes— however, SWG itself has lost some popularity in recent years. The current British standard for such wire is BS 6722: 1986. It is supplemented by various formalised International standards including *Tyler mesh, American standard* and a number of others.

A wide range of finely milled coating material plus chemical additives have been used in processing seed worldwide, some successfully, others not, in both research and field use plus of course concurrent use of liquid adhesives. These coatings include (but are not limited to):

Lime, dolomite, gafsa phosphate, nauru phosphate, gypsum, thermophos, peat, lignite, reverted superphosphate, clay, talc, bentonite, sand, diatamaceous earth, exfoliated vermiculite, perlite, pumice, ex sugar "press mud", alginates, wheat bran and a range of regionally natural (often waste) products (i.e., cork tailings, soybean and lucerne meal, also spent mushroom compost)—either used alone and/or in combinations—preferably available locally and cheaply. Also some more expensive and effective proprietary [prepared] combination coatings now manufactured by larger processing firms which contain modified basic major elements plus some trace elements for localized soils.

Becker Underwood's "Seedbiotics" company sells a blend of macro and micro nutrients for coating grass seed called "N-Rich" which has demonstrated faster establishment than plain seed in tests (not stated as independent tests) and to which fungicides such as *"Apron"* for damping off protection, can be included. These firms have Internet websites with contact details for processors who wish to discuss their commercial products in technical depth.

In addition to the coatings listed above, a range of chemicals to protect, aid, nourish or identify seed including nutrients, inoculants, fungicides, insecticides, nematicides, bird and rodent repellents and colouring agents, have all been included—experimentally at first then permanently if and when proven successful.

Similarly, applied to seed, not as dry particles—but as liquid sprays–a wide range of adhesives are, or have been tried including original and popular gum arabic glue (but soon replaced due to unreliable supply from the Middle East), even, incredibly, for a short time woodworkers hot glue, an animal hoof and horn meltdown gelatine, biologically preserved for commercial use by addition of ortho-phenylphenate, a powerful bactericide!!—employed in the early days by one seed firm when attempting to coat rhizobia inoculated seed[*1]! Included with adhesives has been a range of [hot or cold water] soluble methyl celluloses (of variable strength measured in centipoises (cpi) indicating the level of dynamic viscosity. Also other natural gums, water soluble pva (polyvinyl alcohol) and insoluble pvac (polyvinyl acetate)—these latter two often in specific combination (the insoluble adhesives minimized and heavily diluted by solubles) to achieve a moisture resistant stable coat but which is ultimately dispersible.

Fast setting acrylic spray is applied first to legumes [prior to actual coating] as a wrap around protective film to seal in natural anti-biotic tannins and polyphenols on bare seed prior to inoculation and coating. Importantly, also used are an increasingly wide range of polymers, co-polymers and homopolymers as adhesives, binders, film coatings and "polishers" at various strengths and for finishing of the product to a firm, dust free, almost glossy and free flowing coated surface. The firmness and dust free quality of coating achieved by some processors in recent years is certainly an improvement on the 1970's and 1980's coatings—even Prillcote™ of that time did not achieve as dust free surface as some do today. But, do these glossy finishes have the other technical capabilities of Prillcote™?

[*1]Purchasers of such absurd product would of course find that "seed inoculation does not work"— whereas, it was the method used which did not work—this lack of understanding in relation to viability of sensitive rhizobia has dogged the industry worldwide for many years and sadly has not been confined to commercial processing, but to some research as well.

Other stabilizing and binding agents are used such as pvp(polyvinylpyrrolidone), also pullulan, a natural polysaccharide providing a smooth semi-gloss hard and dust free exterior coat. In addition, there are film forming completely glossy polymers, some fully transparent on hardening and almost "lubricatingly" smooth (but water soluble)—being film formers (not true coatings) which are mainly used on large seed like pumpkin, maize or beet for free flow in precision drill chambers often including chemical protectants, sealed within the film for user safety. All have been—and many still are, used for the inoculation and/or coating and/or precision pelleting and/or film coating of seed.

The chief aim (and greatest technical challenge) of most research into legume seed coating to date has been directed at maintaining maximum viability of *Rhizobium* on legumes, to protect and precisely nourish them plus their host germinating seedlings *in situ* whether dropped, placed or drilled.

Different seed varieties require their own special coatings and the legumes their most favourable inoculant carriers—some of the above coatings such as peat, lignite and diatomaceous earth being specifically for protection and multiplication of *Rhizobium* rather than as a nutrient source for seedlings—other coatings are used simply as a filler for shaping and sizing of pellets—but these solid, dry coating powders all need to be milled to fine particle size before they can be used on small seed (or in the case of sand for precision pelleting, simply sieved to size uniformity and elimination of foreign matter).

In sorting various coatings into groups [related to seed varieties], we start with the legumes.

Coating Materials for Clovers, Lucerne and other Temperate Legumes

The first coatings were developed essentially as a carrier and protector of various strains of rhizobia and *Bradyrhizobia* nitrogen fixing bacteria with no special benefit for seed.

However, these protective coatings had to be favourable not only to survival of root nodule bacteria which are sensitive to pH, heat, sunlight and desiccation, but also not harmful to seedlings; this is a technical challenge as various plants and their *Rhizobium* have differing general and environmental tolerances.

Symbiotic nitrogen fixing bacteria are broadly classified as fast growing alkali tolerant and acid producing (*Rhizobium*)—or slow growing acid tolerant and alkali producing (*Bradyrhizobia*)—the latter being mainly associated with tropical legumes, the former associated with temperate legumes.

pH tolerance is an important factor. Acid intolerant temperate forage legume host plants when lime coated, respond remarkably well to those few grams of finely milled high quality lime around each seed sown on acid soils (i.e., pH around 4.8 to 5.5) compared to bare or just inoculated seed—indeed this miniature lime coat, actually replacing the need for broadcast lime in some situations.[1] Opposed to that, researchers in Australia have found alkaline lime coating of tropical legume seed has actually depressed the viability of *Bradyrhizobia* compared to plain inoculated seed—or inoculated and coated with less alkaline materials. Hopefully, cell numbers were counted and found uniform on each treatment.

In addition, whereas (temperate) clovers do tolerate mild levels of soil acidity in the range pH 5.8 to 6.5 and still grow well as do their associated *Rhizobium,* lucerne (alfalfa) prefers closer to neutral pH (6.5 to 7) at least partly due to the pH preference of its host-specific *meliloti* strains of *Rhizobium.* Inoculated legume coatings are used to support rhizobia survival and to offset naturally hostile soil conditions in each region.

The chief aim of introducing nitrogen fixing clovers onto undeveloped or "native" grasslands on non-arable rugged hill and mountainous country is to improve nitrogen supply as well as forage quality by simple surface sowing [for which there is HUGE potential around the world], originally achieved by hand, simply walking over hillsides tossing seed—or from horseback and truck then later via aircraft; to date usually in temperate regions—notably Australia, New Zealand, USA, Canada and Southern Nations of South America [but with big potential for instance in the temperate regions of alpine Europe and Asia including the Himalayas-Nepal, India, Pakistan, Bhutan, Tibet, Mongolia—and across to Turkey, Russia, The Balkan States—and many more].

The original coating commonly used was lime (calcium carbonate), however, dolomite (magnesium carbonate), also alkaline, was likewise extensively used as was the 50/50 mixture of gafsa rock phosphate and dolomite.

Gafsa rock contains a soluble form of phosphate (typically 28 to 30%) as well as calcium at around 50% [as calcium oxide]; the calcium content can improve phosphate uptake by plants as well as raise pH levels. It also contains trace elements—molybdenum, cobalt, manganese, zinc and copper. Gafsa is a soft material quarried from huge and ancient guano deposits in North Africa, notably Morocco which has more than half of the world's known supply of P_2O_5[1] (being around 8 billion tonnes with [potential] reserves of > 20 billion tonnes), but appears no longer widely used in seed coating having been replaced by other materials either because they are more readily available or they more readily supply P_2O_5 than gafsa in its untreated thus slowly soluble rock phosphate form.

The departure from use of Gafsa may be a retrograde step on reflection because it is (and has shown to be in research), an excellent seed coating material—specially for legumes and, most favourably, in its raw (but finely milled) state. It is however a "soft" mineral, prone to creating much environmentally adverse dust when handling and processing which may be another reason for decline in its use in addition to commercial uncertainty of supply.

---

[1]P_2O_5 is phosphoric oxide of which some 44% is elemental phosphate.

Less favourably, standard rock phosphate—for instance formerly taken in large quantity from Nauru or Christmas islands in the Pacific, is largely insoluble for plant uptake in basic rock form, until processed with sulphuric acid [to become *superphosphate*]; Superphosphate however contains too much free acid for coating of seed, not only being far too acid for *Rhizobium*, but also for most seed itself when in direct contact for any length of time beyond a few minutes even when dry.

One regrettable research paper on this subject published in New Zealand many years ago suggested that from their experience, mixing of bare inoculated seed with superphosphate might be successful—if applied just for the short time the two needed to be mixed for aerial application! That bad advice, originating from the fact that these research workers mixed tiny treatments of inoculated bare seed with superphosphate by hand—on hand sown plots, unwittingly using massively inoculated bare seed (by applying plenty of peat culture from a commercial pack of inoculant intended for 40 lbs of clover seed onto their few grams of seed), in effect *coating the seed with peat culture*) which of course would always leave a few survivors to produce a degree of nodulation despite massive *Rhizobium* mortality. In reality however, 40 times the single dose is an unrealistic rate of inoculation absolutely unaffordable to farmers, unworkable to apply commercially on a large scale and totally impractical in terms of mixing times required to meet both commercial aerial operations where weather alone may postpone a planned aerial application (on the day), for several days, and for which the inoculated seed and superphosphate would already have been mixed in preparation. Even inoculated coated seed should not be left in close contact, compressed with superphosphate for any length of time specially where inoculant has been applied to the surface of the coating (as in over sprayed inoculation, referred to under N.I.T. {*new inoculation technique*} later in this Chapter)—mixing on the airstrip just prior to aircraft loading is, despite potential for delaying rapid aircraft turnaround—technically the best solution.

Provided sufficient rhizobia survive for successful nodule development (<100 cells or more per seed may be adequate in a very good coating which retains viability), and provided biological nitrogen is successfully supplied to the developing seedling, the small quantity of lime around each lucerne seed or gafsa/dolomite around each clover seed was found to be highly beneficial to legume establishment and growth in several research studies as well as many thousands of commercial sowings throughout New Zealand also overseas.

Inclusion of dolomite which is also quite alkaline containing around 59% calcium carbonate (of which 24% is pure calcium) and 39% magnesium carbonate (of which 11.5% is pure magnesium) supplies nutrient magnesium in addition to calcium—both major plant food elements, and in a form supportive of fast growing *Rhizobium* as well as temperate forage legumes.

Gypsum [calcium sulphate–around neutral pH] and thermophos have both been used in legume seed coating research studies, but neither have been adopted commercially as a result. Gypsum supplies sulphur plus calcium but sulphur, a key major element and common soil deficiency is acidic, not rhizobia friendly whereas lime (or dolomite) are better calcium providers than gypsum to both fast growing root nodule bacteria as well as seed and soil. Where sulphur is a significant soil deficiency, application of sulphurised superphosphate as a topdressing fertilizer may be more effective than a tiny quantity of gypsum wrapped around each seed [whereas the reverse can be true where calcium is concerned in many situations].

Thermophos, a slowly available phosphate with fluorine removed, while less toxic, has not been found to be of special benefit in legume seed coating.

The special microbiological coatings—peat, lignite, diatamaceous earth plus some other localized biologically friendly materials [i.e., composts, lucerne meal, etc.] have been used as coatings on legumes, but only in respect of their ability for maintaining rhizobia viability. Materials with protective (microscopic) fissures and cavities with increased total surface areas relative to mass, have been valuable in providing "safe haven" for sensitive rhizobia.

Further to that aim, *peat, lignite and diatomaceous earth* in particular, are used very successfully as chief carriers for legume seed inoculants—being a relatively simple, well documented development [including in this book]. This seemingly simple activity (identifying successful carriers and producing quality legume seed inoculants) is of huge International importance—but sadly limited however by lack of knowledge of farmers, lack of biological production facilities, and widespread lack of equipment such as basic refrigeration in some regions.

If UN's FAO have a single most important message to get out to the food hungry nations of the world, it must surely be the critical importance of bnf [biological nitrogen fixation]

Forage Grass Coatings

Coatings for grasses have a much simplified role—in addition to ballistics for aerial sowing, the coatings provide a once only opportunity to place a small but frequently vital source of nutrient—calcium, phosphate, magnesium or sulphur, finely ground, right where the seed will germinate and grow—if it does get a chance to grow at all in such a hazardous mass surface sowing technique where a result above 50% establishment certainly represents an outstanding level of success under present knowledge.

Reverted superphosphate supplies both calcium and phosphate but requires additional processing by adequately combining lime [25%] and superphosphate [75%] in those proportions for full reversion; the product becomes naturally hygroscopic (absorbs moisture, gets damp easily) yet must be kept dry for roller or hammer milling. If not thoroughly combined in adequate proportions or for a sufficient period to effect total "reversion" [calcium neutralisation mainly of sulphuric acidity]—the product may still contain pockets of sulphuric acidity lethal to rhizobia so is not used for legume coating; it may even be harmful to seedling radicle emergence when moistened. However, when fully reverted then milled it is popularly used in grass seed coating and brassica seed precision pelleting.

Are Lightweight Coatings Good?

More recently, lighter weight coatings but with no nutrient value such as expanded perlite and exfoliated vermiculite [which both greatly enlarge when "exfoliated" (heated) developing platelets like talc, which bind more readily than the raw product], are now used in legume seed coating including precision pelleting manufactured in rotary coaters. These flat platelet-like particles which "interlock" have had the moisture vapourised off and in the case of perlite, has, when heated (exfoliated) developed miniature oxygen bubbles assisting pellets to split open on contact with moisture. Expanded perlite is so lightweight it floats on water having a bulk density of as lightweight as30 kg/m^3, whereas unexpanded (raw) perlite has a density more like 1000 kg/m^3 [1g/cm^3).

The lightweight aerated nature of these natural deposits [mined in Greece, Turkey, US, Australia and China) plus their ability to expand a great deal on substantial heating [>900°C for perlite and >1200°C for raw vermiculite to achieve exfoliation], and relatively low cost, have advantages for coating and pelleting seed. They provide lighter manual handling, lower transport cost (specially aerial lift) plus oxygenation of the seedling when moisture opens the pellet (or coat). Opposed to that, both deposits, being amorphous (not crystalline—not naturally clear and shining) are volcanic silicates—like glass [comprising essentially a combination of silicon (>75%) and aluminium (>15%) plus smaller amounts of sodium, iron, magnesium, potassium and calcium (the latter three below about 3%)] which, while of near neutral pH, are nevertheless of minimal nutrient value.

Consider lightweight bare seed. Forage grass seed generally - but particularly cocks foot being so light, has historically presented the problem of "drifting" even in the lightest breeze when oversown by air, landing if not on the neighbour's property, sometimes several properties away—perhaps out over the riverbed or even out to sea! There is almost always a breeze at 500 to 1,000 ft at which height aircraft normally drop this seed. At 3,000 ft bare seed cocksfoot—even ryegrasses and white clover can drop well off target. Grass seed tends to be more severely affected by this factor of seed weight which nevertheless applies to all surface sown seed.

With its heavier (and nutrient) coatings, earlier Prillcote™ approximately doubled the weight of each seed, increasing its density while barely increasing its volume thus substantially increasing terminal velocity which aids more accurate aerial placement. This also provides a better ballistic value for penetration down through existing vegetation for placement close enough to soil for subsequently successful seedling establishment—even achieving total precipitation to soil when aided finally by hoof movement on well managed oversowings, and also sometimes lifted and dropped by frost—which can cover seed with soil.

These lightweight materials alter the former seed/coat ratios which not only growers, farmers and runholders had become accustomed to, but their advisors and seedsmen as well. If supported by field performance, then changes are justified but to our knowledge these lightweight coatings have not been independently and thoroughly field tested—if at all. It is unlikely they have been tested to the extent that original Prillcote™ legumes were investigated during the 1970's to 80's, so, while they may be successful, the authors are not able to recommend [or oppose] the use of these lightweight substantially inert materials—with the reduced velocities involved—particularly when dropped onto existing foliage by air [where ballistics are important], until and unless good solid field research supports them as equal to—or superior to original Prillcote™ as manufactured in New Zealand in the1970's to 1980's. Those Prillcote™ results are shown in Chapters 2 and 3 of this book as well as in other publications.

While this resume of coatings refers chiefly to legumes, it must be added here that ballistic factors are less likely to affect precision drilled pelleted seed, also now lightweight pelleted, which are not aerially dropped and which can be drilled with nutrient fertilizer either in row—or alongside the pellet drill row. For this purpose, the authors consider lightweight pelleting materials such as perlite provided it breaks down adequately for safe germination, a suitable replacement for clean, graded sand.

Grass Seed Additives

Because there are no sensitive bacteria to be wary of in grass seed coating, these coatings can be used also to carry additives—fungicides and insecticides, bird repellant and colouring agents for identification—and in the case of retailed lawn grass seed—colouring also for attractive presentation in transparent packaging. In due course colouring may well be used for identification of certified seed both internally—and for export, specially when many seed merchant facilities now include machinery for easy application of such colours.

More often than not, coated grass seed needs to be able to withstand mixing with other seed both coated and bare, for oversowing—and to go through a drill. That requires a firmly bonded dust free coat where surface binders are important. There is little point going to all the expense of preparing coatings, applying them—then losing much of it in clouds of dust in the seedsman's store—which will probably settle where it is not wanted. Mixing machinery must not be abrasive to coatings.

The coatings can also be—or include, chemical additives for plant protection from seedling diseases such as—fusarium and verticillium wilts, smuts, rhizoctonia solani, septoria, sclerotinia, white rot, early leaf spot, rust,

powdery mildew, pythium, phytoptera and other "damping off" diseases—
plus many more parasites. These threats can now be controlled—or at least
minimized by inclusion of fungicides as [or in], coatings on seed being more
effective when active at the very earliest stage of seedling development. If
included with a coating such as lime for instance, compatability must be
proven first because some chemicals become de-structured by alkalis—others
by acids. Some may chemically bond adversely with a finely milled coating
rendering the chemical inactive.

Likewise insect predators such as grubs, thrips, whitefly, aphid species,
wireworm, weevils, nematodes, cut worms, leaf miners, leafhoppers and a host
of others can be controlled—or minimized, by selective and tested insecticides
which however tend to be more phytotoxic[1] than do fungicides—indeed in our
experience, some insecticides are generally quite toxic to emergent seedlings
such as the chlorinated hydrocarbon pesticides dieldrin and aldrin.

As seed safe chemicals and associated treatments are constantly being
improved, new users should contact leading suppliers, easily done today by
Internet, exploring via search engines such as "Google"—or others with simple
goals typed into the search such as *"Seed Treatment Chemicals"*—or something
similar where leading websites such as [®]*Bayer Crop Science* will emerge and
can be studied or the manufacturers contacted via those sites.

As discussed in Chapter 4 *"Grass Seed Coating"*, some researchers have not
found coating of forage grass seed to be beneficial [e.g., superior to plain seed]
from surface sowing into existing native foliage [e.g., oversowing] however, it
may be significant that those results came from hand sown experimental plots.

Those very positive results we record in Chapter 4 also came from hand
sown plots; it is difficult to calculate levels of success and failure accurately
from very small applications, unless comparability and practicability are
applied under very tightly controlled conditions. As in commercial sowings,
success or failure is almost certainly dictated to a large degree by ground
surface conditions at time of sowing coupled with post-sown weather
conditions and management—including use of livestock as a tool for seed
trampling right on (and if damp, into) soil surface which factors will play a
major role in % seedling establishment. Importance of drift of bare seed or
coated when aerially sown at 1,000 ft versus hand sown at 2 ft is also important,
as are ballistic effects on soil contact.

In small plot research, specially for grasses only, some technicians may
walk all over plots before—and even after sowings—others may not. That
alone could impact hugely on a final establishment result—so these items
need careful thought before application.

We do know that probably the most potentially negative and misleading
information ever handed down from small plot research in New Zealand
arose from lack of understanding that massive inoculation of bare seed

[1]Chemically toxic to plants.

absolutely ruined any chance of gaining a meaningful result when comparing its performance with identical seed and inoculant applied at the affordable standard rate to coated seed—the latter inoculated professionally at precisely the same loadings of bacteria that a farmer could afford to apply. Thus, meticulous care is required in small field plot research where a seemingly minor variation can cause huge damage (both to understanding—and commercially) when extrapolated exponentially, compounding the error—and that in turn disastrously misleading farm production, Nationwide.

On open ground, largely free of existing native foliage, when trampled by livestock (possibly including by experimentalist staff while working on small plots), it is very likely bare grass seed will establish equally well as [identical seed] coated, particularly where usual fertilizer application has been made at or before sowing. Management of an oversowing is equally, if not more important to a good result—as coating or no coating of grass seed. This is not the case with legumes where successful nodulation is paramount.

If ground surface conditions are less favourable with significant existing foliage (usually lower fertility native species—call it "rubbish" if you like—but better than no feed at all), the seed dropped onto this from 1,000 ft (not 2 ft as in hand sowing), the sowing not trampled by livestock (or anything else), and the area minimally topdressed—or not topdressed at all, then as common sense dictates, it is much more likely heavier fertilizer coated seed will perform better than plain seed—assuming the featherlight plain seed did in fact land on the correct site and did not drift off to neighbours, or riverbed, lake or sea. This ballistic effect is important therefore so are the weights of coatings. Nutrient value is also important, we have seen vigorous establishment of clovers and grass seed on mountainous shingle scree at over 3,000 ft altitude from use of standard inoculated Prillcote™ coatings without fertilizer topdressing[2] which however, needs follow up topdressing—eventually, to ensure survival in relatively low fertility locations—but the seed coat achieved establishment and when that is successful, then the (often huge) cost of flying on superphosphate (or other fertilizer) is much more justified and applied with less risk.

Of the coatings applied to grass seed, reverted superphosphate has been the most widely used in New Zealand giving each seed a good chance of phosphate, sulphur and calcium uptake pending availability of nitrogen when clovers become established and nodulated. Then, when there is sufficient growth, short term grazing by a large mob of livestock to "inject" (e.g., transfer nutrient via livestock from other established feed) a further dose of animal manure nitrogen into the newly developing grassland. This is vital in securing a well established sward, as well as to consolidate the seed bed (specially after severe frost, pre-sowing burn-off or other soil disturbance). More recently, some processors are using lime alone as a coating on grass seed for oversowing— better than nothing (and cheaper) but not as potentially valuable in lower fertility situations as reverted superphosphate—in New Zealand.

Coating grass seed simply for ballistic effect may be worthwhile in some situations, in which case inexpensive clay coating—or perhaps lime coating would suffice, but it would seem more sensible where the cost of milling the coating must be met plus the cost of coating and handling itself, that a more nutrient oriented coating be used—the nutrient should of course aim at the most deficient element on that soil type of the five major elements—calcium, phosphate, sulphur, potassium and magnesium, as identified, usually by soil test in that locality wherever it may be.

There are actually 6 major plant food elements, but nitrogen, not a natural soil component like the others, is basically "deficient" in most soils (although is available as a combined chemical in natural deposits by way of potassium or sodium nitrates or ammonium chlorides).

The soils of the world are surrounded by N_2-elemental diatomic gaseous nitrogen, many billions of tonnes of it comprising 78% of the earth's atmosphere—yet not one ounce of that gas is available to plants until converted to ammonia by nitrogenase enzymes. Electrical storms also "fix" some nitrogen in the atmosphere which comes down to soils in rain—but that source too is not a product of soil or an elemental deposit and the volumes available from it are small.

Coatings with explosive capability. The importance of nitrogen is simply this—palatable grasses will eventually die without it. However, there are three good reasons for not including nitrogen as a coating on grass seed.

The first reason is that nitrogenous fertilisers such as sulphate of ammonia (pH 5.5) or Urea are much too concentrated (acidic and/or evaporatively toxic) to be compressed right around and in close contact with seedlings at emergence of a germinating radicle; secondly, the small quantity of nitrogen involved for each seed would soon be lost in rainfall dilution and vaporization. Thirdly, some forms of ammonia are potentially explosive in any quantity when combined with other materials. For instance when combined with fuel oil and any one of several types of intentional—or unintentional detonators (such as static electricity) ammonium nitrate can become a powerful explosive and is not permitted in the bulk bins of topdressing aircraft in New Zealand where alternator electric generation, fuel, oil, vibration, dust and heat (in close proximity) combine to make a dangerous environment. Other combinations have explosive qualities such as potassium nitrate (saltpeter) used in the manufacture of gunpowder.

Sulphur fertiliser is also prohibited in topdressing aircraft because of its explosive potential—nor welcome in seed processing facilities and much less welcome in roller or hammer mill factories where explosive potential is heightened by unavoidable generation of ignition dust—particularly static electricity ignition of unstable micro-dust particles. Milling of electrically combustible dust forming materials can be quite hazardous despite factory engineered containment even where explosive doors are fitted to release pressure upon sudden unpredictable detonation, particularly where men are working nearby. Fine milling of coal is also notoriously explosive-likewise grimy, adding to cost, thus, apart from lignite for *Rhizobium* protection, coal has not been used commercially in seed coating.

Explosiveness of sulphur, coal and ammonia products can however be minimized by maintaining dust levels at less than 20% of atmospheric volume. Mixing with other fertilisers can achieve that, but less satisfactorily, by mixing with lime.

Miscellaneous additives. Trace elements are safely included in seed coatings for situations where soil deficiencies are known. Molybdenum can be included in the adhesives liquid [as sodium molybdate]—or sprayed on, for inoculated legume seed coatings, it being important both to good nodulation and healthy plant growth, specially for more acid soils. It is also used in brassica coats for protection against molybdenum deficiency which, when serious can produce "whiptail" whereby plants retain the central stem (resembling a "tail"), but with little or no foliage. Any foliage remaining will typically present the same pale green/blue colour, or light brown, generally indicative of the unthriftiness normally associated with nitrogen deficiency.

Boron deficiency, fortunately rare in New Zealand soils, causing symptoms such as hollow centres, watery areas or heart rot in the turnip crop can be minimized if not remedied by application of borax (sodium borate) in the seed coat; however, if severe as it may be on very alkaline soils where it is "locked up" (less available to plants under high pH), a topdressing of borated superphosphate—or use of borated, reverted superphosphate drilled with the crop—will achieve a more positive result than in a seed coat.

Other trace elements, once cleared of phytotoxic tendencies when included in the seed coat (a test which a technician can quite satisfactorily check out in the laboratory or glasshouse) can be added but if we eliminate the animal-essential minor elements such as selenium, cobalt, copper, manganese, fluorine, iodine, zinc and iron, the actual range of micronutrients essential to plant health is small and those deficiencies fortunately quite rare around the world.

Colour dyes are also used with and without other coatings and are generally applied as dry powder mixed with coatings which when wetted out during coating produces that colour throughout the coat, or it may be applied as liquid sprays not harmful to seed except that of course any moisture applied must be dried out of the product before it can be bagged or stored. Traditional suppliers like I.C.I., Ciba-Geigy and BASF have offered a range of colours for many years but there may well be new suppliers now—again, a web search should identify competitive and locally available options.

Even very tiny seed such as New Zealand "Browntop" (*Agrostis tenuis*) has been coated and coloured for identification and introduced to the market by Coated Seed Ltd many years ago [1960's]. It included bird repellent and fungicides in the coat and remains today an excellent fine turf seed—double the bulk when coated and much easier to sow uniformly by hand to create a new lawn grass area—not least because the pattern of coverage is more easily seen via the colour. With such small seed, the particle size of milled coatings applied must be smaller than the seed itself—otherwise coating particles can become the core of a coating build up—with seed randomly attaching to the core! This would produce a most unsaleable product.

Precision Pellet Materials. For many years, Prillcote™ pellets usually employed only sieved coarse sand—not salty seaside sand, which however could be used if the salt were washed out of it. Riverbed and other sand deposits were used—specially plasterers sand being a good source because it is partially processed, clean of sticks, stones, plant material and other debris whereafter (for seed pelleting) the sand must be further sieved to a regular particle size with dust discharged through the sieve and oversize removed over the top of a two layer sieving action.

This sand is relatively heavy (compared to (say) perlite which is preferred now). There is little ballistic or other reason that pellets should be "heavy" because, sown through precision drills into prepared seed beds, a little extra weight could be helpful in achieving a more positive drop from machine to soil–however, advantages of other materials like perlite, vermiculite and diatomaceous earth (not least being local availability, pre-processing and cost)—will easily outweigh a small ballistic value for this product (if any).

The three main advantages of new materials are likely to be 1/availability locally 2/cost and 3/product convenience. However, if [the author] were starting a precision pelleting plant somewhere in Oceania, Asia, Middle East or Africa, our expectation would be that sieved sand is most likely to be more readily available, more convenient and less costly to use [with a talc or clay finish]—at least initially and until better or cheaper materials were able to be proven in small scale experimentation.

Because sand grains are sharp by comparison to other materials (specially after others have been milled), they attach to each other frequently at extremities leaving micro pockets of air/adhesive between, which is essential in assisting the pellet to break down under moisture, as well as creating a source of oxygen to the developing radicle.

The following figure, while diagrammatic, explains the benefits of sand grain attachment. Over many years of commercial production and in dozens of glasshouse investigations, there have never been problems of slower or impaired germination or seedling development from use of sand as filler with Prillcote™ pelleted seed compared to identical bare seed.

On the contrary, as with coated seed, under identical and normal conditions, sand pelleted seed usually germinates a little faster and more completely than identical bare seed. The reverse could be true under certain conditions, but that has not been observed in any trials except occasionally with coated lawn seed.

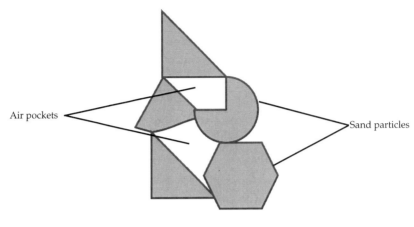

Air pockets

Sand particles

Fig. 61

Sand Particle Attachment

["magnified" diagrammatic image]

Adversely however, because sand is sharp, it is also abrasive (and for instance is industrially glued to stiff paper to make what we know as *"sandpaper"*), not a particularly welcome participant in the moving seed chambers of precision drills where an occasional pellet may be crushed (silica sand can be very abrasive to any mechanical part it is in contact with). All fillers of a crystalline nature are likely to attach similarly to sand.

To alleviate this abrasiveness and create a smooth, largely abrasion free surface on these pellets the product is finished off with a coating of talc—a smooth, almost "lubricant" soft dry powder giving a silky surface to these otherwise abrasive pellets (and exactly as used by ladies in scented cosmetic preparations). Abrasiveness is less likely with other fillers and finishing talc may be replaced by smooth binding agents like pullulan.

Brassica Coatings (Turnips, Swedes, Rape, Canola, Kale, Broccoli, etc.)

Because most brassica seed is drilled into prepared seedbeds and fertilized adequately in or alongside seed rows, there has not been a great demand for nutrient coating.

Direct sowings (aiming at reduced cultivation, or for instance on "light land" to avoid lifting stones), into existing vegetation are more widely used now and in that situation or in very low fertility soils, brassicas could be coated with reverted superphosphate; In cultivated seed beds, there is also the opportunity for drilling precision pelleted seed where nutrient coating is not required as the chief aim is to achieve a uniformly sized pellet for precision placement and then, being pelleted, can be drilled with reverted superphosphate (*supplying calcium, phosphate and sulphur*) which is safer than superphosphate alone. Potassium, sulphur and/or magnesium can be included—the latter by using dolomite instead of lime in reversion of the superphosphate—and the former by using potassic or sulphurised superphosphate instead of plain superphosphate. If drilling fertilizer in the same row at same depth as pelleted (or coated) seed, reversion of the above superphosphate combinations either by the farmer himself (specially if he has a mixer), or by purchasing "super" reverted with either of the two alkalis (lime or dolomite) will always be more prudent in avoiding the possibility of seed damage. Time in contact [with no rainfall or irrigation] is likely to be critical where acidic fertiliser is used.

There is value in New Zealand however for brassica seed protected with both insecticides and fungicides against a host of predators which sometimes do considerable damage to the seedling crop soon after emergence—indeed even to seed itself. In a cultivated paddock, a seedbed drilled to a crop with no other vegetation, when succulent and highly proteinous brassica seedlings begin to emerge, becomes a smorgasbord of rich food for a host of insects which concentrate on these juicy seedlings for their survival.

Available commercially for many years in New Zealand and overseas have been various plain seed treatments such as "Poncho" and "Goucho" for control of some of these pests, however these are not coatings but simply chemical treatments, sometimes coloured for identification and sown at the same rate as bare seed because there is no appreciable weight gain once the product is dry.

Firms in New Zealand such as PGG/Wrightson offer farmers their Superstrike™ brassica seed, providing protection from insect attack—chiefly "springtail" in New Zealand—a notorious brassica crop insect predator. There are others, as well as protection from some fungous diseases—particularly seed rot and the "damping off" group including pythium, fusarium and verticilium wilts (brassica oleracea v. botrytis) & spore free rhizoctonia solani. These very small chemical treatments may save a crop from various predators/ diseases—or may not do so—depending on so many field and agronomic conditions. Considering all the work and expense in preparation of a field ready for a brassica crop, it would be good management practice to employ such cautionary products where they are available—or by the farmers [who have adequate safety gear] preparing their own seed from this book and other references.

For locations with greater risk of {potentially} serious problems in New Zealand, PGG Wrightson offers Ultrastrike™ brassica seed—again not coated but protected against the same predators as Superstrike™—in addition, also against aphids, nysius fly and argentine stem weevil—plus carrying molybdenum to rectify that deficiency which in serious cases can lead to "whiptail"—as explained earlier here. There will be many other insect predators in various regions of the world including for instance diamond backed moth (now worldwide), soil grubs and nematodes which localised research may show are able to be controlled or modified by insecticides included in seed coatings where the combination of *precise placement* and *early timing* is often important as it not only limits the quantity of chemical required—highly desirable environmentally, but increases its powerfulness by way of that precise and early intervention.

Young brassica seedlings are nutritious, readily attracting small soil insects (as well as fungi) making these varieties ideal for seed protection treatments. Even if coatings are not widely used, the best brassica crops sown are likely to be those using high quality seed precision drilled to achieve optimum spacing of plants therefore optimum crop production due to minimal inter-plant competition; some are also likely to be drilled using precision pellets carrying locally deficient nutrients plus locally targeted insect predator and fungous disease chemicals (in very small quantities—remember, the chemicals are protecting the seed—not the paddock thus not saturating the environment). Precision drilling requires minimal seed/hectare, therefore needs top quality seed (which is expensive) but using less seed is more affordable—not only of economies of seed, but also the cost of precision pelleting, however, it must be very high interim germinating seed.

New fillers will be developed as science marches on—however, to get started on a satisfactory product, the coatings above have been found quite successful in addition to which every processor should have its own formulation testing facility, with a small laboratory and associated glasshouse—or climate control house, all essential in sorting out coatings and formulae best suited to each agronomic region for maximum commercial use.

No commercial seed processor should be in the business of selling seed treatments without adequate facilities for thorough and ongoing quality control. That is of course common sense—no serious manufacturer (of anything) hoping to stay in business keeps on churning out product—without frequently checking its quality. If a quality control technician finds a fault— don't blame him or her!—reward them—then fix the problem smartly and before continuing with production.

Once a successful range of treatments has been identified locally then a demonstration of those treatments to local farmers (a field discussion group), is possibly the best and most effective way to promote successful treatments and have them adopted commercially. Precision drilling can be quite adequately simulated by planting pellets alongside a ruler or tape measure with small tongs (marked to a depth) to achieve the spacings and depth which a precision drill will do either in the laboratory or (where predator attack is required to assess the merit of various treatments), in the field.

Small plot treatments must be kept absolutely free of trespass or interference by animal or human threat, so in the field, one of the most critically important facilities is secure fencing of plots—preferably using close-weave wire netting. Where animals or children might try to climb that fence—or large animals knock it over, an electric fence simply run off a car battery using a pulsator is quite effective—auto-turned on at night (by a timer) when unattended, may be an essential precaution as well as double fencing and even some security wire. These measures are important in obtaining *accurate* results of such experimentation. It is serious business (often treated much too casually in the past).

The essential point being that it is critically important for researchers or technicians to be able to run closely natural, uninterrupted field studies from small plots in the best possible interests of the greater community as well as the farmers who will adopt superior products if they are satisfied they work.

If birds are a problem, it may be necessary to carry the small weave netting right over the top of small plots—this is not silly or impractical—it is ESSENTIAL that the best treatments be identified ACCURATELY.

If for instance an elephant, giraffe, pig or sheep accesses a farmers field, tramples or chews part of the crop—it will no doubt be most unwelcome to the owner of that crop, but not a total disaster for that whole community—yet **it will be** where a field investigation is ruined. That community must then farm on year after year vulnerable to pest and disease until a worthwhile remedy is found—simply because pilot trials could not be conducted without interference.

Adhesives

The widely used original adhesives for coating seed—gelatins, gum arabic, methylcelluloses, syrup and sucrose solutions have long since been improved on by the rapid technical development of polymerization. For coatings, there is nothing wrong with using the tried and true earlier adhesives such as methylcellulose—a benign and harmless product proven to be entirely compatible with seed, inoculants, additives and not least, factory workers.

It is the intention of this book to record the ongoing development of seed processing from its early days, not only of adhesives, coating methods and formulae used, but of every aspect of the industry because new users of these techniques need to understand how and why all components were adopted—to build up a better capacity for continued success and above all to encourage use of less expensive methods and materials just to get started. From such success, better, faster and more sophisticated products can be developed.

When we introduce key factors such as cost, ease of use, storage shelf life, suitability for new coating methods and machinery, firmness of coat with a dust free finish, a suitable response to moisture breakdown once sown in the field—then, among adhesives, polymers do tend to rise above the others—furthermore are under constant improvement. Nevertheless, various materials and equipment will be more available in some countries than others—so we must relate all materials and equipment used successfully and explain why others are not successful. This will save months, even years of research time, and will avoid waste conserving cost.

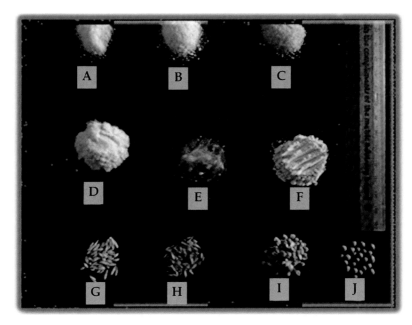

Fig. 62 Adhesives used in the past for seed coating and precision pelleting plus examples of resultant products are: A. Insoluble [after use] latex powder. B Soluble [after use] acrylic powder. C. Gum Arabic. D. Methyl cellulose (Diacell 2,100). E. Polyvinylpyrrolidone (pvp). F. Superfine talc. G. 50/50 seed/coat [by weight] reverted superphosphate coated cocksfoot seed. H. Identical plain cocksfoot seed. I Plain tomato seed. J. Precision pelleted tomato seed. The ruler shows centimetres alongside. NOTE: The PVP (E above) has absorbed atmospheric moisture and within hours changed from a white dry powder into a hard crystalline compound.

Already, some of the early polymers have now gone from the market—often due to a better or cheaper replacement and sometimes due to disuse by some other larger purchasers such as ink or paint manufacturers where their disuse has rendered that product uneconomic for the chemicals companies to produce. Compared to some, seed coating is a small user—but growing.

Products used initially by the leading seed coating manufacturers included:

Syrups, molasses, hoof and horn gelatin, even woodworkers pva glue have all been tried—unsuccessfully for various reasons—most reasons already explained. Glues which have been successful, though some now also replaced are:

- Gum Arabic. There are various grades and sources of this natural gum, however the standard product imported into New Zealand was usually applied as a 20% solution for both precision pelleting and earlier coating, first dispersed in hot water (then cooled). Depending on quality of the gum and type of seed being coated, the strength of solution would ideally be varied between 15% to 25%. Where available it will still be a useful adhesive for any processor wishing to get started on testing products and their manufacture with a coating bowl.

- Methyl-cellulose (of various qualities, the brand "Methocel" being popular). This was low viscosity adhesive in the range 200 to 1,000 cps which although easy to use and fully compatible with seed, inoculants and coatings (indeed is almost chemically "inert"), did not saturate and adhere the coatings sufficiently tending to produce loose and chalky coated seed not well suited to handling—specially not to machine mixing with other seed in conventional seed mixing machines prior to sowing, in fact not well suited to commercial conditions generally as a sole adhesive.

Then began the polymers:

- P. 316 LV a synthetic latex (designed to substitute for gum Arabic).
- Wallpol 03192, a polyvinylalcohol homopolymer emulsion. This is a water soluble synthetic polymer. This product as described by Wikipedia Encyclopedia exactly meets the requirements of a pre-treatment film former (quote) *Polyvinyl alcohol has excellent film forming, emulsifying and adhesive properties. It is also resistant to oil, grease and solvent. It is odorless and nontoxic. It has high tensile strength and flexibility, as well as high oxygen and aroma barrier properties. However these properties are dependent on humidity, in other words, with higher humidity more water is absorbed. The water, which acts as a plasticiser, will then reduce its tensile strength, but increase its elongation and tear strength. PVA is fully degradable and dissolves quickly.*
- Primal AC 234 (originated from famous Rhoplex 33 the original latex polymer of Rhom and Haas, USA)—Primal AC 234 a latex co-polymer was used extensively for coating grass seed—and as a primary sealer of legume seed. It was later discontinued whereby Rhom and Haas (acquired by Dow Chemicals Co in 2009) produced ●Primal E-1764K as a replacement—and A.C. Hatrick in New Zealand substituted Primal with Aeropol and Hetropol.
- Aeropol CA 146 and ●Hetropol AE 320 were among a wide range of proprietary polymers tested for suitability—a demanding job but usually well suited first to laboratory investigation, then factory trials, and only field tested when optimum polymer combinations were identified—not all polymers were successful. These two for instance may also no longer be available, being pure acrylics were much slower to break down under rainfall than unplasticised acrylics.
- Acropol CA 146 a vinyl acetate acrylic latex. This is a water insoluble latex, as used in paint. However, Acropol 63079 for instance of 800 cps strength is remoistenable after application and drying.
- Synthemul 63973 is a pure acrylic emulsion.

- Viscopol C500. A vinyl acetate homopolymer of about 4,000 cps, comprising 50% solids, being water dispersed; it was a stock item of Nuplex in the 1990's effective on a wide range of substrates. This is clearly an excellent product in the polymer adhesives range, however, for use with inoculated seed coating it is a primary example of an adhesive which would require thorough testing—for two main reasons. It has a pH of approx. 4.3—too low "as is" for *Rhizobium* though this can be lifted by inclusion of an alkali; and it contains 0.16% [by weight] of formaldehyde which is toxic to both *Rhizobium* and seed though a trace may not be significantly harmful. While it also may no longer be available from Nuplex, Viscopol 500 is available at time of writing being a vinyl acetate close cousin of C 500 of 3,500 cps, but also carries a pH of 4.3—so would need fully checking before commercial use.

Cold and hot water soluble Methyl-celluloses (with a wide range of viscosities) have been used—initially [and alone] as low viscosity celluloses of around 400 cps but later a high viscosity [around 4,000 cps] product was used sparingly as a very successful dispersant of milky latex and wet acrylic formulations. When mixed together they give better coverage, suspension and dispersion through the seed. High viscosity allowed application of reduced quantity, less expensive where all grades of dry powders were similarly priced by weight.

All these polymers and cellulose adhesives are supplied as granular (powder) or liquid concentrates (there is no point users paying for transportation of water from manufacturer to processor, possibly over great distances—some internationally). So they are usually dry powders or liquid concentrates mixed with clean water to make up the required level of both saturation (i.e., to saturate chalky coating powders thus overcome dust activated by abrasion)—and adhesion, but also of a viscosity which will spray conveniently—even atomise, for maximum coverage. Obviously, to minimize cost, the smallest (but adequate) volume of adhesive solution, and lowest (but adequate) viscosity and bonding strength need both to be identified—specially true with large scale commercial production where 1 cent of wasted cost per kg of seed can mount up to thousands of dollars lost per annum from just one supplier.

Combination Adhesives

The levels of dilution therefore vary—for instance Aeropol CA 146 and Hetropol AE 320 concentrates (developed to replace Primal AC 234) for grass seed coating has been successfully diluted to 50% with water as for Primal itself—but in all cases applied mixed into 1 to 2% of methyl cellulose 4,000 cps before application, as a spreader and stabilizer. These adhesives have been applied in bulk into a revolving drum process in earlier processing techniques—not sprayed, so this is an important consideration as well, because spray applications are more demanding of adhesive suitability.

Unfortunately, a large volume of substandard inoculated and coated seed was sold in the early days in Australasia—worse, opinions and field results were frequently determined on the basis of these poor products. The seed was usually good and some of it would always grow making it difficult for farmers to really understand good v inferior quality coatings. But, as shown in this book, the differences in practice can be enormous.

TYPICAL OF A SEED TREATER WHICH CAN SIMPLY APPLY A CHEMICAL BUT CAN ALSO COAT SEED.

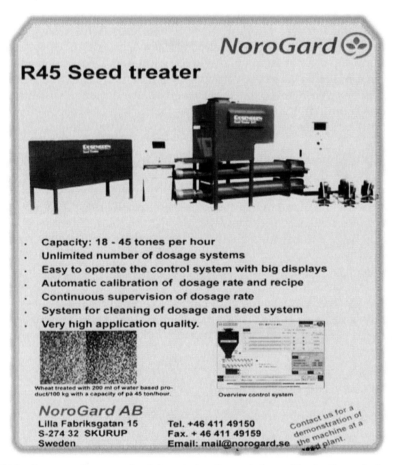

Fig. 63 The above page is a copy of an Internet advertisement of a seed treater. It can also be used to coat seed. It can apply many dosages and can be thoroughly cleaned between seed varieties, which is essential.

In revolving drum coating, an improvement to stability can be achieved by application of a light spray of water (via a hollow pipe axle fitted with fixed nozzles) into the revolving drum during the fixed beater rolling phase (after beaters have done the mixing while the drum remains stationary) which water spray further dilutes the adhesive slightly—but when applied at that stage assists saturation from the outside of the coating into, and combining with, the adhesive.

A water spray of only 3 to 4% of seed weight was required to achieve this improved coating—but both the open bowl and closed drum equipment is superceded in terms of efficiency by the fast new spinning disc rotary coating machines available today. Nevertheless, the widely used revolving drum is still a good method to get started into reasonably large production at minimal cost.

If a polymer such as Viscopol 500 has a viscosity rating of 3,500 cps and is diluted to 50% strength, then of course its viscosity will drop at least to 1,750 cps being approx. half that of the concentrate; and whereas a polymer of that viscosity *could* be supplied by manufacturers, it would then have to be applied at full strength, but because the cost of such polymers is likely to be similar per kg, the higher cps product may therefore cost around half that of the lower cps rated product.

For legumes, an application rate of 25 to 30% (of seed weight) of a polymer adhesive concentrate diluted to 50% of original strength can be successfully used in the revolving drum application but always pre-mixed with Methylcellulose 4000 cps concentrate at typically 1 to 3% of seed weight. Where a combination of a water insoluble latex and soluble acrylic polymer are used being soluble and insoluble after application and drying, the combination would be typically applied at about 14% of seed weight (of the 50% diluent) for the insoluble latex—and 16% of the (50% diluent) for the soluble acrylic—a lower total than the single adhesive and with a moisture resistant but eventually soluble quality.

Single Polymer Application. More recently, and with the introduction of more automated equipment than a revolving drum, rotary coating machines employing spinning disc application of adhesive has demanded the introduction of individual polymer adhesives which require no other support. Polymer research has substantially perfected new proprietary single adhesives simplifying production—such as Becker Underwood's "Flo Rite 1085" polymer which is also water based, of neutral pH, low viscosity and can be used in continuous flow or batch treating systems both in slurry or direct injection equipment and is claimed to be fully satisfactory as a sole coating adhesive. Polyvinylpyrrolidone (PVP) was found to be, not just a polymer strongly protective of rhizobia but also with adhesive qualities which can be used for coating itself though more expensive than some alternatives. Special storage and use is required with products like PVP which, left exposed to ordinary atmospheric moisture is so hygroscopic[*1] it will solidify from powder form in less than 12 hours.

Large Firms now manufacture a wide range of such synthetic adhesives. *Nuplex Industries Ltd*—a leading global manufacturer of polymers and resins with operations in ten countries on four continents manufacture products currently sold in over 80 countries—now easily contactable via the Internet (just Google, *Nuplex*) where their International contacts are given. Their chemists are constantly improving products with a tradition of working out suitable formulations for specific purposes and for substantial users of their resins.

Other Firms such as Germains which has been involved in the seed processing industry for many years sold its Polycote™ range of polymers under the *Spectrum*™ brand suitable for rotary or conventional pan or roller coating plants for a wide range of treatments and of various viscosities. Some of these polymers are available in a "finish" range from matte to high gloss, also where required in a wide range of colours for seed identification, or treatment identification [in addition colour can be separately included in or onto any coating].

Croda International (www.croda.com) sell their Atlox™ SemKote polymer range ; *Precision Laboratories* of Waukegan Illinois, USA, offer their "Sentry" seed coating polymer and Suboniyo Chemicals and Pharmaceuticals Pvt. Ltd. of Maharashtra State, India, sell their "Smart Cote" polymer for seed coating.

There are now literally hundreds of firms manufacturing polymers worldwide, however, only a handful of these are specifically supplying the seed industries of the world. Again, one should simply use an "Internet search engine" to identify suitable contacts in various International locations.

Polymer use is no longer secret and its availability is widespread. Manufacturers of rotary spinning disc seed coating machinery, such as "SEED SUPPORT" [contact details shown in Chapter 5] who supply seed coating machinery of a range of sizes and types from small pan coaters to large rotary spinning disc machines—also supply coating materials and polymer adhesives to their clients and importantly demonstrate their successful use to new clients.

Innovative technology is now creating advanced polymers such as utilizing thermo-sensitivity whereby the polymer excludes absorption of moisture into the coating at low temperature but when temperatures reach a safe growth level, the polymer changes its physical structure allowing moisture to be absorbed for germination.

In the 21st Century, polymer resins have become an essential ingredient in a huge range of products used every day worldwide, from basic necessities, through consumer durables to luxury goods.

So, you wish to get into seed coating?—then to obtain an ideal adhesive— probably a synthetic polymer resin, or maybe a range of adhesives suitable for the variable seed and commercial conditions you as a new supplier will

---

[*1] Hygroscopic means readily absorbs water which changes the nature of the product.

be involved in, you would be wise to approach one or more competent and experienced polymer chemists—preferably someone working for an existing proven and substantial manufacturer and set out the following qualities which a seed coating polymer should meet. It should be:

1. Water dispersable and supplied as granulated or a liquid concentrate.
2. Packed in secure, moisture proofed and manually manageable containers. (e.g., >25 kg).
3. Storeable for all seasons (at least >1 year) in dry conditions and in the temperature range of (−10°C to 40°C).
4. Water dissolvent (or dispersable by swelling or slumping) when wet after polymerization and drying of the coated seed.
5. Of pH range 5.0 (which can be raised with an alkali) up to about 8.0.
6. Free of formaldehyde and all other chemicals toxic to seed and/or *Rhizobium* and *Bradyrhizobium*, such as copper or sulphur or chlorine.
7. Readily sprayable in solution at application to seed.
8. Of satisfactory adhesive strength—probably not less than 2,000 cps nor more than 4,000 cps—the latter quite stiff when made up—e.g., similar consistency to jelly, but thinned if combined into a water/glue mix where combination adhesives are used.
9. Priced at a level where its application does not make the finished product too expensive to use. For example, where a product concentrate costs (say) $30/kg and is diluted to appropriate adhesive application strength, lets say 20% (i.e., 100 kg of concentrate becomes 500 kg of aqueous solution costing $3,000) which is then applied at (say) 25% of seed weight then the cost of adhesive per 100 kg of seed (using 25 kg of solution) would be 25 kg x $6 = $150 or $1.50 per kg of seed coated (in 2012). That is an acceptable cost in relation to the potential benefits of a higher % of seedling establishment.
   However, these costs and application rates are hypothetical and will change over time—hopefully, better technology and increased competition will facilitate an easing of prices around the world.
10. Available at short notice in commercial quantities and of a consistent quality.

New Zealand's Coated Seed Ltd pioneered the move away from early bio-adhesives (which either contained anti-biotic preservatives, had a short shelf life, or were difficult to obtain) also away from insufficiently robust low viscosity cold water dispersable methyl-celluloses (as a sole coating adhesive which had limited binding properties)—into the world of polymerization where it was quickly discovered that many polymers were compatible with both seed and inoculants.

Importantly, some polymers are formaldehyde free *(this is a bactericidal gas from which commercial solutions are used to create formalin—a powerful preservative of organics also reported to be carcinogenic and of course absolutely toxic to seed and totally negative in association with inoculated seed).*

Encouragingly, various polymer formulations had suitably median to high pH levels; they transported, stored and mixed well and in solutions suitable for seed coating (usually diluted in clean water) were of acceptable cost in addition to providing a better bond on chalky surfaces than earlier glues.

This move into polymerization was formerly secretive commercially—CSL called its two main polymers by code names "rsa" (acrylic) and "rsx" (latex) for the soluble and insoluble dry crystalline product supplied and made up in the factory to specifically formulated aqueous solutions. While superior polymers now exist removing the need for above "combination" adhesive mix approach which is now outdated, this coding was important during pioneering days for retaining a leading role where for instance factory staff could (and did) leave CSL and join other competitive processors who however, lacked the technical resources required to up skill autonomously into polymer technology. CSL employed technically qualified staff who's personal skills brought this technology to that Company.

Principles in Adhesive Use

Two major problems presented themselves from the outset in the inoculation and coating of legume seed (chiefly clovers and lucerne).

The first problem being the oily and waxy (slippery and sometimes even shiny) surface of such seed (specially white clover and lucerne) which made the adhesion of coating materials difficult. Early coatings were of poor quality because any abrasion or handling at all tended to "throw off" the coat leaving half or more bare seed exposed and substantially defeating the purpose of inoculation and coating.

A second problem was the presence of polyphenols and tannins present on seed surfaces [contained in those oily and waxy substances] which protects seed in nature [when it lies on the ground] from soil borne infections. These seed exudates are quite lethal to *Rhizobium*. To overcome the adhesion and toxicity problems (but not remove Nature's protection of seed), these anti-biotics can be kept clear of inoculants by being bound onto seed (see "pre-treatment"—described above) and sealed off from inoculants by this thin but tough film of rapid air-drying [primal] polymer mixed through the seed while rotating in a drum [or bowl or pan] in which the coating process was completed. After rolling clover or lucerne seed in a solution of primal polymer (applied in bulk—tipped into the coating drum) it was given usually just a 10 second blending, followed by a wait of approx. 3 minutes allowing the film to polymerise (change from a liquid to a protective film). This polymer application around each seed created a vital thin plasticising latex film coat aimed at blocking off the toxins after which application of peat inoculant (which followed) was thus protected from that natural seed toxicity. In addition, the next application of coating and adhesive were better able to adhere to this pre-treatment film than to oily seed.

Though the primal film is technically water insoluble upon hardening on seed, being such a thin coat it is removed by break down of the coating material after sowing.

So, the natural seed toxins were contained but importantly not removed, and the oily surface was "keyed" for subsequent bonding of coatings and adhesives. This pre-treatment was not necessary for grass seed coating because the seed surface is not so oily and there is no inoculation to protect.

A further principle employed by CSL at adoption of polymerization, and for some years after was clever and quite original, that (described above) of combining a percentage of water, dispersible but *subsequently insoluble* (after drying) polymer with a similar percentage of a water dispersible—and *water soluble* polymer which in combination achieved a tough bond able to minimize dust and providing a reasonably stable coating. This combination of soluble and insoluble polymers would not normally be considered seriously by most manufacturers because insolubility of [part] adhesive on seed (which must obtain water and oxygen to germinate and grow) would appear harmful; however this well investigated combination eventually slumped off quite readily (as insolubility was weakened by solubility) and eventually dispersed under rainfall and/or soil moisture.

In a diluted solution, "insoluble" polymers eventually disintegrate—rather than solubilize.

While actually designed for the paint manufacturing industry or as timber adhesives or packaging glues plus other uses unrelated to porous, dusty or chalky surfaces involved in seed coating, the seed coating industry nevertheless adopted this polymer technology to its advantage.

Limitations of Commercial Exclusivity

Adhesives and coating methods for grass seed were more straightforward than for legumes. Despite that, when author Bennett resigned from Coated Seed Ltd after 14 years service the latter 5 years released by the Board of FCC Ltd from the provisions of his employment agreement [except as to confidentiality of technology which the board claimed in its opinion remained operative during those 5 years] the two companies optimistically sought yet further protection for its seed coating formulae in the New Zealand High Court in the 1980's including grass seed coating. That application was denied because the Court did not consider grass seed coating was "owned" by FCC Ltd—or CSL—or even sufficiently exclusive as a product to warrant protection of any sort. Grass seed coating is simple, an obvious concept, certainly not "exclusive". Author Bennett was entitled to consult in grass seed coating publicly, continuously and totally legitimately from the day he resigned from CSL. In fact, all technology of any ongoing use to such consultancy had indeed been published and was available publicly[3]. It was not owned by CSL or FCC Ltd [both of which no longer exist].

Nevertheless, despite a further period of restraint on CSL's seed coating methods [which Bennett was not interested in], all confidentiality restraint [between CSL and author Bennett] ceased "by consent" within Australasia in May 1983 and throughout the rest of the world on July 29th 1990–this was, as requested by CSL/FCC, as unopposed by author Bennett, and as formalized by High Court Order.

As of that date, having faithfully observed confidentiality of CSL technology over many years Agricultural Consultant Gerald Bennett [*Associate of the New Zealand Institute of Agricultural and Horticultural Science*) was free to consult without further harassment in all aspects of the industry and finally via this book. It was an unreasonably excessive period of both pre and post-employment commercial restraint which succeeded only because author Bennett's legal advice and support had proven to be unreliable. He was compensated for that by a $30,000 professional indemnity payment in 1980. This whole matter was an absolute legal shambles which brought no credit to those involved. Legal people are not technically competent and their simplistic decision ruling that everything not published was therefore "Confidential" was ridiculous. There are important lessons to be learned here by technically qualified employees.

Seed/Coat Ratios

Grass seed more easily accepts adhesives and a coat, has therefore traditionally and conveniently been coated with its own weight on a 1:1 seed/coat basis (when dry) thus 100 kg of bare grass seed made 200 kg of coated—that applied to the ryegrasses, cocksfoot, timothy, fescue, browntop—all non-leguminous grasses. Whereas the early coaters of legumes in New Zealand could only manage to get a limited weight of coating to "stick" on that oily seed—but had to maintain commercial consistency—so opted for a ratio of 4/7ths as seed and 3/7ths as coat (when dry) thus 40 kg of bare seed made up 70 kg coated (To work out bare seed from coated weight one simply divided by 7 then multiplied by 4). In effect, it was the limitations of adhesives which dictated seed/coat ratios in the earlier days.

Grass seed coatings are applied as one material without any need for pre-treatment—these single coatings initially being gafsa phosphate, or dolomite, or lime but best results in the field were obtained later from lime reverted superphosphate—milled to 300# BSS of course [a very fine powder]. The adhesives were also simplified—gum arabic then methyl cellulose were originally used but when those coatings were found too dusty and insufficiently stable they were also replaced by polymers giving a tougher—yet easily soluble coat—and better protection for factory staff (less dusty) where chemical additives (fungicides or insecticides) were included.

Today, due to the freedom of superior adhesives, seed/coat ratios can be variable, but growers need to know what they are buying. Manufacturers all have their own seed/coat ratios today, and they should advise purchasers what their recommended sowing rates are to achieve best results for each product.

Adhesives need to be able to mix harmlessly with inoculants. A major change in adhesives/inoculant use arrived with development of a new method of legume coating.

Technical Breakthrough (A "ONCE IN 50 YEARS" Development)

A major breakthrough in technology was achieved when a new method of inoculation and coating of legume seed was proposed in a Report to CSL directors by Development Manager Bennett as a result of research with a sample of PVP [polyvinylpyrrolidone] which he brought back to New Zealand from USA. He had earlier arranged to supply Invermay Research Centre (Dr. Bill Lowther) with processed treatments (including pvp) for comparative testing—being part of the ongoing search for optimum seed inoculation methods for benefit of New Zealand land developing farmers including The State via its Lands & Survey Department. The results of these tests were momentous.

Bennett had arranged for treatments to be processed by Technical Manager {contributor to this book} John Lloyd at the FCC Ltd laboratory which process subsequently gave markedly superior *Rhizobium* survival and nodulation results in Invermay's independent laboratory and field tests. And more than that.

This new process was in fact a "reverse inoculation"[(4)] technique whereby legume seed is simply coated first using any compatible adhesive (but not at that stage inoculated) is simply warm air dried [no rhizobia to safeguard from heat, etc.] and safely stored pending orders to supply, then rapidly run through a spray-on inoculation procedure incorporating polyvinylpyrrolidone (PVP type NP—K30 at around 2% of seed weight) which, it was discovered, provides exceptional microbial protection.

That coated product does not need to be dried after inoculation (because only small quantities of water are involved)—and it is supplied to the Industry freshly inoculated giving the best field and research results of any inoculated product we know of. This eliminated the tortuous procedure of attempting to disarm or coverup natural seed toxins to protect rhizobia—and the need for partial coating with protective peat—and the need for expensive vacuum refrigeration (cool) drying which, all discarded, saved much time and cost of processing. Thus in one giant step, the need for high class adhesives, additives (except pvp) and procedures such as vacuum/refrigerated drying, became obsolete!

So the adhesive for this new inoculation technique (N.I.T.) can be any polymer which is compatible with seed and rhizobia, which is inexpensive, easily applied, readily available, provides a strong stable coat providing minor absorptive qualities (to accept the inoculant later—on top of and into the coat, also a binder if used as the final coat), which stores well and breaks down in the field after rainfall. This was a major leap forward in the industry—and, we suspect, in worldwide commercial seed inoculation production. Author Bennett chose to introduce this major leap forward to CSL, on the eve of his resignation and did so in good faith. It is the author's opinion that this will have saved CSL (and farmers) a fortune in processing cost henceforth, quite on top of superior field performance.

To achieve the highest quality inoculated and coated product possible however and as a precaution, the authors recommend *continued use of the pre-treatment with N.I.T.*, whereby seed is first wrapped in a thin polymer film which not only seals off seed toxins (e.g., natural seed exudate protection which is toxic to rhizobia) but also gives a better oil free "key" for subsequent stability of coating materials. If shown to be unnecessary in studies of *R* survival and coat stability, then pre-treatment can be discarded.

Binders

A further refinement which can "polish" the final product is the application of a finishing binder such as pullulan, a natural water soluble polysaccharide fermented from sucrose which is tasteless and odourless therefore used in pharmaceutical tablets as a semi-gloss firm and stable outer coat. Sucrose supports rhizobia viability so pullulan is not likely to be harmful to inoculants and as a finishing binder would normally be applied as the outer thin coat over the top of inoculant application + pvp, applied as "N.I.T.". However, the author has little experience of use of a binder over the finished product to keep it firm, dust free, flowable and mixable (with grass seed, lime or fertilizer). So, processors will need to identify whether a binding film should be applied over the coated seed before it is inoculated (with inclusion of magical PVP— or a successor)—in which case the binder will need to be able to absorb low moisture inoculant + pvp overspray without a drying step; or whether a binder should be applied over the inoculant + pvp where it is possible a drying step will be required adding to cost, unless a polymer is available which vaporizes on exposure to air. It is not good practice to dry inoculated product with hot or even warm air, so it may also be the best option to abandon a binding film (such as pullulan) if its use will demand forced drying, reducing viability of the inoculated product.

Cost is always a factor, but with pullulan applied at just 1.4% of coated seed weight and at (say $40/kg in 2012 ex China), it will add just 56 cents per kilo of coated seed—acceptable where we also sow a lot less seed because it is effectively coated, effectively inoculated and has a higher % establishment threshold than plain seed—including plain inoculated (see Chapter 3). That is for forage clovers oversown—but a finishing binder on all coated and pelleted seed will give a much more stable, mixable, firm, semi-gloss (or satin) finish which even appearance wise—is superior to chalky or dusty products. For precision pellets, an additional cost of 50–60 cents per kg is easily affordable in relation to the value of most crops and the machinery it is sown through, which are not designed to withstand the effects of loose sand (from sand pelleted seed) in contact with steel bearings and other moving metal parts.

Pullulan is readily dissolved on contact with soil moisture. Other binders are available such as Becker Underwood's recently developed Flo Rite 1197—however, the authors are not aware of any studies showing compatibility or otherwise of such binders with *Rhizobium* which proving may nevertheless have been done and of course must precede their widespread commercial use.

Formulae

For newcomers to inoculation and coating, and to get started with legume seed samples for field research prior to committing money to expensive machinery (which is required where larger volumes are needed), the earlier methods of processing, though more basic are reasonably satisfactory for farmers and are easiest. Simply prepare cold water soluble methyl cellulose (around 300 to 400 cps is satisfactory—dispersed in hot water first then cooled) {note: gum arabic is better if you can get it}, and when the methyl cellulose granules have dissolved in water to make a semi-transparent gel, peat based inoculant is mixed with the methyl cellulose adequate for the weight of seed being coated and then thoroughly mixed with that seed. When all the seed is "tacky" with glue/inoculant, mix the appropriate quantity of coating powder into it gradually (to avoid lumps). The resultant wet coated seed is then set out to dry in a cool place—ideally with a breeze, but out of the sun—then when dry enough to handle safely without rubbing off the coat, for optimum nodulation results, it should be sown as soon as possible. But remember, this allows seed toxins access to the inoculant—so it is technically inferior to the product you could manufacture as the "new inoculation technique"—or "NIT".

For combining these materials, a simple concrete mixer is satisfactory, however the agitator bars inside a concrete mixer do not allow any hand tools to be used inside the bowl, internal bars are dangerous in terms of trying to hand mix the materials—or for instance to break up any lumps of product. To avoid that danger and have more personal control of the process, a revolving drum, just like a concrete mixer—but without the internal agitators, is preferred (*see bowl alternatives under "The New Zealand Experience" in Chapter 5 at page 196*) and is also satisfactory for manual precision pelleting—which a concrete mixer is not.

Formulations which have been used successfully are:

1/ A 20% Gum Arabic water based solution at 10 to 15% of seed weight is mixed with legume seed in a revolving drum or pan coater to which the recommended rate of peat based inoculant is added followed by finely milled gafsa phosphate or lime (calcium carbonate) or dolomite (magnesium carbonate). More gum arabic can be sprayed onto the product to firm up and stabilize the coated seed surface—but must not be overdone—adhesive will sweat up through the coat as rolling continues so too much glue and the product will begin to attach and form lumps. As in baking and cooking, skills are acquired with practise

The following formulae are suggested as a starting point for new processors to develop their own precise formula depending on which of the ingredients are readily available locally at reasonable prices. This is not the N.I.T. method of coating and inoculation which is optimum for a larger scale commercial processor, but it is a formula proven successful in principle via Govt. research studies with Prillcote™.

Due to complexity of *Rhizobium* inoculum manufacture, adhesives, seed, coating technology, storage, handling, application methods, climate, soils and many other factors outside control of the author, contributor and publisher of this book, no responsibility will be accepted by them for any loss, failure, cost or other detriment claimed to have been caused by the following suggested experimental formulae or any procedure or formulae in this book.

Legumes {To make 1:1 seed/coat ratio by weight—not 4/7ths seed, 3/7ths coat}

Ingredients based on 100 kg of bare clover or lucerne seed made up in a revolving drum process are: (NB only clean water is used—not chlorinated).

- 100 kg seed
- 3 kg pre-treatment concentrate. [Primal AC 234 and methylcell or equivalent.]
- 15 kg of 20% gum Arabic [in water] (or any one of several suitable polymer substitutes at their recommended rates mixed with 1% [of seed weight] methylcellulose 4000 cps).
- Peat inoculum culture recommended as suitable for 100 kg of seed. (increased by up to 5 times single rate for those wishing to better ensure good field nodulation). The water dispersable inoculant is usually mixed into the adhesive prior to application onto seed.
- 115 kg of either gafsa phosphate/dolomite mix [50/50], or calcium carbonate (limestone) for lucerne. These coatings milled to 95% passing a 300# BSS—a very fine dry powder.

This combines more than 235 kg of materials to make 200 kg of inoculated coated legume seed.

Where has the 35 kg gone?

These weights allow for moisture loss at drying plus a small % of dust and possibly some oversize lumps of seed, coat and adhesive. These unwanted by-products of seed coating—dust and lumps, are difficult to avoid in roller drum coating of large quantities per batch where internal beaters in the enclosed drum are relied on (without manual assistance) to separate all individual seeds with a coating of adhesive which then wrap themselves into a cover of superfine principal coating during the revolving action.

Dust and oversize are removed from the bulk of the product by 3 level screening after coating, having first been dispatched to a drier. However, very little seed if any is lost in dust and the loss of product from oversize (by this method of coating) is usually less than 5% of the total weight—of which the bulk of that is coating material and adhesive—but there is often also some loss of actual seed. Where oversize quantities are unacceptably high (quite possible during early trials) the lumps can be broken open when dry, carefully washed to clean the seed, dried again promptly and re-coated (with more care!). Oversize which is to be washed clean and re-coated would of course be more easily broken down and washed **prior** to any drying.

More recently (and a lot more expensively) rotary coaters with spinning disc adhesive application technology which now permits continuous production (ideal for grass seed) able to be observed in action, avoids major lumps and dust and can be cleaned more easily between seed varieties—which is important. That technology is using lightweight coatings like perlite both for pelleted horticulture seed as well as inoculated and coated seed. This however, eliminates nutrients in the coat and may present other changes, is therefore not the same product as the thoroughly researched 1970's and 80's Prillcote™ which performed exceedingly well in independent field testing and as reported in Chapter 3 of this book.

The author has not seen any independent field test results of more recent product, or what it consists of and is therefore not in a position to endorse or discourage its use. It is claimed that one or more New Zealand processor sells coated legume seed for oversowing without inoculation at all claiming *"New Zealand is now inoculated"*! There remain large areas of New Zealand hill and high country which carry only low grade inefficient rhizobia and much of it, no *Rhizobium* at all. This condition applies similarly throughout the world. Some legumes (i.e., Caucasian clover, plus the lotuses) have host specific strains not found at all in many areas of New Zealand hill and high country.

For new processors who need to become established commercially before branching out into sophisticated machinery such as continuous rotary production, the process can be carried out first in a very small way (even in a laboratory beaker for research samples) or as small scale production (a pan coater for coating and pelleting) on up to large open mouth revolving mixers and to horizontally revolving enclosed drums fitted with internal spray and agitator beaters which revolve inside the drum.

For newcomers, a more technically advanced coating for inoculated legume seed [than the formulae above] is the inclusion of a coating layer of finely milled, pH modified peat (or lignite) onto the adhesive/inoculant layer. This is an advancement on the above formula from point of view of increasing viability of rhizobia by providing a layer of host supportive peat or lignite into which the microbes can shelter, protecting them from mortality [This formulae also to make 1:1 seed/coat ratio].

- 100 kg legume seed.
- 3 kg polymer pre-treatment concentrate{Primal AC 234 and methylcell or equivalent}.

- 15 kg PVA at usually around 15% concentrate as an aqueous solution plus 1% of 4,000 cps methyl cellulose, or gum Arabic [or plain methylcellulose solution], with inclusion of sufficient inoculant for 100 kg of seed (and up to 5 times single rate where certainty is pursued).
- 25 kg of finely milled pH modified peat—or lignite.
- 115 kg of coating material (exactly as in above formulae).

Grass Seed [As a 1:1 seed/coat ratio—being the standard for grass seed coating in New Zealand for many years]

- 100 kg Grass seed (ryegrasses, cocksfoot, danthonia, dogstail, timothy, prairie grass, tall fescue, chewings fescue, browntop as well as several new proprietary cultivars).
- 28 kg of Primal AC 234 solution (made up as 50% concentrate + 1% methylcellulose 4000 cps, balance clean water) or a replacement polymer.
- 110 kg of finely milled (95% passing 300 mesh BSS) reverted superphosphate, or gafsa phosphate, or lime for acid soils where superphosphate (or other deficiency) is being topdressed.
- Water sprayed onto the coat where it is considered too dry or dusty—but any water applied must be removed by drying, so to conserve cost, water application should be kept minimal.

  Precision Pelleting (Processed in a revolving pan coater or mixing bowl).

- Selected weight of seed related to seed size and bowl capacity. Say 10 kg of onion seed per batch (This will "grow" to >8 x 10 kg = 80 kg pelleted).
- Fine spray application of gum arabic (or substitute polymer) onto seed at approx. 20% aqueous solution with sufficient "tack" to pick up sand particles onto seed, but not so strong it will join two or more seeds

Fig. 64 Woven wire mesh screens manufactured by Endecotts, of UK, a world leader in sieve design and construction which supplies their product with a certificate of compliance for National and International standards where required.

together. The % solution of adhesive may need to be varied for different varieties of seed as well as variability of sand weight or size.

- Alternate applications of clean [salt free] dry riverbed sand (ideally 50# to 100# BSS) sprinkled onto seed—glued with adhesive (above). This is a skilled operation whereby the operator develops a "feel" for the correct balance of sand and adhesive, gradually building up the pellets until they are—rounded, enlarged, uniformly shaped (hollows filled) at which point he stops pelleting—until hand held sieve tests show no more build up is required.

- To test for optimum size and shape, the operator measures a sample from the bowl to determine the median size of pellet in small (raised sides) hand held test sieves [as illustrated above].

  When optimum size for that seed variety has been reached, he applies a thin coat of smoothing almost "lubricating" talc as dry powder onto the tacky pellet surface which, as a filler, seals off abrasiveness of the sand leaving pellets with a silky, tack free surface. The quantity of talc applied is at discretion of the operator—usually not more than 2 to 3% of pellet weight. Allowing adhesive to permeate through the coat, he then tips the pelleted seed (with a wet interior coat) into a drier. Pellets change colour when dry and are then transferred to a 3 tier screener to take off any lumps (sometimes a little "offal" which came with the bare seed—i.e., small sticks, stones, vegetable matter which of course also get pelleted). The screener retains the main product for bagging onto platform scales but allows any dust to fall through to a dust bag.

Above is the formula for precision pelleting as it used to be carried out in New Zealand (which was never "confidential" but simply adapted by FCC/ CSL from published expired patents describing revolving drum confectionery manufacture, i.e., boiled lollies), and is likely to be the least expensive, safest and most guaranteed method (in terms of commercial success) for any new operator in that business. However, it is dependent on availability of a skilled pan coating operator who, however, can practice those skills on reject bags of horticulture seed which have returned substandard germination tests or for any other reason is not fit for sale—provided such pelleted seed is dumped immediately after processing—no matter how good the pelleting quality may be.

Currently however, horticulture crop seed is being precision pelleted in rotary coaters with automated spinning disc application of adhesive plus calibrated application of appropriate pelleting fillers. Sand has been replaced with other non-nutrient pelleting materials—particularly lightweight perlite or exfoliated vermiculite which are "oxygenated" with tiny air bubbles formed while in the exfoliation (heating) process, making the product itself so light it can usually float on water. These lightweight pellets have quite different seed/pellet weight ratios and are not usually costed by weight gain—as are sand pellets. Alternatively they are generally costed by weight of seed pelleted.

However, there is no longer a "National standard" for pelleted seed and each operator must advise their client growers how they should handle and drill their pellets for optimum results. For instance, belt drive precision drills (such as Stanhay Mk 2) are gradually being replaced in popularity by pneumatic (air suction) drills (manufactured by many engineering firms—including Stanhay) for which these lightweight pellets are ideal.

Other pelleting materials can be used—there are no fixed parameters for this, just provided the product is able to be evenly applied, is not harmful to seed or equipment used in drilling it, can be dried and stored; also provided the pellet disintegrates and is oxygenated when moistened after sowing—then any material which can be sieved or milled to the range around 50 to 100# should be satisfactory.

Additives

Fungicides

Because the main aim for legume coating is to safely introduce sensitive rhizobia to the rhizosphere after sowing, there has been little attempt until recently to complicate the product by introduction of chemicals such as fungicides, insecticides, trace elements or nutrients other than are contained in standard well tested coatings—such as calcium, magnesium and natural phosphate. Some manufacturers in New Zealand have recently abandoned inoculation in favour of chemical applications claiming the latter is more important. However, where host specific strains [and probably later, new more competitive, more èlite strains] don't already exist for those legumes it is unlikely anything else will be as important as effective rhizobia introduction with legumes. The situation in this field is subject to constant change however and some chemical use will no doubt be justified.

CSL successfully introduced fungicides such as Thiram and Captan for control of pythium (captan), verticillium wilt (thiram) and related "damping off" diseases incorporated into the coatings of grasses and precision pelleted horticultural seed (but not inoculated legumes) by inclusion in the adhesives water diluent at >350 g of 75 wp [wettable powder equivalent]/100 kg of seed [75 wp means 75% active ingredient of both captan and thiram but large users may be able to access the concentrate for application at 260 g/100 kg of seed]. These two fungicides have been used for many years most effectively provided they are used in combination. They have never shown phytotoxicity (harm to plant tissue) in research or commercial production to the author's knowledge. But, used alone can be disastrous where only one spectrum of seedling disease is controlled leaving the other free to "bolt" having had its natural competition removed.

Other fungicides such as the combination of benlate (mildews) and demosan (pythium and damping off) [each at 1% (formulation) of seed weight] have been proven safe when incorporated into the coating of, for instance inoculated and coated or precision pelleted alfalfa (lucerne) seed—a crop which is more likely to meet these threats in higher quality soils than oversown inoculated clovers on newly developed raw soils—so this combination would require proving with inoculated coated clover seed. However, as we are advised benlate is no longer available, this combination needs to be replaced.

New fungicides are constantly being developed in various regions of the world for control or reduction of localized seedling diseases—all of which can quite readily be tested in the laboratory for compatibility. In other research a wider range of control treatments are being investigated—not just a wider range of chemical fungicides, but environmentally friendly and organically developed disease resistance plus genetically bred resistance; they are all being advanced.

Research conducted by Messrs du Toit, Derie, Brissey, Holmes and Gatch of Washington State University, Mount Vernon who in 2006 exhaustively investigated the control of seed borne verticillium wilt (*Verticillium dahliae*) in spinach crops via seed treatments, found among other treatments, that "control" seed carried over 20% infection while treatment with Topsin M 70 wp (thiophanate-methyl) produced no growth infection at all—neither did the combination treatment Farmore D300 + Mertect 340 F.

That's always the primary job done (identifying effective, non-phytotoxic seed treatments).

The next step for seed processors in USA is to find if these chemicals remain viable incorporated into a seed coat—or in such application harm the seed. A coat such as perlite which is virtually inert chemically, is unlikely to damage the chemical structure of such fungicides which, for that matter may well be satisfactorily applied to seed alone—without a coat. But for protection of handlers from dust or direct contact, and just as long as the seed is going to be factory processed anyway, the value of incorporating a nutrient coat (also conserving fertiliser use environmentally by precision placement) plus the fungicide or any insecticide, may well be worth investigating by US spinach growers. A simple film coat overlaid on the chemically treated seed, can also protect workers.

Some seed Companies do supply chemically protected legume seed which is coated or uncoated—including inoculated clover with protection from clover root nematode. But most chemically protected coated seed sold commercially today is insecticide protected rather than fungous disease protected. In most of New Zealand the range of insect predators usually presents a greater threat to seedling establishment than do fungi. That may not be the situation elsewhere.

Insecticides

Bare grass and clover seed has been processed for many years with Bayer Crop Science chemicals "Gaucho" and "Poncho" against a range of insects which attack proteinous seed and/or succulent seedlings. These can of course now simply be over-coated as well (provided coatings do not chemically alter the insecticide structure) but in addition there are now available a range of effective seed treatment insecticides which were not available to CSL in the 1960's to 1990's.

In those days chemicals like Lindane and DDT were available (though seldom used on seed) having now long since been withdrawn from all farm use (where DDT was mainly applied as topdressed prills) due chiefly (like effects of radiation) to unacceptably prolonged toxicity. Some of these toxins were so persistent in soils they also engendered a build up of insect resistance to those chemicals). Selection of suitable product was quite limited by the fact that many of the organo-phosphorous and chlorinated hydrocarbon range of insecticides, while very effective on a range of insects, were usually found phytotoxic [quite harmful to seed in close contact], were not systemic and due to severe human and animal toxicity plus unacceptable residual viability of some in soil, have been steadily withdrawn from agricultural use—and are now largely confined to domestic and industrial insect and rodent pest control.

However, we now have a range of insecticides important to a well controlled agriculture which are not phytotoxic, on the contrary are systemic (work inside the plant) and are environmentally friendly because very small loadings are required when applied to seed coatings. Because of precise placement with seed wherever it goes, plus the [early] timing factor can have a large impact by "nipping an insect outbreak in the bud" which, if left unattended would require full on spraying or granulation application at much larger volumes of chemical required to save the whole crop—a procedure usually not good for the soil, or indeed worms and other soil life such as beneficial bacteria.

Application in small quantities with seed is an insurance policy well worth its cost in locations prone to specific, known insect problems. These chemicals can also be organophosphorus in formulation such as "Counter 20G™" manufactured (formerly by BASF) more recently by Amvac Chemical Corpn. A synthetic pyrethroid, it is not phytotoxic, quite safe with seed, ideally short lived in soil—but safe and persistent once inside the plant where chewing or sucking insects are killed by all three, direct contact, a fumigant effect and by stomach ingestion.

Again, it needs to be tested with preferred coatings to ensure chemical stability and that coatings can both host the chemical and wrap it up more safely for dust free handling. Seed varieties such as grasses, brassicas and pelleted horticultural crops can all be protected with "Counter 20G" which is partially or fully effective against insect pests commonly encountered in New Zealand agriculture—including *grass grub, argentine stem weevil, nyssius huttoni flies, springtails, some aphids and importantly clover root weevil.*

A carbamate liquid insecticide from Bayer Crop Science called "Thiodicarb" is also claimed to be safe for use on seed—not phytotoxic, on the contrary is also systemic in action. It is importantly effective against the [New Zealand] North Island's black beetle as well as some of the same insect range destroyed by "Counter 20G". A combination of both chemicals provides a broad spectrum of insect control for New Zealand conditions—may also control or deter other predators in other regions, but as with all new additives including other non-phytotoxic chemicals, they should all be carefully compared in new trials before commercial use.

Tests for chemical stability (both as mixed chemicals as well as mixed with other coatings) need to be carried out in whatever combinations the processor considers best supplies a nutrient deficiency in its region plus seedling predator control.

The Thiodicarb supplier in New Zealand ("Nufarm" [www.nufarm.co.nz]) requests that all seed treated with insecticide (coated or bare) should be dyed a distinctive colour, preferably a colour which would deter young children or animals from attempting to eat it. Well done Nufarm, good advice.

In addition, New Zealand hazardous substances regulations (Hazchem) require that these chemicals must be under the personal control of an *"approved handler"*. When not in use, the chemicals need to be securely stored out of access to anyone who is not an approved handler.

Further, treating seed with toxic chemicals such as this requires full safety protection for processing staff as well as safe storage and handling of the finished product. It is a matter of commonsense that these precautions should apply internationally.

Other chemicals are and will become available for insect predators in various regions of the world however, the treatment rates for the above two chemicals are:

Application rates

Counter 20G: for control of seedling pests and grass grub. Apply 3 to 4 kg of (20% terbufos granules) per hectare (as in sowing rate of seed) and if applied dry should be well mixed with pasture species seed (but not inoculated lucerne—or clover–to date, as clover inoculation is usually not required on higher fertility soils anyway). It is also mixed with cereals, brassicas and most horticulture crop seed. *This is for sowing with seed (but not coated onto it).*

If applied in solution, e.g., mixed into adhesive water for application to coated seed, two factors need consideration. Firstly a processor of such product in any reasonably commercial size operation would be better served to apply for approval to access terbufos concentrate from the manufacturer either in soluble granular form—or as a water dispersable liquid concentrate. That reduces the weight (and will improve the suitability) of product which needs to be applied with coated or pelleted seed. Secondly, precise placement with each seed (not just along the drill row) means that economies of chemical can be exercised (the seeds are being treated—not the hectare) thus slightly smaller quantities of chemical will normally be required for an equal and even better effect. An application rate of chemical to actual seed weight (plus polymer film sealant) which applies closer to 2½ kg/hectare would essentially provide much stronger protection per seedling than 4 kg of 20% granules spread along the drill row where the dilution factor (by surface area increase) is significant.

Thiodicarb: for control of black beetle, argentine stem weevil and other seedling pests. Apply 26 ml (of 375 g/litre of active ingredient, thiodicarb suspension concentrate) per kg of grass seed (including clover not inoculated) and double that rate for pelleted horticulture seed, forage brassicas and possibly forestry seed in some districts where problems exist.

The same chemical use approvals, safety precautions, approved handler rules, withholding periods and need to cover seed drilled into cultivated soils away from bird poisoning risk—all apply to this chemical (with a hazard class of 6.1) as well as to *Counter 20G* (above) being hazard class 6.9. Both chemicals are claimed to have a shelf life of approximately 2 years.

There are now other chemicals which are systemic in action, can be mixed with fertilizer for drilling in or near seed rows but should not be mixed directly with seed, such as Phorate technical 90% and 75% manufactured by [or were available from] American Cynamid Co., USA; United Phosphorus Ltd., UK; also W.R. Grace & Co., USA, Boots International, UK; Cyanamid India Ltd., Mumbai, and probably others—at time of writing.

- *Phorate 20G* granules are sold in New Zealand by "Crop Care" for conventional drilling in rows mixed with fertiliser. Much smaller volumes of 90% concentrate from manufacturers might be suitable for application on the exterior of already coated or pelleted seed, but this or any other new chemical would require extensive testing to ensure this caused no harm to seed (and no harm to well protected handlers by application of a polymer sealing coat over the chemical). This product has a [claimed] very wide spectrum of use when applied with fertilizer as dry granules at drilling. The author is not aware of any research attempting to place it on coated or pelleted seed. It is promoted for control (or suppression) of argentine stem weevil, grass grub, lucerne aphids, springtail, nysius fly, aphids, soldier fly, stem weevil, black beetle, wireworm, carrot root weevil, carrot rust fly, mites, root lesion nematodes and white fringed weevil.

The range of seedlings protected include: grass, clover, lucerne, brassica, cereals, maize, potatoes, carrots, vegetable brassicas, strawberries, ornamentals (nursery stock dahlias, lilies and roses), squash and other cucurbits. Use of this chemical must also be under control of an approved handler and carries withholding periods of—from 6 weeks for maize or for livestock on pasture on up to 8 weeks for most vegetables and 13 weeks for potatoes.

Only the most well equipped factory with experienced processors should attempt to use these toxic products able to be used with seed and only after official clearance for such use plus some field evidence that they are effective as claimed—before launching into commercial production. Small test samples cost very little to prepare and most chemical manufacturers will supply test samples of product for such proving experimentation on condition they are reciprocally provided with the details and results.

Other Additives

Trace elements

Molybdenum [mo] is an essential trace element important in legume seed nodulation.[*1] Mo becomes less available to plants as soil pH drops, so on acid soils where lime (calcium carbonate) application would be too expensive, it will be an inexpensive insurance policy to include molybdenum in the coating of inoculated legume seed. Better class soils where vegetable legumes [peas, beans, lentils] are grown are usually easily limed, not acid and will seldom benefit from molybdenum—even though its inclusion when using precision pelleted or film coated seed anyway, would be a simple and inexpensive management tool.

Brassicas can suffer molybdenum deficiency disease which in severe cases causes the stem to shed its leaves. As earlier described, this can lead to "whiptail" of the central stem.

Sodium molybdate is readily soluble so can simply be added to adhesive water for inoculated and coated legume seed processing. Sodium molybdate crystals are available from chemical distributors and an application rate of 5 g of sodium molybdate per kg [.5%] of clover or lucerne seed is a standard rate of application. Where mixed into the correct quantity of adhesive water (without increasing the actual water involved—or marginally so), each 100 kg of clover seed will get 500 g (½ kg) of sodium molybdate where that ½ kg is dissolved into the adhesives water for 100 kg of seed. With N.I.T. (new inoculation technique) the molybdenum is applied with the coat—not mixed in with inoculant.

Other Trace Elements

Using inoculated and even just coated seed as a vehicle for carrying and distributing trace elements has limitations under present knowledge.

- Copper deficiency can be remedied by application of 7 kg/hectare of $CuSO_4$ (copper sulphate), but is lethal to rhizobia in concentrated form. Cattle and deer suffer more readily than sheep from this deficiency which induces *enzootic ectera* a nerve disorder, as well as *osteoporosis* (weak bones). It is however mainly associated with peat soils and is at its worst (least available to plants) when wet and in mid to late winter. It is traditionally applied with fertilizer because, in this case we DO want

---

[*1]Molybdenum-containing enzymes are used as catalysts by nitrogen fixing bacteria to break the chemical bond in atmospheric molecular nitrogen (N2 → ammonia), via biological nitrogen fixation.

to treat the whole hectare, not just using the seed as a carrier, even if it were not toxic to rhizobia or seed when concentrated.

• Boron deficiency is rare but can cause stunted growth through cell wall dysfunction (also damage to seed formation) and is more acute in high pH soils. It is an essential element of both animals and plants but in plants can become toxic if concentrated or applied to excess. For these reasons application of borax granules (11.3% boron) at 5 to 10 kg/hectare—or by use of borated superphosphate is likely to be more simple and less harmful than attempting to apply it as a component of coated seed.

Other trace elements may be deficient in plants but only to the extent that they are essential to animals—not the plants themselves which simply "carry" those elements. Prominent among these are *selenium, cobalt and iodine* which should all be applied [where required] separately from oversown or drilled seed because we need ALL plants to pick them up for maximum animal health.

Bird Repellent Treatments: It is difficult to deter birds effectively—and not kill them. Mankind could destroy all of Nature's birds relatively easily, a lot more easily than pests which live underground like rabbits or concealed in thick growth like venomous snakes or stoats and weasels. Birds are easily seen, always hungry and highly mobile which is exactly why mankind needs to take care with use of poisons, traps and baits. Of course an epidemic of predatory birds out of control and causing substantial damage or threatening the survival of farm crops must be dealt with positively in a return to balanced Nature. But for everyday protection of valued product, mankind simply needs a deterrent to steer birds away to their traditional food sources—and away from—for instance the fine turf seed recently hand broadcast onto a laboriously well prepared new lawn area where such proteinous seed offers an attractive meal—but leaves an unattractive patchy new lawn.

The two fungicides which protect seed from various soil pathogens, Thiram and Captan do tend to deter birds when included in the grass seed coating of fine turf seed. There is evidence birds detect it without swallowing which is good because they then leave it alone—we have seen no evidence of sick or dying birds from use of these two chemicals in thousands of coated lawn seed sowings. The coating itself on CSL's fine turf seed (traditionally coated in reverted superphosphate) also helped to disguise that seed at least until the coat is removed by rain or irrigation; colouring is also considered to be moderately repellent so fine turf coated seed (for lawns, reserves, parks, bowling greens, putting greens and cricket pitches) has for many years been colour coated since it was first introduced onto the market by (Lloyd and Bennett of) Coated Seed Ltd. in the late 1960's.

Colour Dyes

CSL's colour dyes "Rhodamine red" for chewings fescue (*festuca rubra* ssp. *commutata)*; "Brilliant green" for browntop (*agrostis tenuis*); and "Auramine yellow" for crested dogstail (*cynosurus cristatatus*) are readily available from Chemical Company dye suppliers including I.C.I. Ltd. A much wider range of colour dyes are available now from dye makers like "Abbey Dye" or "Becker Underwood" (Ames, IA, USA) as part of its ("Color Coat™") range of liquid seed conditioners. These colours are usually added to water in the adhesives mix, however, some larger size seed (corn, beet, soy) is simply spray-on coloured for identification (either as varieties or because they have additives) then clear polymer film coated, to seal in additives, specially toxic chemicals, or simply to smooth, "lubricate" and standardize the irregularity of shape for better performance in precision drills while remaining transparent for colour coding. These are not "coated" seeds—but transparent polymer film glazed or encrusted seeds usually over a colour earlier applied directly to the seed.

Above colours (as used by CSL) have been shown harmless to seed of course and the time may come when commercial colour standards will be set—even internationally, for coding the increasing multitude of seed varieties, multiplied by a range of well established chemical additives, multiplied by genetically modified enhancements, multiplied by a range of different coatings—in other words hundreds if not thousands of alternative seed products—all requiring identification.

Perhaps these colour codings will eventually include mottled and speckled multicolouring, i.e., standardized double and even triple colours per single sample which can be applied quite readily by a succession of spinning disc and drier combinations and perhaps the addition into seed lots of other similar sized products—such as chemical granules for a specific treatment for direct drilling.

MACHINERY

The method and scale of processing determines what machinery will be required. There is a huge difference between small scale operation and large commercial production of inoculated and/or coated and precision pelleted seed.

Although already partly described in Chapter 5 of this book (precision pelleted seed) the range of machinery which has been used for coating and pelleting, from simplicity to complexity from the beginning to the present offers many options for new processors as a first step on the processing ladder and from small on-farm batches to more sophisticated modern equipment of today which is capable of processing large volumes either semi-automatically or continuously.

Firstly for very small scale processing, perhaps for a grower or farmer in a newly developing locality preparing his/her own seed (well able to do so with the assistance of this book)—or for research samples, a hand turned bowl requiring two people to operate—one turning the bowl handle and the other operating the coating process can be used. That is about the most simple machine available and a machine like this is marketed by "SANLI" of Hebei,

China at website http://www.alibaba.com/*handle seed coating machine*. For cleaning, this small handle turned bowl can be emptied into a drain or bucket. Earlier described in Chapter 5 because it is equally able to process precision pellets as well as coatings is the following machine.

Fig. 65 Suitable for larger quantities than the small manual bowl (described below) and electrically or motor driven but still operated by manual input of all ingredients is this standard coating [and precision pelleting] bowl similar to a small concrete mixer but with some important differences.

Because this is a most suitable size bowl for new processors to make a start as well as being adequate for a larger scale grower keen to process his/her own seed, we have given the specification for building this mixer in Chapter 5. It has given excellent service both in inoculated/coating and precision pelleting. Driven ideally by electric motor, but where electricity is not available, a small petrol (lawnmower type) motor though noisier, will work quite well. With electric motor this bowl size can process about 2 tonne/day of inoculated and coated legume seed, and even more as simple coated seed subsequently inoculated as in N.I.T.

It has a tipping handle which allows emptying the bowl either forward—or right over to the back. Full tipping capability assists not only emptying a full load into a bin—or onto a conveyor feeding the drier, but allows washouts (hopefully into a collection sump and drain) for thorough cleaning of the mixer—which is very important in this business, as contaminant seeds are not welcome in the seed industry and can cause much harm (no owner of steeper hill country wants to introduce even two or three tiny seeds of something extremely virile and seriously noxious such as *gorse, nassella tussock* or *sweet briar*, which in future years could cost a huge sum of money plus "toil, sweat and tears" to get rid of.

Fig. 10. Preparing pelleted clover seed in a concrete mixer.
Photo: S. A. Rumsey

Fig. 66 A well known and favourite photograph reproduced from the (former) DSIR [Department of Scientific and Industrial Research] Information series No. 58 being a Manual titled *"Legume Inoculation In New Zealand"* published 1966 shows Technical Officer (Inoculation) Mr. Athol Hastings of the Plant Diseases Division of DSIR demonstrating inoculation and coating of clover seed in a standard concrete mixer. While today there have been huge advances on this technique it has nevertheless played an important role in the early history of legume seed inoculation and coating in New Zealand.

Note: Farm Advisory Officer, Department of Agriculture for Northern Southland [Author Bennett] sent several sacks of rhizobia free soil from TeAnau to Athol Hastings at DSIR Auckland which confirmed the need for clover seed inoculation earlier suspected by Department of Agriculture staff. This was later proven in field experiments laid down near TeAnau in Southland assisted by Lands and Survey Department (the developers) who supplied cultivation equipment. Results of these experiments were published in the New Zealand Journal of Agricultural Research, Vol. 5 No. 3, June 1962, 278—93, by Messrs C. During, N.A. Cullen and G.M. Bennett. Subsequently, the more important huge success of oversowing [much less expensive than by cultivation] became of paramount importance in development—when made possible by the availability of effectively inoculated coated seed.

Any machine which cannot be immaculately cleaned in this industry, should be sterilized of viable seed between varieties, by the best means possible, including steam cleaning. Equipment [specially coating bowls] do need cleaning thoroughly between seed varieties.

For larger scale commercial coating of seed there are now many manufacturers producing a wide range of types and sizes of machinery for the full range from simple treatment with a chemical or colour right through to film coating, encrusting and multiple coating—but some of these machines are not suitable for precision pelleting with recent exception of the rotary coater now also producing precision pelleted seed under skilled operator control.

In pioneering days, elongate blending drums fixed horizontally were used and made with a lid enclosing the contents—some with internal beaters and water spray nozzles fitted into the central axle whereby the adhesive and seed are mixed by the beaters inside the stationary drum, the lid opened for receiving a dose of coating powder, closed again, blended again while the drum itself remains fixed (not rotating). The lid must be opened for inspection of the product and if necessary (i.e., too dusty and not wetted out) a fine water spray added. When blended satisfactorily, a bottom trapdoor discharges the product to a rolling and compacting drum and then to the drier.

This blending can be improved by making both the drum and the beaters inside it able to revolve independently so that beaters can achieve rapid mixing and then the drum (with the beaters now fixed) able to compact the product by revolving together—or alternatively discharge the load to a simple rolling drum for compaction (sometimes compacting two or more batches for added weight). That basic equipment required the expense of floor height for installation of blending and then rolling cylinders so they could discharge their loads down to the next floor; that is—the top floor is for initial *blending* of adhesives and coatings (fed by elevator for seed, and by pallet lift for coating materials, with inoculant and adhesive mixing capability available requiring water supply and drainage on that floor). The blended seed/inoculant/adhesive/coat product then discharged down to the next floor—rolled for *compaction*, then discharged down another floor to the *dryer*. After drying the product would either need to be dropped down again to a weighing and bagging device, or more likely elevated (acceptable for dry product but would be damaging to wet product) into a holding bin or hopper (fixed or on wheels) under which the bagging off facility is positioned. That requires a three or four storey factory, elevation equipment and pallet lifting gear. Today, that has all been made obsolete by equipment at factory floor level whereby seed, coatings, adhesives—and most other ingredients are fed either by auger, blower, elevator or pump to processing equipment which discharges finished product directly into warm air driers then receiving hoppers for bulk or bagged dispatch—all executed on or just above ground floor level, reducing factory design cost considerably as well as improving convenience, use of fork lift vehicles plus drainage and cleaning.

Not all modern equipment for seed treatment is designed for coating, inoculation or pelleting. For many years simple application of inoculants (for soybean), or of fungicides, insecticides and colouring agents has been achieved with continuous flow treatments at a simple volume control rate via machinery such as for instance "Gustafson", available for various volumes of application from "Bayer Crop Science LP", 4895 12th Avenue East, Shakopee, MN 55379, USA (www.bayercropscience.com/)—who have been involved in commercial seed treatments since 1914. These are not coating or multiple layer pelleting machines but simply seed treaters which can handle large volume throughput in districts where substantial areas of corn, cotton, sugar beets, soybeans and wheat are grown.

Also available from Bayer Crop Science are ancillary items of equipment which both the coating/inoculation and pelleting—and plain seed treatment factories require—such as inoculant mixing and application pumps, driers, conveyancing, screening and holding plus storage equipment.

Wherever product is fed into a machine, there has to be a level of feed (or supply) above that machine (or a means of lift up into it)—and wherever product is discharged from a machine, there needs to be a receiving facility under it (or a means of conveying product away to the next facility). So, historically, seed coating plants have been built into multi storey factories whereas today, machinery design does it all.

While all of these machines (above) have played a valuable part in the development of seed processing—and will remain invaluable for those starting out in newly emerging agricultural economies, the basic rolling drum or pan coater has been replaced for large scale coating/pelleting production in developed agricultural and horticultural systems by the introduction of the Rotary automatic coating and pelleting machine.

Rotary coaters can process fixed batch quantities of seed or be programmed for continuous production; more recently have also been adapted for multiple coatings as required for precision pellet build up to specific sizing.

Fig. 67 Rotary coating/pelleting in operation. [This excellent photograph is repeated for good reason in both Chapters 4 and 5—this machine can both pellet and coat.]. (Photo by courtesy of "Seed Support " of The Netherlands.)

What used to take 2 to 3 hours to process by 3 assistants in the factory using roller drum processing can now be achieved by one person and a rotary machine in less than half an hour. Precision pelleting in an open mouth bowl required considerable operator skill in assessing how much adhesive was required before a certain (operator judged) quantity of coating material was applied—then the adhesive again, then more coat and so on until (in another skill judgement) the operator considered the optimum sizing had been achieved, checked it on a test seive and finished it with talc. This pan coating has also been commonplace for much longer in confectionery and other similar build-up manufacturing—which is where CSL obtained its initial processing skills from—via patent publications—expired.

The major element of successful coating and pelleting by earlier methods [and in the rotary coater too], is the need to apply adhesive to the seed (and coated seed)—NOT onto the bowl or cylinder which contains it.

If adhesive is sprayed onto the holding vessel (whatever it may be), seed (specially smaller seed) will then attach to that equipment which of course will "grab" coating powder or sand and specially lightweight materials such as perlite or vermiculite.

A major consideration with all such equipment is the need for immaculate cleaning between seed varieties—even between orders of the same variety but with a different seed test analysis.

Some rotary machines have a removable lining which overcomes abrasive effects damaging the structural central mixing chamber itself, also allowing separate thorough cleaning and rapid turnaround when a clean replacement liner is fitted.

There have been so many patent applications for seed processing methods that it is not clear who in fact engineered the first rotary coater. The principle may not have been applied at first to seed but that use is now well established and fully described around the world—thus at last, commercial secrecy surrounding seed inoculation, coating and precision pelleting, is now largely gone. Some manufacturers will have confidential methods and formulae but if they are selling the process and machinery to make it, the secrets will not last long due to staff movements and termination of confidentiality in employment agreements. These products must also be fully described to buyers.

"Germains Seed Technology" is a long standing highly experienced seed processing Company who market their "Spectracota" systems which offers accurate application of ingredients for those processors preferring batch systems where precise ratios of seed, coatings—and particularly chemicals are required keeping costs to a minimum. Germains market a full range of seed treatment machinery including rotary models plus polymers for coating/pelleting as well as for encrustment and film coating.

This book cannot illustrate all the equipment and materials available or even list all of the Firms involved around the world in seed coating, inoculation and pelleting, however several of the larger and longer term suppliers are included here for further enquiry by interested parties. Most are easily found via search engines on the Internet. The authors have no commercial interest in any of them.

"Norogard AB" of Sweden also manufacture seed treaters such as their R15-2 and R20-2 Seed treaters shown (below).They both manufacture and market these machines and their contact details are:

NOROGARD AB
LillaFabriksgatan 15
S- 274 32 SKURUP
SWEDEN
Email: mail@norogard.se

R15/ R20/ R24 NoroGard continuous Seed Treater

Fig. 68 The machine in the picture is extra equipped with powder feeding system and a system for handling very fragile seeds. The manufacturer's description is:-
A machine suitable for seed plants with a need for:
• Treatment of cereals, corn, peas, soybeans, OSR, cotton, sunflowers etc.
• Capacity: R15: 6 to 15 tons/h
              R20: 8 to 20 tons/h
              R24: 10 to 24 tons/h
• Easy dosage calibration in a closed system
• Easy and quick cleaned
• Accurate dosage and application
• Compact and reliable construction
• Minimum of service required

Norogard also manufacture their Seed Treater R1 for liquid treatment (including simple inoculation of legumes) and for "dosing" cereals, corn, peas, soybean, sunflower and other staples with a capacity of 2 to 5 tonnes/hour.

Cleverly, this can be purchased as one unit or as a wide range of individual processing support items such as a 100 litre liquid hopper, the actual treater, liquid mixers on a wheeled frame, a second dosage system, dust evacuation, double bagging units and dry powder feeder with a 3 litre liquid hopper—which all connect to-gether to make a compact small treater. Skilled design and very practical.

Driers

In addition to machinery designed to convert bare seed into various conditions of the inoculation, coating, pelleting, encrusting, film covering or simple dosing "spectrum", all of which are wet processes quite harmful to seed if bagged or stored wet, is the need to remove moisture, so must be dried back[*1] to natural seed moisture level which for most forage seed is around the 6 to 8% mark.

Over the years of development, CSL proudly demonstrated its seed coating and pelleting factory and described its processes to hundreds of seed firm and other visitors including use of vacuum/refrigeration dryers [pictured on the front cover of CSL's "Technical Manual"—from a photograph taken by development manager Bennett and published in December 1972] which removed moisture from inoculated/coated seed while keeping it cool. This was during the period (e.g., prior to N.I.T."—"new inoculation technique" which dried coated seed—then inoculated it with minimal moisture)—a period when legume seed was pre-treated, then adhesive covered including inoculant, then peat covered, then coated—each step introducing moisture, and sometimes wetted further (particularly grasses) to secure the coat—but at that stage, quite wet. Immediately after coating the product was free-flowing but moisture content of the coating quite high thus rapid drying was essential to prevent seed absorbing moisture from the coating. If that occurred the seed would swell then shrink upon subsequent drying causing the bond between the coating, adhesive and seed surface to weaken.

Though free flowing, undried product was certainly much too wet to screen, bag or store in that condition. Drying with heat (blowing hot air) even in winter temperatures is potentially harmful to *Rhizobium*—hence the use of vacuum/refrigeration which extracts the moisture by condensation (more efficiently when in a vacuum) then removes it without heating the product. The principle of these machines thus ideally removing the need for heat.

Fortunately for CSL, engineering firm Cuddon Ltd's objective was [and is] to achieve "*innovation in design, excellence in quality and efficiency in service*"—exactly what CSL required.

---

*1except N.I.T. inoculation which is applied on to coating, not seed, employs minor moisture and is sown as soon as possible after inoculation.

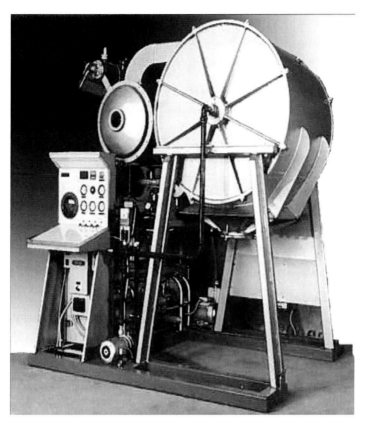

Fig. 69 A Cuddon vacuum–refrigeration drier [excellent for efficient drying of heat sensitive inoculated and coated seed].

Photo by permission of Cuddon Ltd. [World renowned expert in Engineering Innovation [Ed.]. www.cuddon.co.nz

Cuddon Ltd specialises in implementing world leading lyophilization technology in their freeze dry machines and associated freeze drying machine equipment.

Manufactured at that time by W.G.G. Cuddon of Blenheim, New Zealand, CSL had three such units manufactured because the whole production including grasses was vac/refridge dried.

Warm air driers are technically quite satisfactory for grasses and uninoculated product—however, it would require large factory space to install both types of drier. The Cuddon driers could do both inoculated and plain seed whereas warm air driers were not suitable for inoculated product—so Cuddon driers were relied on solely and gave excellent service.

These efficient Cuddon driers were first introduced to CSL by Technical Director Geoff Taylor M.Agr.Sc., and were very much part of the reason for such superior rhizobia survival on Prillcote™ coated seed compared to other processes, and plain inoculated seed. It was one of the many important and innovative technical contributions to New Zealand agriculture made by Geoffrey G. Taylor {deceased}. However, much of the finally successful design was achieved by W.G.G. Cuddon [now Cuddon Limited] skillful design engineer Don Kinnaird, assisted by FCC's Technical Manager [Inoculants and Seed] Mr. John Lloyd (contributor to this book) who worked closely together on a very successful solution to a biological manufacturing problem—first in the world to adopt this technology in managing inoculated coated seed to our knowledge.

Conventional warm air driers simply consist of fine mesh screening fitted along a travelling conveyor, gently vibrated to invert the product on that air bed on which the damp coated seed travels (or can be kept stationery for a period if more drying is required)—the whole conveyor is enclosed underneath and inflated (pressurized) by a fan blowing through heating coils up into the conveyor enclosure passing warm air through the travelling conveyor bed which will carry dampness into the atmosphere; however if there is any material nearby which could get damp, specially hygroscopic items such as reverted superphosphate, then the damp air may need to be collected and discharged outside—or the vulnerable item removed. Driers are used extensively in a wide range of manufacturing and come in all shapes and sizes. Long tubular driers are common as they contain warmth and dust and can be fitted with viewing points (windows) to allow operators to monitor progress (coated seed usually changes colour as it dries).

Screening

After drying, product which has been coated or pelleted should be run over a two tier screener of which the top screen retains oversize and any lumps which are run off into a receptacle, either discarded (if minor) or washed back to clean seed and returned for further processing. This top screen of appropriate hole size allows the main product to pass through onto a lower screen which retains and runs it off to the bagging-off gate; the hole size of the lower screen being too small to allow main product to pass through, but allowing the passage of dust which is discharged to a rubbish bin.

Screeners are also commonplace in many industries—and very much so in the seed industry however, the machine needs to be easily and quickly cleaned in this industry.

A number of screens must be made up to fit to this machine (see wooden frames at back of the screener shown in Fig. 70 awaiting the fitting of punched steel plate)—mesh size should be prominently written on the wooden frame of each one [and a note of which screens need to be fitted for each pelleted seed variety and treatment]. This note of sizes needs to be held as a reminder, even for the best operator who can easily make a mistake. The range of sizes required to make up suitable screens for sand pellets is shown in Chapter 5 "Precision Pelleting" in the Table shown on page 205 of seed varieties pelleted, in the

Fig. 70 Built for quite small production, more clearly demonstrates the way these screeners work. This illustration is repeated here as in Chapter 5 because it is also suitable for precision pellet screening].

column showing metric diameter of finished pellets. New manufacturers may wish to establish their own pellet sizes—also for seed varieties not mentioned in that Table.

The screening part of the process is recoverable if a mistake is made (provided main product has been kept clear of any contaminant seed), because if not screened correctly, it can be done again. Mistakes with wet product however, or boxing two varieties of seed which should be kept pure—are not so easily remedied. Such errors can be costly—even to the point of having to wash the whole coating off, dry the seed promptly, process and dry it again. If this involves a tonne or two, it could be very expensive indeed, as could dumping "boxed" seed and starting again. The operator needs to be wide awake on this work. The most seriously deceptive error however, is forgetting to inoculate.

Pre-mixers. *[Mixing adhesives and other products ready for coating].* An electrically driven inoculant/water mixer and dispersal cylinder is essential for peat inoculum culture (whatever method of inoculation is to be applied). This consists simply of a cylinder (stainless steel is superior) with a small electric motor fixed on top and an "egg beater" and/or (milk shake type) vibrator shaft down into the cylinder. Water is connected (preferably permanently but not fluoridated)—and with a measuring device while inoculant packs from the manufacturer are cut open and discharged at appropriate loadings into the now water charged cylinder. Total dispersion may take 5 to 10 minutes for a thorough result—maybe longer as research has shown better rhizobia counts where inoculants are thoroughly dispersed.

For a reasonably large operation, there should also be a similar motorized mixing vat or cylinder (a lot larger than for inoculant) with measured water fitted for adhesive concentrate dissolution, plus addition of inoculant and some of the additives (but not those even remotely able to harm seed or inoculant in concentration) with hose attachment for discharge directly onto seed in its pan or drum—or by discharge into measured buckets then tipped into the mix.

Miscellaneous Equipment

There needs to be provision for measuring, weighing and sometimes mixing additives for inclusion into the coat or pellet. Where certain items are regularly added to the process such as peat inoculant or specific polymers or colouring dyes, a separate dedicated mixing and dosing pump/applicator (whether sprayed on, or tipped in) should be provided for these regular items to avoid constantly cleaning for other additives.

Commercial production requires weighing seed into the process (unless bagged weight can be relied on) as well as weighing or volume measuring all other ingredients accurately. So, for seed alone under commercial conditions, an input seed hopper might be hanging on scales fed by elevator—the hopper able to be discharged into the mixer, both copiously and down to very sparingly, to get the right weight.

Likewise, outgoing processed dried product (coated, pelleted, encrusted or film coated) needs to be weighed—probably into (hessian, jute or multi-wall paper) bags fixed to a bag holder under a discharge hopper—the bag holder only holding the bag open—not taking any weight and the bag sitting on accurately calibrated weight scales.

These (preferably pre-stencilled) bags are then securely and rapidly sewn with a weight-counterbalanced electric bag sewer suspended over the bagging off point—or by a hand held electric sewing machine, then stacked carefully on forklift pallets where they must be clearly identified and run out to a truck or railway wagon—or just into store.

Certified Coated Seed

Seed which was "certified" when bare, remains certified when inoculated and/or coated but approval from the Certification Authority to break the seals on bare seed bags for processing then to replace them—or new ones—or carry out whatever arrangement is made when the seed has been processed, will need to be arranged being very much a matter of trust which should be earned, not lightly approved. Because trust already existed between Department of Agriculture in New Zealand and well established seed merchant half owner Wrightson™, CSL had no problem obtaining approval to break and re-seal bags of inoculated, coated and certified seed, never deviating from total adherence to the rules.

Handling Care

Unfortunately, transport industry staff have traditionally tramped all over pallets of bagged seed, also happily stacked loaded timber pallets on top of exposed coated seed bags, slung ropes over loaded trucks and ratcheted them tight until bags are almost cut in two, occasionally even use hooks or steel capped boots to move bags around—but this is certainly not acceptable for coated, pelleted, encrusted or film coated seed—indeed is usually not good practice for bare seed either.

Bags (and process staff) should indicate these three instructions clearly:

→FRAGILE←
HANDLE WITH CARE
No hooks, boots or tight ropes

DIY Handling

For a grower wishing to inoculate his/her own lucerne seed to drill a field—or clover seed to oversow a hill block—maybe coated grass seed to mix with it, the only equipment actually required is a sturdy tarpaulin laid out on a concrete or timber floor—and a flat wide mouthed shovel. Slow and physically demanding—sure, but cost efficient.

Measured quantities of seed, inoculant and coating powder are required. The seed can be coated first, dried on the floor in sun and breeze (turned over occasionally)—using a heater/blower if there is no sun or when temperatures are low, then when ready to sow, can be sprayed with half the inoculant (which has been thoroughly mixed over several minutes) as a fine solution—applied in a clean sprayer using clean water—(not chlorinated or fluoridated), applied very sparingly over the legume seed (not to get it wet again), turned over and sprayed with the other half inoculum culture so it is well covered—slightly damp, but not wet. This can be left to dry also—in a breeze of cool air—but not in the sun. When dry, a measured weight of grass seed can be mixed with it (by shovel) coated or bare, then bagged or loaded ready to be sown.

That is the most simple application of the most up to date technology.

Ideally, this post-inoculated product will be sown alone, not mixed with superphosphate or any other acidic fertilizer which (again, ideally) will have been applied earlier. Where contractors are to be engaged for application—specially aerial application, it may be necessary to relax part of the ideal procedure to the demands of practicality, whereby mixing (coated then N.I.T. inoculated) seed with superphosphate immediately prior to loading the aircraft is unavoidable.

The following Table of results of research conducted by Technical experts Taylor and Lloyd as delivered in a paper to the New Zealand Grasslands Association Inc. indicates that only *very effectively inoculated* coated seed can be mixed with Superphosphate—and then only for a limited time prior to application.

Table 18. EFFECT OF SUPERPHOSPHATE ON NODULATING CAPACITY OF INOCULATED WHITE CLOVER SEED.

Exposure time to Superphosphate [hrs]	← % nodulation →		
	Uncoated	Gafsa-dolomite	Peat plus Gafsa-dolomite
1	nil	34	100
2	nil	10	100
4	nil	9	91
8	nil	3	84
24	nil	nil	63

From this important study it is clear that mixing inoculated seed with superphosphate *for 1 hour* totally destroys the rhizobia (author suspects mortality begins the instant they are mixed). Inoculant damage is also severe when "Super" and inoculated/coated seed are mixed together for 1 hour.

But if there is a shielding peat layer included in the coat of coated seed (as was the case with Prillcote™ over many years) providing rhizobia with that special protection, then this product can be mixed with superphosphate for a couple of hours with minimal damage to rhizobia—but no longer, *thus it is safer by far to apply the inoculated and coated seed separately from superphosphate topdressing.*

On a larger commercial scale, the better quality and range of processing equipment, the faster and more consistently professional the finished product is likely to be. For any organization wishing to commence this industry but preferring to engage a Consultant who is fully familiar with every aspect of production—including of inoculants, Axis Associates Ltd of 25 McPherson St., P.O. Box 3414, RICHMOND 7031, NELSON, NEW ZEALAND. Phone (03) 544-2379 and E-Mail: olympus46@xtra.co.nz, are able to offer this service worldwide at time of writing which may even involve contributing co-author Mr. John Lloyd, considered the most expert *practical seed enhancement processor* in New Zealand's history {and quite possibly around the world}, with over 50 years experience in microbiological, seed and chemical processing, actually visiting a proposed new facility conditionally.

Machinery Manufacturers also favourably offer to demonstrate their equipment to new clients but well beyond that, Axis Assoc. Ltd are aware of the science including of inoculant production, the practical details, the coat/pellet technology, the many pitfalls and tricks which exist within the industry and in particular the critically important task of quality control analysis which any processor, new or old, will choose to ignore at their peril.

One of the most important ingredients to smooth flowing factory production of processed seed, is the necessity for an alert and aware person (a secretary) to receive, record, organize and safely dispatch orders correctly processed and on time for the range of products that Company offers. Not only telephone expertise, but (ideally) instantly accessible computer records, an understanding of seed, the seed industry and the range of products processed plus a firm grip of transportation in and out of the factory and the need to issue changing prices for inoculated and/or coated seed as the price of seed fluctuates—which it does internationally and frequently.

Maintaining current seed price is only necessary for clients buying the whole product—if a client has sent its own seed for processing, then cost of seed is not relevant—only the cost of coating/inoculating or pelleting is involved which, unlike seed cost, does not fluctuate as readily.

Procedure. There are some "do's" and "don'ts" in this industry. The whole business of processing orders of seed would fill a book because there are so many alternatives. In general terms, the following are items to watch out for. These items refer to processed seed only.

For all seed varieties

1. Quality of Seed

Commercial processors should set minimum standards for seed quality—particularly interim germination standards, but of impurities as well. Only seed of the highest levels of interim germination should be sent or received for inoculation, coating, pelleting or filming. Why? Because the seed owner is going to have to pay for every seed to be processed whether that seed is dead or alive. Seed of (say) 95% final germination will mean 5% of that seed is going to be processed at a cost despite that we know it will not grow—indeed, when oversown, much more of that seed order will not survive—but the 5% of "dud" seed has NO chance of growing. Of the balance, if a legume and if the interim germination were (say) 80% and 95% final—then even the 15% of "hard" seed which eventually will germinate, may also possibly die if, by the time it does germinate, the rhizobia inoculant has expired, or with its coat washed off having been sitting there for a long time, birds have found the bare seed.

So it will be *excessively expensive* to use (say) 75% final germination seed because we know for sure that handling and processing 25% of the 75% [final germ.] order has to be paid for when we also know *it hasn't got a chance of growing*.

Equally, if a line of seed has a favourably high interim germination count but carries a high "foreign seed" or "inert matter" content, then that rubbish too will become inoculated and coated at some cost—wasted cost. Furthermore, where weed seed is present, it is not good practice to boost its chances of establishment and growth by giving it both protection and a fertiliser coat. Poor quality seed may give resultant coated seed a bad name.

2. Quality of Inoculants

As discussed in Chapter 2, even when using high quality inoculants (with elevated numbers of cells per litre of peat culture) *there is no guarantee* that inoculation of seed will result in successful field nodulation. There are many

Fig. 71 Tubes of sterile agar jel are implanted with legume seed then inoculated by serial dilution. Resultant nodulation levels indicate [both] the approx. numbers of viable rhizobia (by mathematical calculation) as well as the correct strain. At current knowledge, this laborious test is the only recognized cell count procedure available.

reasons that the technique may fail—while the principle remains robust. It is essential that the processor obtain periodic rhizobia cell counts to *ensure the loadings per seed meet a reasonable standard (as discussed in* Chapter 2, Page 40, etc.)—but also that these cell counts be complemented by serial dilution agar tube infection checks as well because cultures can get minor contamination of "foreign" *Rhizobium* not host-specific to the seed that this culture was intended to inoculate.

The threat of contaminant strains can be exposed by (above) "*Most Probable Numbers*" rhizobia cell counting technology. For instance, *Rhizobium trifolii* which nodulate white clover do not nodulate lucerne, so if a Mother culture of *Rhizobium meliloti* intended for inoculation of lucerne is being multiplied under most favourable cell development conditions to be multiply split later for commercial culture supply—but has picked up a few cells of *trifolii* during preparation (easily done in a laboratory supplying many strains), it is a fact that *trifolii* can grow faster than *meliloti* thus insidiously changing that culture from a lucerne inoculant to a clover inoculant—plenty of viable rhizobia cells of course!—but mainly the wrong ones or even if 50/50—then that means the lucerne culture contains only half the rhizobia required. Even the agar tube infection test must be carefully carried out because a culture supposed to be for lucerne which in fact has become (say) 95% *trifolii* (for white or red clover) and only 5% *meliloti*, will still contain adequate lucerne cells to nodulate lucerne seedlings specially under more favourable laboratory conditions. But under rigorous field conditions, such a weak culture will fail.

Thus it is essential to run serial dilutions down to a very low level of cell numbers so that a weaker dilution may show no infection at all when in fact the cell count is "adequate". We then know it is a contaminated culture.

Inoculants should always be frequently tested for quality—e.g., purity, infectiveness and strains integrity.

3. Finished Product Quality control

Not only *Rhizobium* counts and strains validation checks are required in factory size production, but a small sample of every single order processed should be taken by an appointed staff member (who may be paid a small tariff for each sample taken, on top of wages—to encourage 100% sampling, also deducted by a much larger sum for any order overlooked!). These samples of both bare seed [as it was received at the factory]—and processed seed [as it was despatched]—each numbered (as per invoice) and dated, should be securely stored for (say) 2 years, then discarded.

From those samples, two or three of them per week should be randomly selected for a full quality control check in a suitably equipped laboratory. If the Company concerned is actively engaged in research, it may have suitably qualified staff to carry out these quality control checks, if not, they should be sent off to an independent laboratory which does have the expertise.

In addition to growing them in suitable media and under natural conditions to ensure normal germination viability (of both bare seed and processed) the samples should be checked for seed/coat ratios, coat stability and integrity (i.e., total seed coverage) plus effect of additives on seed performance (if any)—and the usual nodulation check for inoculated legumes (as a separate exercise from MPN tests which will be less frequent).

4. Handling

Where seed arrives from various sources—some for "in house stock" being (say) popular clover or lucerne seed lines to be coated then stored awaiting orders for inoculated product then, conveniently, that stacked product already coated, can be broken out, rapidly over-inoculated and supplied hopefully the same or next day. Also seed from firms supplying their own seed for coating and return as inoculated—or perhaps coated grass seed, or from farmers or growers including horticultural seed for pelleting—whatever, with all orders, a careful system of identification, labeling, stacking for processing in rotation, for trustworthy and fair revolving of processing and return of that same seed appropriately treated, is essential to avoid major foul ups which if delivered to customers clearly in error, will very soon prove to be commercially fatal.

The temptation is to store bare seed as it arrives on pallets starting on the floor and piling subsequent lots on top awaiting processing. This however ensures that either there is a lot of unproductive shifting around required by forklift truck—or, worse, the most recent deliveries (on top of the heap) which are most readily accessed are processed ahead of earlier deliveries at the bottom left till last in that stack, which is inefficient of course and likely to cause trouble.

Factories all over the world have this problem with various goods and many will have worked out a scheme that suits them best to overcome it. One such idea for an established seed processing firm where orders are seasonally urgent, substantial and where strictly rotational turnaround is important, is to provide steel frame pallet racks whereby pallets can be inserted and retrieved individually from the rack free of any pallet above or below it. As part of the identification and labeling system, those racks be boldly numbered, painted (say) green for bare seed waiting to be processed and (say) red for processed seed awaiting dispatch with the processing invoice also identically numbered. On arrival in store, the computer cannot allocate that number again until first, the factory part of the invoice has been entered, the computer record completed then the secretary enters final details of that order (including date, time and mode of dispatch) so that when despatched off the premises, factory worksheet finalisation will in turn generate a charge-out invoice.

Factory Production

1. Location

Naturally, to save transport cost and facilitate rapid processing of orders, a seed processing factory should ideally be located in the heart of a Nation's seed industry—or at the heart of production of a main seed variety—such as soybean or beet, and even more ideally (but less likely) located near where the processed orders will be sown.

New Zealand, a small country, is a classic example. The largest quantity of both grain and seed is grown in the Province of Canterbury (45,346 km^2) and today, that is where most seed processing stores and factories are located. Pelleted seed production for horticulture is however located in the South

Auckland region centred on Pukekohe which is where the largest volume of vegetable production is grown.

But Canterbury is not the location of greatest use of inoculated and/or coated seed, neither is Pukekohe the sole location of horticultural production—though it probably is for some varieties such as onions.

It is not easy anywhere in the world to satisfy centralization for both seed production and seed use. This is undoubtedly where lightweight coatings such as perlite have an advantage because land freight is calculated mainly by weight (with volume usually more important for shipping).

Careful consideration of these factors, as well as others such as ready availability of quality staff, of a reliable supply of clean water (or its availability if an investment such as a deep bore or piping from a river, lake or reservoir is possible) of electricity (not only for factory machinery, but also for refrigeration of inoculants), nearby rail freight services, safe, dry and cool on-site storage, suitability of the land for multi-storey construction (so seed can be dropped from one operational level to another during processing)—or suitability of multiple elevator use which will be required with single storey production—all these and other factors need careful consideration before a decision is made as to location of a seed processing factory.

Location of demand (for the product) is paramount.

2. Staff

It is easy to exceed the most efficient numbers of staff required for processing, often because the manager responsible for that decision has never worked on the production line himself. The production line should be under the control of a foreman who personally works full time on it and is solely responsible for it; there should be an assistant equally capable of taking charge during the foreman's absence. It is better to have two hard working good quality reliable staff and pay them well, than have three or four doing the same work but none of whom are well paid and therefore not as dedicated to the job. Office administration should be under control of a secretary who should be solely responsible for that aspect—recording, invoicing, office administration, wages, payments and transport organization—are all part of that "admin." where safe keeping of accurate records is also important plus regular independent auditing, tax calculation and a graph of production.

If the Company is involved in research and development plus sales promotion including seed firm liaison, there will need to be a formally qualified or exceptionally skilled and capable technical person doing that work. Being an activity small in scale [as in most cases at commencement] there will also need to be an appointee able to undertake development plus general management duties and who, in order to competently address farmers, seedsmen and researchers as well as advising a board of directors, will need a thorough knowledge of both the technical and commercial details of the industry. That person should be general manager and an executive director, the board's "right hand man" to keep all parties well informed and facilitate periodic 3 person meetings with factory foreman and secretary to ensure smooth running of the business.

The foreman and his assistant need to have a good understanding of seed, the processes, the machinery, materials (specially sensitive ones like inoculants—and dangerous ones like insecticides), of safety and sound staff management, of the need for close co-operation with administration and overall management. They should periodically be given an opportunity to see some of the results of their efforts out in the field.

Former experience in a similar situation would be a valuable qualification, but while the manager does not need to be a mechanical engineer, agronomist in seed, a microbiologist in *Rhizobium* and a master of business administration all wrapped up in one person, nevertheless a good knowledge and understanding in each of those fields of interest will be valuable—indeed essential. It would not be easy to find someone so trained for a new processing facility, so a period of training of someone with the basic formal qualification and hopefully some of those additional eperiences might be an acceptable starting point and an essential investment to eventual success. A commercial risk in commencing this industry is for newly appointed and inexperienced directors to believe it needs "an expert" in each of the fields of expertise, which is unlikely to be affordable. Another error is to appoint successive ambitious "stepladder managers" in overall charge who are popular identities, maybe well publicized sports people, or have pleasant personalities, but know little about the industry and have no real interest in it; they can do considerable and often unrealized damage to the company before moving on.

The overall manager {technical, development and administrative}—is the key appointment to growth and success who will be more secure in his position when formally qualified in agriculture (i.e., B. Agr. Sc. in agronomy and microbiology), also well read in engineering, skilled in public speaking, commercially aware and will need a strong level of board support and reward.

Inoculant production, if also undertaken by the coating company requires an additional range of skills and substantial laboratory commitment thus separate management will probably be desirable. Contributor/author John Lloyd has proven this role does not demand qualifications beyond exceptional personal ability and wide experience—however he is an exception to the rule. These roles are critical: we have already seen well structured seed processing firms stagnate and eventually fail due to incompetent and/or inappropriate staff appointments plus unaffordable costs.

Inoculant production may however be available from an independent source which, provided quality is maintained and costs contained would be a preferred way to get started. If production justifies it later and staff plus facilities are affordable, a processor may wish to commence production of its own inoculants for its own use, as well as for sale to farmers, seed companies and other processors.

An essential pre-requisite to commencing inoculant production, of course, is to be able to access a reliable source of Mother culture—of proven èlite strains of *Rhizobium* and *Bradyrhizobium* for the whole range of legumes the factory is likely to be asked to process—or that growers or seedsmen may need to purchase. Not all legume seed which needs inoculation also needs coating—mixing inoculant with seed in the drill box right at sowing is going to be adequately successful in some situations—but definitely not for oversowing onto vegetated soil surfaces.

3. Packaging

The worldwide seed industry has consistently sought the most suitable and cost efficient materials for containing the wide range of seed products available—including coated and pelleted seed. Earlier last century, jute (and its more coarse parent, hessian fabric) were widely grown in Bangladesh (United Pakistan) and shipped to UK, principally to Dundee, where it was milled for manufacture into bags and sacking—or carpet backing in the case of hessian. Cotton has also been widely used for bag manufacture—specially for flour, but not often for seed as it is flimsier than jute, will not tolerate hooks nor heavy weights.

Today with the arrival of synthetic polymers, principally nylon and thermoplastic polymers, principally polypropylene, also multiple layered toughened paper, seed sacks are now widely made from these lightweight and relatively inexpensive materials (but to the detriment of Bangladesh). Grain, cereal, pulse and oilseed crops are shipped in such huge quantities that container transportation in bulk has been used to good effect, but for relatively small orders, often batch produced, individual bags able to be handled manually without endangering workers spinal columns (OK >25 kg each) are the preferred container and widely used for inoculated and coated seed.

For precision pellets however, multi-wall paper bags have been considered more suitable—less vulnerable to impact damage, more versatile as to pre-labelling, size and stacking and can be further protected by loading into cardboard cartons.

Normally, as bags of coated seed are filled while sitting on (bag weight adjusted) platform scales, the bags pre-stencilled as to identity and date of processing (which supplying seed merchants may prefer to be left out—but to which information the buyer/farmer is absolutely entitled—particularly with inoculated product), they are electrically sewn when the target weight is reached (including ticketing and certification tags where appropriate) then carefully stacked on a pallet awaiting forklift dispatch to appropriate storage—or vehicle.

4. Storage

Most orders for inoculated, coated and pelleted seed are made just prior to the seasonal onset of ideal sowing time. In temperate climate New Zealand

Fig. 72 Earlier coated seed was damaged by truckies in hard boots climbing over bags while loading their vehicles, whereas pallets packed flat loaded gently onto other pallets also loaded flat spread the weight acceptably and did little damage provided sharp objects were avoided. Today's more stable coatings from use of binders or other techniques allows mixing and handling with much less damage but also requires care.

(cold frosty and wet winters, with warm to hot often windy dry summers), the ideal oversowing time by aerial application of forage clovers and grasses, is August (on lower *mild winter* hill country), to late September (*colder* snow risk *high country*) so processing for that market is at a peak in June, July and August. Precision pelleted seed is also mainly sown in spring to early summer, but some crops (i.e., onions) are substantially autumn sown, winter grown and spring harvested. Lucerne and brassicas may also be autumn sown into cultivated seed beds.

These orders usually go straight out to seed company clients (or growers) when processed, so storage by the processor is not required, apart from bare seed awaiting processing of course. However, with the development of N.I.T. where legume seed may be coated first—then inoculated later, just prior to use, the processor and/or seed companies may wish to have clover or lucerne seed on hand (coated) awaiting orders for inoculation when product already coated can be run through the inoculation only process promptly and despatched to the seed company client. That would *seem* simple enough and is convenient all round, because the seed was already held pending sale in the seed merchant's store, will keep quite safely in any dry storage for many months and can be pre-coated in off-peak times when the factory needs work to keep staff employed.

However, it is unfortunately not quite as simple as that.

Seed Companies are not certain which lines of clover or lucerne seed (grasses can be processed well before time too) the buying farmer is going to select for coating. In addition, if legume seed is to be stored awaiting orders and then inoculation, it would seem sensible to leave it stored where it is going to be inoculated. In the case of CSL who's factory was not designed for N.I.T., this would have meant adding more storage, organizing more insurance and adding a cost for that to processing invoices. In a case such as at CSL, jointly owned by former seed merchant Wrightson™, but also by former FCC Ltd, the latter not a seed merchant, to be fair, some arrangement would have had to be made to compensate FCC for convenience of extra storage required by the seed partner, but where non-processing seed merchants have their seed processed then want it stored awaiting orders for inoculated product, an additional charge for storage would seem justified.

This however, increases the cost of the product.

The obvious solution to this, is for seed merchants themselves to select lines of most popular forage legume seed (high interim germination, no noxious weed seed, modest price) and in return for having it processed "off-season" the processor should offer a small reduction in cost, making that product the most attractive a farmer can buy.

Then, in respect of having it stored at the processor's factory where inoculation facilities exist, either each seed merchant rent its own storage on that site—or if not possible there, or nearby (within forklift access), should consider retrieving the coated seed itself into its own store, and, because N.IT. is now so simple (spray on with minimum water and no drying), a facility for this could easily be set up by individual seed suppliers, quickly inoculated and re-bagged as and when required.

This does mean double handling {coating then later inoculating}, but N.I.T. itself will often mean two stages of handling and while factory managers have been aware for centuries of the downside of "double handling", in this case where it provides an important technological and timing benefit, then it changes from being negative—to being significantly positive.

Let the processor do all the technically demanding coating, pelleting, filming, encrusting, chemical treatment, colour dyeing—and so on, but the inoculation process (the process with the short shelf life which offers so much reward when well done), can now be handled relatively easily by the seed merchant itself—even by the grower—even on any airstrip. All that is required on the airstrip are a PTO [power take off from say a tractor] or petrol driven revolving drum, where a reliable spray device can be provided, where some insulated ice packed containers are provided for the inoculant and a clean water/inoculant stirrer can be made available [maybe by a battery driven electric drill].

"On the airstrip" may be too demanding because the revolving drum for overspray and the stirrer for water/inoculant mixing will require power—electric or fuelled. The clean water might require a tank on a truck and the dumping of coated clover seed from bags on a truck into the mixer then tipped out [when lightly sprayed all over] and then if mixed with grass seed then into an aircraft loader—would seem—rather complicated.

Whereas, the power, clean water, refrigeration for inoculants, conveyor to load the overspray mixer with coated seed, spraying it then addition and mixing in of grass seed then a bin under it to offload back into bags neatly re-sewn full of inoculated, coated, pasture mixed seed all ready to fly on (or sow), are facilities and activities not beyond the average seed merchant back in their own store. Overspraying lucerne [as N.I.T.], in the barn or on-farm (with no grasses to be mixed in) would seem much more available to the farmer as an on-site project, just prior to drilling (but still a good idea to drill the inoculated coated seed separately from superphosphate, potassic or sulphur super, etc., though quite OK to mix with lime). Remember, the N.I.T. inoculum must have PVP added.

Seed Industry

In New Zealand the earlier success of Coated Seed Ltd (and the technological benefits passed onto clients), were due in large part to the recognition and strong support of Seed Industry leaders—not just Wrightsons™ but also Pyne Gould Guinness Ltd, Donald Reid & Co, J.E. Watson Ltd., Farmers co-ops, Newton King Ltd, Arthur Yates & Co., Webling and Stewart and a number of others. The clients of these firms all benefitted from the experience, knowledge, research and development of staff of both CSL and Fruitgrowers Chemical Co. Ltd. (Taylor, Lloyd and Bennett as to formulation, proving and application of the products) and of Engineer George Bowman, Leading Hand Les Smith, Production Manager Bob Morton, Precision pellet expert Norm Hill—and again Geoff Taylor and John Lloyd, as to pioneering of factory and machinery design, processing and micro-biological development and handling facilities.

Supporting firms did well to utilize that Prillcote™ technology without themselves getting involved in factory establishment costs plus development and wages costs plus expensive research—much of the latter undertaken by Govt. Agencies, but arranged and supplied with treatments by Coated Seed Ltd. Without financial risk, these seed companies earned a profit margin from their sales and their clients benefitted substantially from superior technology—well done by them!. Had the concept of inoculated and coated seed and/or precision pelleted seed failed at any stage (and there was significant Departmental opposition to it originally)—then those firms who used the technology by supporting product sales without actually investing in it would have been well placed simply to abandon those products with no loss of investment and no obligation to Coated Seed Ltd.

Coated Seed Ltd organized many one hour, two hour or half day meetings with Seedsmen of these Companies with conducted tours of the factory after which the technology and field use of the full range of products was fully explained in the Boardroom by development manager Bennett, supported by his library of colour slides of FCC Ltd laboratory, propagation house, field research and actual farm results from around New Zealand and overseas. As stated, some of these classic but now aged transparencies obtained from various sources included those taken by the author for benefit of CSL when he considered a subject may assist CSL's promotion. For those, cost of film and development was re-imbursed to Bennett. But others created for his own professional records are illustrated in this book for benefit of readers. Many are well worn and much used so an apology for poor quality of some is warranted—however they are also irreplaceable.

Details of seed/coat ratios, inoculant care, field application, research and commercial results, costings, supply and transportation arrangements added to the confidence these sales technicians gained in introducing and confidently selling their own seed professionally Prillcote™ processed, to their clients. Farmer clients of Seed companies, specially former PGG Ltd benefitted from the specialized knowledge gained by these Seed Specialists serving many districts in New Zealand.

On several occasions, Coated Seed Ltd had to run its factory production in three 8 hour shifts working 24 hours round the clock—and for weeks on end seasonally, to keep up with demand.

There is an important "flip side" message in this for other parts of the world. In New Zealand there was the example of one or two other firms which (placing pride ahead of rationalisation) refused to use CSL's products [because seed competitor Wrightson was involved] tried to run their own small coated seed ventures without adequate formulation technology, agricultural, engineering, microbiological and laboratory expertise, or the financial strength required to reach the standard necessary for success. The size of their businesses in this specialized field would never justify the required and essential installation cost for success. They supplied product which, in the author's opinion did not consistently do justice to their clients (i.e., samples often compared quite unfavourably with Prillcote™ in many tests and *rhizobia* cell counts carried out over the years by independent authorities (including the New Zealand *Inoculant & Coated Seed Testing Service* which set minimum standards for acceptance of Govt. tenders). The farmer clients of such firms suffered the loss of product quality—but not all "other" inoculated and coated seed was inferior—as stated some product had higher rhizobia cell counts than Prillcote™ however, it was subsequently found that rhizobia survival was not as satisfactory.

The Strength of Combined Resources

For many Nations, it will be absolutely sensible and quite astute for individual participants to unite some resources—certainly seed merchants, maybe mineral and polymer suppliers, hopefully a microbiological laboratory and perhaps one or more chemical manufacturers, to jointly establish a facility, appoint a competent technically qualified manager—an executive director reporting to a board of directors comprising representation from substantial contributors (seats on the Board reflecting not only the level of investment by that participant, but critically, knowledge and understanding by each board member of at least some part of the industry).

By establishing a single processing facility to process all seed delivered to it, including from growers [and farmers] to ensure commercial success of the venture it will be much more able to meet required standards and supply inoculated products that in fact DO WORK. CSL only processed for members of the seed industry in New Zealand whereas *Seed Enhancements Ltd* (in Pukekohe, New Zealand) which bought out the pelleting business from Wrightson™, now process both pelleted and inoculated/coated seed for all.

Rules as to minimum seed quality and any other item felt necessary can be laid down by the board to maintain integrity of the processes.

In New Zealand, to achieve commercial success in supply of inoculated and/or coated seed, it is important if not essential to be established in (or have members of the facility who are established in), the seed industry. With a team of travelling seed sales technicians on the road who now don't just "sell seed", but are now often qualified agricultural technicians or graduates who, albeit with some commercial bias—are able to advise farmers about new grassland and crop cultivars, methods, protection, timing and establishment regimes for these products—including the place of coated and inoculated seed as a management tool. It would be difficult for any new processor to get past this grip on farmers' interests by the industry. That situation may however not be the case in other countries.

If a manufacturer developed an important new product (say other than seed related) and were able to show farmers that they really do need that product—(say) by some expensive TV advertising or circulars, then they may well find enough people will buy the product that its sales "take off" because there are enough of these important units out there (whatever they are) for other farmers to see the benefits and gradually "get the message"!

But where that product is seed and is to be purchased from a seed merchant anyway (who may be supporting that farmer financially—for reciprocal trade), if that seed firm does not offer the new product, specially if there is a competitive product that they do—then the chances of getting a sale of the new and independent product remain slim indeed. Few products are so obviously "revolutionary" that farmers see instantly that they must all have one, regardless of supplier though invention of the chain saw must have come close. "Selling" the idea or product becomes more difficult where only the very experienced eye can detect a good result compared to an average one with a grass seed sowing which, to further cloud the evidence, will soon be grazed after establishment. In short, the unfortunate facts are that the benefits of *good* result versus *ordinary* or *bad*—with coated + inoculated seed are not easily recognized unless careful analysis is applied. Cold calling on farmers is not easy in most countries even if a company has something as revolutionary as a newly invented chainsaw to sell. But sales technicians cannot carry actual results of seed establishment around with them—except in photographs which a brochure can do—if it is read, understood and trusted. Sadly, it is not a simple matter to demonstrate true comparisons and best results of seed establishment as shown in this book.

Costing and Invoicing

In well developed regions of the world, huge quantities of seed are traded Nationally and Internationally by a communication network among seed traders. They purchase that seed from primary growers (as certified or not), then run it through their seed dressing plants (a skilled job whereby maximum foreign material, dust, dirt and weed seed is dressed out of each lot with hopefully only minimum good quality seed being lost), then send a sample away to an independent authority (The Department of Agriculture's Seed Testing Station—in New Zealand, now called "AsureQuality"[5], a Government owned commercial company) obtain a *Purity and Germination Certificate* (P & G cert.) whereby, armed with that supporting evidence, the merchant may offer that seed to the international market, or locally. Other local seed merchants may not have that variety, or sufficient quantity, or of adequate quality of seed which is required for their own clients—so they buy. If they have surpluses of other varieties or grades—they sell. Sometimes seed is bought simply because it seems "a good buy" and (by arrangement) left in store (where it was purchased) until the original buyer can find a new buyer who will pay more for it than he did. Occasionally, a bulk quantity of seed may be bought and sold several times without physically moving at all—good business for traders no doubt, but unnecessarily expensive for the final purchaser and grower. Being largely speculative, it is not safe ground for small underfinanced companies or anyone with "gambling" tendencies to be involved in.

Such trading has historically been quite dramatic, in Australia for instance where small fortunes can be gained "overnight" in good seasons for some—[specially with poor seasons elsewhere around the globe] but has also been absolutely disastrous for others. This has driven more than one Australian seed merchant out of business in poor seasons locally when, perhaps due to prolonged drought, no one is buying (but most seed steadily losing viability, costing storage and insurance, vulnerable to rodents (specially during drought) and in many cases the seed being of average or inferior quality in the first place.

As a result of ongoing sales, prices keep changing—e.g., as they do for shares on the Stock Exchange. So it can be risky to send off a bulk line of seed at [whatever] current price, get it coated—then hold onto it in store awaiting orders for purchase—as coated, to be inoculated.

The astute seed company will:

- Identify two or three varieties of clover seed most commonly purchased for oversowing forage grassland—or lucerne for hay and grazing, or any other legume to be inoculated over its coating which carries a good P&G test and is most minimally priced.
- Arrange with the processor a discounted processing cost for seed supplied off-season. If available, arrange the same for popularly sown grasses as well (most clover and grass seed will keep quite well for some months if soundly stored—but, for instance *phalaris tuberosa*, chewings fescue and parsnip seed do not). If a discounted cost is not available, seek alternative processing, or (now that the technology has been explained in this book), install its own processing facility. A rotary coater operated by two technicians off-season, may be a good investment for any seeds company where sales will justify it.
- Where processed by a contract coater, store the coated (but not inoculated) seed carefully (circulating dry air, not too hot or cold, vermin proofed (not just a cat left loose in the store), germination tested regularly—yes to support sale—but also to check on any unwelcome deterioration.
- Instal its own inoculation facility (if needed, *see Axis Associates Ltd, Consultant expertise available anywhere in the world* on page 274 of this Chapter), and over-inoculate the coated legume seed when sold, mix it with grasses if required and get it out to the farmer smartly—but not to inoculate until it is quite clear the grower has everything ready to oversow (forage species) or drill (crop species). "Weather permitting" needs to be kept in mind. Mobile phones and weather satellites make all this possible, it was much harder to organize such close co-operation 50 years ago.
- This is not an impossible programme for an efficient seed supplier who has or can develope the business to justify and the potential to persevere with it.

A. Costing

Because seed prices fluctuate, seed merchants would be rewarded for thinking ahead, as stated, assigning some ideal lines of seed to coating, purchasing and coating them off-season (because they can be stored)—all aiming to keep costs down. This will create very saleable product when the season arrives. Because *processing cost* is much less variable than seed cost, at onset of the sowing season it will be possible to circulate in plenty of time a coated and inoculated price for some good quality seed lines which will satisfy the demands of that seed merchant's clients who must also plan ahead.

Cost of processing various seed species will of course be established by the provider of that coating and inoculation service for the contracting merchant to choose either:

1. The whole finished product (coated or pelleted).
2. Seed and coat supplied by the processor—but not inoculation.
3. Coating only, of seed supplied to the processor by the merchant (Includes pelleted seed).
4. Inoculation only, of already coated seed—as a separate cost from coating (Including re-inoculation of expired product).
5. Coating plus chemical or other additives supplied by the processor on seed supplied either by the processor—or by the independent merchant, or a farmer or grower [where it is known the chemical will not damage seed viability during storage].
6. Inoculation of seed coated with additives as separate from the cost of plain inoculation (above in 4). It may be necessary to apply an extra polymer film over the whole product (after over-inoculation) as a safety measure for handling by staff, e.g., for protection from toxic dust during drying (because a drying step may be required). Also protection during mixing with grass seed, when weighing and bagging off, during transport, on the farm when sacks are opened, in the aircraft loader and certainly when being flown on, blown on, spun on, manually tossed on—anywhere where toxic dust may adversely impact on staff of the combined operation. If drying is required after inoculation for any reason, then vacuum/ refrigerated drying will be superior to warm air drying, the latter not recommended at all.

While a safety polymer film would normally be applied over the top of any finished coated seed containing toxic chemicals in or under the coat, where such coated seed is yet to be inoculated over the top of the coat applying a film barrier before inoculation may not permit the inoculant to "seek shelter" inside the exterior coating. This would probably fail as an inoculation technique as well as create a moist, maybe sticky coating surface—hopeless for bagging, and dangerous for placing in an aircraft hopper. This would require more thought as to the best solution.

To establish actual cost of processing, the processor needs a formula of contributing costs which may be difficult to establish at first—or at least a lot of work to do so, but once assembled, any fluctuation in a component cost could be promptly corrected in a computer programme and if that cost shift was significant then a processing price change would be made.

Cost of the inoculants, coatings, polymers, other adhesives, additives, dyes, bags or boxes, total cost of running the factory—wages, power, maybe water, rates (property tax), insurances, depreciation on machinery, forklift cost per hour multiplied by X hours per 100 kg of X seed process, cost of accountancy, advertising and promotion, research and development, legal services, director's fees, management, manager's vehicle and travel, office costs plus "general and unforseen", all need to be costed out, related to successful processing of (say) 100 kg of each seed variety, method of processing, return transport and mixing charge (if any).

Once these items are identified, listed on computer, costed in relation to that process on each variety of seed, all totaled up, then it is a relatively simple matter to adjust one item (say wages up 5% this year) adjust that accordingly, re-total the lot and if it makes more than (say) 2% of change in the total cost, then circulation of a new cost per 100 kg of bare seed—or coated may be required.

The Company will add to the list the very last and most important item being a profit margin (without which there is no incentive to continue in business) and can factor in those fluctuating seed prices promptly (where the processor is providing the seed).

Of course these costs will vary greatly all over the world, so there is little point in trying to identify them here for one region only—and likely to be well out of date by the time this book is published.

B. Accountancy and Invoicing

An honest and reliable accountant is usually essential—not just to keep a record of payments and receipts, but to follow up on any overdue accounts. A raft of these can send a business into financial strife quite quickly no matter how technically successful its products are.

In addition, tax rules, dispensations, depreciation, Bank accommodation, wages and holidays calculations, ACC payments (Accident compensation Commission—New Zealand only), compliance with Occupational safety and health—are all specialist subjects best relegated to the accountant or specialist. Make sure the fee for such service remains reasonable and affordable. Once an accountant has sorted out all the detail for a business like this, annual adjustment will be relatively quick and simple. We consider an accountant's services should not be costing more than a maximum 5% of that Company's gross income but on a sliding scale—5% for small companies sliding down to more like 4% and even 3% for larger gross incomes. If the Company secretary is assembling all the data required for the accountant, and maybe preparing an annual summary and/or a tax statement for checking and approval of an accountant, then above costs may be considerably reduced.

No matter how good the product may be, farmers and growers will find another way if the cost [even] <u>appears</u> to them to be excessive, let alone whether it actually is or not. If it seems high but is not in terms of improved results, then the supplier must identify and demonstrate true cost reduced by better results to buyers. The astute firm can do this by several means—technical manual, fliers, Radio and TV plus important lectures (with a socializing element, i.e., after a promotional dinner) to farmer groups in their home locality.

Fig. 73 New Zealand Grasslands Association Conference in an appropriate field day setting among the tussocks above beautiful Lake Wanaka, Otago New Zealand. Use of inoculated and coated seed was endorsed at this gathering by independent Government employed grasslands scientists.
This was a special moment for the developers of Prillcote™ inoculated and coated clover seed. Many farmers attend these conferences where they know the unbiased facts of new developments can usually be relied on.

Internet

Where farmers these days have computers (as many do in New Zealand and elsewhere), most keeping key management records, storing valuable information, completing annual budgets as well as staying up to date with increasingly complex technology, one of the best and least expensive ways to get new information across, is the use of the Internet where a superior website can convey valuable information to a large audience on a daily basis. This can include videos of seed processing plus agronomic advice all published on-line plus downloadable brochure information accessible to users and students worldwide. It provides a really valuable service to clients, is readily updateable by its webmaster conveying vital technical information for farmers, growers, technicians and contractors, all essential to sustain "lifeblood" sales.

Naturally, no company can afford to educate the whole world, or even its competitor's clients, and some details do need to be kept confidential.

Where today, technology is becoming increasingly complex, is frequently changing, where the clientele is widespread over rural areas, not easily contacted, where the information needs to be readily accessible to, and downloadable by the farmer, where pictures literally replace a thousand words—certainly the case in agriculture, a very good, well presented website on the Internet is, *without any doubt whatsoever*, the most technically effective and cost efficient way for a seed firm to get its message across to the modern farming community—where future progress takes place.

Ideally, such website needs an emailing "question and answer" facility loaded, to clear up any questions or concerns—also so that farmers may place orders for product offered on-line because they can do that so much more efficiently at home in the evening even during bad weather where all necessary information is in front of them. That is better than having to interrupt a busy daytime activity, with their mind elsewhere, no support information at hand, suddenly interrupted by a telephone call, a seedsman's visit or a neighbour's advice—with nothing recorded in writing as a consequence and a reluctance to defer the practical job on hand—then have to ask about the technology again later.

So, an important item of *"machinery"* in the coated, inoculated and pelleted seed world, these days, is **a computer, connected to the Internet.**

References

(1) Lowther, W.L., Scientist. 1976. Invermay Agricultural Research Centre, Ministry of Agriculture, Mosgiel, New Zealand in *Proceedings of the New Zealand Grasslands Assocn.* a paper titled *Factors affecting the response of clover establishment to inoculation and pelleting*) [Volume 38, number 20, p. 176]—and others.

(2) Bennett, G.M. 1970. Farm and Station. Published in Christchurch "Press". New Zealand. A Newspaper article explaining to New Zealand Runholders promising early success of Prillcote™ inoculated and coated clover seed and reverted superphosphate coating of grass seed successfully established (photograph included) without other fertilizer on shingly scree at more than 3,000 ft altitude on the large 90,000 acre "Muller" Sheep and Cattle Station in New Zealand's Marlborough Province. Published on Feb. 6th 1970

(3) Hale C.N., W.L. Lowther and J.M. Lloyd. 1979. Effect of Inoculant Formulation on Survival of *Rhizobium Trifolii* and the Establishment of Oversown White Clover [Trifolium repens]"The New Zealand Journal of Experimental Agriculture" 7 [1979] 311–14.
So even the use of PVP as an enhancer of *Rhizobium* survival had been published (but was not available to author Bennett) when his use of it was disputed by CSL—thus it was ruled (May 1980) in the court as "confidential", in error, by the Judge to whom the publication had not been made known. That item alone was the only ingredient [incorrectly ruled "unpublished"] required by author Bennett to fully demolish the exclusivity claims of CSL. which would have cost it a heavy penalty as required under the terms of the ex partè injunction issued by CSL. Beyond all doubt, use of pvp with inoculants had been published see p.46 Chapter 2 in above research study dated 1979 certainly published prior to the court ruling [1980] that "it was *not published* and therefore not available to Mr. Bennett"!. While not the fault of the judge, a serious miscarriage of justice was a feature of this event and has never been remedied. This 1979 publication stated: *"Hale and Mathers (1977) showed that the survival of R. trifolii on white clover seed could be increased by either ....or treatment of seed with phenolic absorbents such as polyvinylpyrrolidone (pvp)....."*

(4) Thompson, J.A. 1960. Inhibition of nodule bacteria by an antibiotic from legume seed coats. Nature 187: 619–620.

This Australian scientist and *Rhizobiologist* published a research paper in which he recognized natural seed coat anti-biotics, by introducing a method called "reverse inoculation" to overcome rapid *Rhizobium* mortality. He coated the seed first then inoculated it in reverse of the previous standardized method. Reverse inoculation was "published"—also prior to 1980. So was use of PVP as an inoculant enhancer. These in combination constituted the best process, both were fully published and certainly not confidential. This perfectly legal process was not owned by CSL or FCC and should have been fully available to Consultant Bennett.

(5) AsureQuality. 1st October 2007. Internet Dissertation. http://www.asurequality.com A commercial company owned by the New Zealand government. Offering worldwide testing services it is stated to be the leading provider of testing, inspection, certification and verification services to the Australasian seed and grain industry. One of its many functions is the service which was provided by the Department of Agriculture's former Seed Testing Station at Palmerston North, New Zealand.

7

Present and Future Considerations

Food Growth Considerations

Three key elements feed the world, seed, soil and water. Excluding water of the oceans which of course produce seafood, these three exist in almost every nation, but of variable quantity and quality. Some of it being readily accessible and in use as our current agriculture, some usable with modifications (irrigation under low rainfall, reclamation from desert, jungle, swampland or the sea), some regions adapted to forage grassland, tree crops or timber forest (on stony soils, hill and high country, also acutely acid or alkaline soils). This leaves a substantial percentage not economically usable under present knowledge (being mountains above the snowline, icelands inside the arctics and rockland from small islands to The Grand Canyon and from the Fiords of New Zealand to the Fiords of Norway). Sensibly, mankind has farmed the best first, being the easiest and fastest to produce food and earn a profit from. To develop more difficult land, Governments must initiate and lead the way.

As world population grows, it is becoming increasingly essential to develop more marginal land despite the higher initial cost—not just cost in money, but in planning, effort, sacrifice and determination. The most powerful limitation to such progress is a lack of understanding and a lack of widespread knowledge.

If, for instance, everyone from Heads of State to the most humble goat or sheep herder knew and understood all of the technology discussed in this book, the world would be a very different place. There would be millions of tonnes more food produced, simply as a consequence of harnessing larger volumes of atmospheric nitrogen through grass and crop which would be growing acceptably well in areas hitherto considered marginal. Drought prone seemingly arid and barren regions would, with this knowledge, grow food from human effort and determination by better understanding. Then as wealth improves, so will health, education, political stability and engineering achievements from which yet further agricultural development will follow.

To accomplish this, water will be conveyed to many dry locations as has already been achieved in miraculous feats of engineering around the world. Much of it by government initiative via huge dams which also generate electricity, but some also by thousands of smaller irrigation schemes distributing water from a wide range of sources.

Fig. 74 [Illustration by courtesy of Wikipedia Free Encyclopaedia and: the GNU Free Documentation licence.]

The mighty Aswan high Dam [right] which separates the waters of Lake Nasser from the Nile River Valley releasing around 55 km^3 of water annually of which a massive 46 km^3 is diverted into irrigation canals. In the Nile valley and delta, almost 33,600 square kilometre benefit from these waters producing close to 2 full crops per year. Due to the absence of appreciable rainfall, Egypt's agriculture depends almost entirely on irrigation.

In China, the Hwang-Ho [Yellow River] flows through many provinces comprising a 795,000 square kilometre river basin irrigating an enormous 12 million hectare of arable land. However, because it is a river, huge silt and sand runoff remains a serious problem.

India has one of the most intensive irrigation systems in the world and ranks third for dam construction after the US and China.

It is estimated by authorities that well over half the increase in food production in the 20 years prior to 1980 has been due to irrigation.

Thousands of large dams and hundreds of thousands of small dams have now been built worldwide—most of the large ones since World War 2.

Not all of the large dams have been built for irrigation purposes, but about half have included irrigation, contributing significantly to the estimated 270 million hectare physically watered around the world since the beginning of the 20th century. As 100 hectare = 1 km² this is just 2.7 million square kilometre of an estimated 209.32 million square kilometre comprising the Earth's land surface—about 1.3% [irrigated].

From this information we may safely conclude that there remains huge potential for much more irrigation.

"More irrigation" does not have to be solely via large projects at a national level. A whole range of methods for shifting water from A to B have been developed around the world, from river and lake outlets via main race reticulation to multi side channels and dykes across graded land "bordered" to direct and contain the flow: to deep bores with submersible electric pumps powering filtered water up to large "rainmaster" overhead irrigators on motorised wheels which travel uniformly across large fields: to simple pump extraction from rivers and streams feeding overhead sprinklers and by trickle irrigation via alkathene hose from header tanks by gravity feed to drip water at the base of horticultural crops, grapes and fruit trees—all play a vital part in the global irrigation network.

Large areas of the world ignored for agriculture or forestry because they are traditionally too dry in summer, too cold in winter, too isolated from amenities such as electricity, too infertile or too vulnerable to animal and/or insect predation, actually can be developed where there is the determination to do so. Economists may declare a potential scheme "uneconomic" in terms of its productivity v cost and while that may be true of the first 5 years, it may not be true if calculated over the forthcoming 50 years or even less. At that point (Hopefully not artificially defeated by charging fictitious interest rates against its capital cost) such project can become hugely profitable into the future. Usually, only the national purse can afford such long term investment however and good governments will pursue the "big picture".

These marginal lands represent another "barrel of productivity"—except this more marginal barrel will have larger holes lower down in each barrel which we must "plug" [satisfy] before we can make progress "filling it". Undeveloped or wilderness areas may however have only one or two major inhibitors (to production) which, when identified, allow that land to flourish. The inhibitors may be easily fixed or overcome—establishing a small team of effective researchers can be one of the most productive investments any developing nation's government can make.

In New Zealand, government through its *"Lands and Survey Department"* commenced development of a large (over 140,000 acres) area of glaciated moraine soils growing fern and scrub including a windswept desolate plateau frequently snowbound in winter, hot and dry in summer, locally called "the wilderness".

Fig. 75 New grass, all dead, not by drought (see water lying in puddles), simply a classic nodulation failure—no *Rhizobium* in the soil and none inoculated onto the seed before sowing. First the clovers died—then because there was no nitrogen supply from clovers, the grasses also died.

This is an historically important photograph taken by the author shown twice in this book for full realisation of the reality associated with outright establishment failure.

Fig. 76 Closely located to the nodulation failure shown (above)—except: here, all the clover seed was Prillcote™ inoculated and coated; also the grass seed was coated, and no lime was applied. Up to 12 cwt of superphosphate was applied per acre in the first 18 months and most importantly, many areas like this were no longer expensively cultivated but simply oversown through native fern and scrub with the outstanding results shown here. In addition, there was eventual disappearance of scrub [shown standing upright in the photo] plus timber—ultimately providing clean grass fields. This has worldwide significance.

Contour was good—from level to rolling hills, rainfall was good—around 1,000 mm per annum reasonably well spread throughout the year, a servicing centre not far away with electric power, petrol, diesel, fresh and frozen food, plus schooling (by bus).

To this limited base, the government extended road, electricity, and built a house per economically sized farm, a barn/workshop, some shared sheep shearing sheds and cattle yards—but when the bracken fern and scrub was first cultivated, clover and grass seed sown, alarmingly, the clovers failed to grow, grasses then also failed due to nitrogen starvation and the whole cost of development (heavy cultivation, seed, fertiliser, lime, fencing, stock water)—all became at risk until it was found how to make the grass grow.

Of course, in hindsight, that research should have come first. But the developers were simply not aware of such classic establishment failures.

Having experienced such a monumental failure, the next step obviously—was to investigate every aspect of establishment—but chiefly of clover establishment failure.

That investigation[1] by The Farm Advisory Division, Department of Agriculture discovered during the next couple of years that:

(a) there were no rhizobia nitrogen fixing bacteria in these ex-glacial soils at all—and,
(b) lime was barely deficient (despite large volumes traditionally used nearby in Southland Province)—and,
(c) Superphosphate was required in large quantities due to the high content of ionising aluminium and iron oxides in this soil which negates phosphate and sulphur (i.e.—is not available to plants)—all of which changed the situation dramatically.

There are now some 56 very successful farms in this area all purchased by adequately experienced young farming families who are now all paying annual tax—its capital costs long since recovered.

Clover inoculation and correct fertiliser use are only two of many possible barriers to successful grassland or forestry establishment. Most undeveloped locations will present variable challenges—some more complex than others. Mankind, however, has the knowledge, the wit and the skill to solve them. The "difficulties" are no longer difficult when identified, but they will often require qualified, skilled technicians of agricultural and other sciences, well equipped, to study the details.

There are also opportunities for those who keep a sharp eye open for them.

Tramping through the middle altitudes of Himalayan Nepal, the author was astonished to discover an isolated orange grove—a welcome sight after three days without food.

Fig. 77 This orange tree (centre foreground) in full fruit is one of several in a grove of good quality oranges at mid Himalayan altitude in Western Nepal. Author's apology for the poor quality of photo which at least survived the rigours of a long trek in a backpack through a wide range of temperatures, months later being developed in India then eventually transported back to New Zealand—in 1960.

More importantly, they were very good quality oranges obviously well suited to the soils, climate and environment which could almost certainly be developed as an industry with vast potential. Obviously with amazing freedom from natural biological predators initially because these orange trees would have received minimal care.

This must surely be one of few places in the world where healthy orange trees can be photographed in the same frame as snow and ice (see mountain tops in the distance).

Combined with widespread and superb mountainside terracing also found in Nepal [and elsewhere] irrigation could so readily be introduced to some terraces from small to large reservoirs created on stable ground or in gullies filled from higher altitude creeks and rivers [by simple alkathene hose] plus monsoon rains in summer to provide more reliable water supply to various tree crops, corn and other grains through dry periods—including the dry winter when orange trees would be picked.

Development is Essential and Critical

A basic recipe for such development (anywhere) is as follows:

1. State funded research to identify any local problems of plant growth, of profitable farm or forestry production plus the technical, economic and social requirements for success.

2. Where production has been found feasible, State provision of a nucleus of essential services—maybe a "frontier" town centre for a rural Region. A State run demonstration and improvement property.

3. When proven in essence, State allocation of an economically sized area of land is made to pioneering and proven skilled growers—either by subsidising their purchase (State loans) or gifted after a period of successful production, or leased with the right to purchase, all on condition they farm the land successfully for a specific bond period following the procedures laid down in 1. (above).

4. Included in the programme/s identified in 1. are likely to be: State provision of houses, sheds, water, electricity, gas or other power, storage of [or access to] water for irrigation where summers are dry, trees for shelter in exposed situations and shedding for vulnerable animals where winters are cold. There needs to be all-weather roading and bridging where isolation applies, State loans for [electric] fencing where wild animals are a threat (in conjunction with animal control measures), biological and/or chemical control of insect predation—as identified in 1. (above).

Water for irrigation can be a big challenge, but government or community resources can explore sub-surface ground water opportunities (deep bores), if not right on site, then nearby where rainy season run off from mountains, hills, winter streams and ice thaw are evident, piped [i.e., buried alkathene running 24 hrs a day to a reservoir], from a river. Engineer designed creation of reservoirs to hold rainy season water for summer irrigation (such reservoirs using as natural as possible geographical features), outlets controlled and ponding beds lined with impervious clay—(i.e., bentonite),*¹ smaller ones lined with polythene for drip fed irrigation, maybe evaporation loss controlled by a thin polymer surface film (with access excluded including to birds by providing an adjoining drinking facility)—the list goes on.

Governments need to realise that even if they have to borrow money to undertake such development schemes, once in production, those farms, forests or vineyards will or should be paying taxes to the government forever more, the production of goods keeping not only those who produce it busy and content but creating more employment for others who then also pay tax. Every under-developed nation should have such bold schemes making certain they are not derailed by corruption, imposters, incompetence, violence or mismanagement. Independent auditing being a must, particularly in the development phase.

To dive straight into a development programme however, without careful thought and professional advice as to the full consequences of it—successful

---

*¹A natural clay with variable main elements, e.g., calcium, potassium, sodium or aluminium some of which swell when wet. Used as drilling lubricant in oil and other exploration it also seals reservoir basins from major water loss. Other natural deposits have similar properties. Bentonite is quarried in New Zealand as well as in Greece, Australia, India, Russia, Ukraqine Turkey and China. Other deposits no doubt exist.

or not, would be most unwise. There are some already well known major hazards in such developments which can be controlled and should not be ignored. Some of these are:

Irrigation Warning

As stated above, irrigation can contribute significantly to food security as well as economic progress—e.g., more purchasing power for essentials like housing, better roading, wider power distribution, health care and education plus much more—but there are warnings to heed—such as the World Bank paper prepared by Lire Ersado [*Poverty Reduction and Economic Management Sector Unit, Europe and Central Asia Region*], this paper being part of a larger effort in the Bank to achieve global food security and poverty alleviation. He has reported that some irrigation schemes have a history of facilitating transmission of vector borne disease, particularly malaria and dengue. Other vectors include arthropods (organisms with an exterior skeleton such as centipede, scorpion, butterfly and crab) but some vectors such as ticks also include wild animals in their cycle of blood feeding which makes them almost impossible to eradicate.

Remedies like vector parasites may assist.

So alongside irrigation expansion, control measures need to be provided where disease lurks including personal protective measures, barriers, house screens, bed nets, appropriate footwear and generally suitable clothing plus insect repellants and face nets. Unrealistic? Ask a malarial or yellow fever sufferer.

There are vaccines for some, i.e., yellow fever, tick borne encephalitis and japanese encephalitis while malaria vaccine remains under development—we understand there are others. Chemical and biological control to destroy arthropod larvae [and adults] plus environmental control of breeding areas will also become essential to prevent parallel expansion of disease in some irrigated localities. For example, flooding of rice fields by surface irrigation can create multiple shallow ponding which malarial mosquitos find ideal for breeding.

Due to insect chemical resistance, some of these diseases have sadly been on the increase in recent years because some vectors carrying pathogens have learned to tolerate that method of control—chemicals such as DDT, Lindane and Dieldrin have been effective insecticides but long since considered environmentally hazardous, and are now widely banned in agricultural use so the search for other chemicals and for biological control must continue.

Not only do we need to beware of biological predators when expanding irrigation, there are physical factors to take into consideration as well.

On the "Yellow River" in China, so named because of the huge volume of pale sandy silt it carries, there is an average 1.6 billion imperial tons of mud and sand washed into the river each year. Conservation measures are as vital as irrigation itself in such situations.

In other areas, soil saturation, salination and sunlight weakening from constant fog in windless locations are factors which need pre-determination and remedy where possible. Progress will require ingenuity and education of communities.

In many of mankind's endeavours, major steps forward (like electricity, the motor vehicle and the aeroplane) have been tempered by adversity (electric fires, motor car accidents and aeroplane crashes) which, cannot be allowed to stop those major positive developments, but require heightened awareness of those "downsides", demanding extensive safety measures and preventive action.

When this irrigation ball "begins to roll" however, it gathers momentum, and, with adequate nutrition and better health has come modest wealth, then education, social discipline, more stable Government funded by fair taxation, improved facilities and a much improved quality of life for many millions of people. Why don't all developing nations simply "go for it"!? After all, as we see (above) like the Nile valley many have done so, very successfully.

Sadly, progress is slow. In a very sluggish and frequently painful way where bombs and guns are sometimes the preferred path to change, instead of education and honesty in introducing social discipline, mankind is nevertheless making some progress—yet millions of people in less well developed societies continue to live in appalling poverty, hunger and fear.

Clean water to drink plus harnessed water to irrigate must be at the forefront of International assistance programmes and of localized endeavor.

Seed and Soil

From many years of plant breeding, supported by associated research, a substantial international seed trade has developed which buys and sells large volumes of proven stabilized seed varieties—now a major world industry.

Hardly surprising considering the miracle of seed and its importance to mankind.

In addition to plant selection and breeding, more recent developments include new genetically modified seed providing important advantages like tetraploid growth enhancement. Also disease resistance or predatory insect resistance from endophyte inoculation of ryegrasses, carbohydrate (sugar) enhancement of grasses and other features pursued by plant breeders either by plant selection or chromosome and even DNA modification.

These refinements show important potential for the established, well developed agricultural economies of the world, but the basic seed varieties—clovers, alfalfa, grass, cereal crop and horticultural seed remain important for developing Nations which should be pursuing a programme of importing and growing the most suitable range of seeds from which selection of the best performers should be made for multiplication alone, as well as cross-bred with promising indigenous varieties which have adapted to local conditions but are in need of genetic upgrading for better production, resistance or palatability.

While plant breeding has made an enormous contribution to increased production of world grain, rice and crop seed there remains huge potential for further enhancements in yield with better disease and insect resistance from further intensive plant modification.

The following facts underscore the need for promoting urgency and excellence in plant breeding:

Reported by FAO, August 2009.

Annual global level of lost food production = $106 billion caused by pathogens and $57 billion caused by insects.

In UK: disease resistance alone saves about $205 million/yr in crop protection products.

Quite small samples of "pilot" seed which costs little, may well be made available via aid programmes and will allow agronomists to commence programmes of breeding (utilising localized genetic values), selection and modification—it does take time—yet there is no time left in Nations with starving children.

These programmes are urgent.

Basic seed imports need to be good quality varieties, proven as to production capability in their places of origin, of course chosen for hardiness, disease resistance, harvest yield and germination performance for sure—but importantly obtained from regions as closely similar in climate and soil fertility as possible to the locations in which such seed will be grown.

Seed and Grain can feed the world—but it must be good seed, it must have moisture to grow and it needs reasonable soil fertility. We can grow it almost everywhere with a determined input of *investigation, planning, expert management practice, adequate finance and effort,* Government led with leaders aware of rural conditions—not led by desk bound financial office holders.

Undernourishment is not just a matter of insufficient food in the world—but of distribution coupled with extreme poverty.

The same insufficiency of food applies to livestock.

In bizzare contrast, there are actually now more obese and overweight people in the world than the 800 [plus] million who are undernourished, as recently reported from the US. Obesity epidemics have developed during the 21st century in many of the more wealthy societies—certainly in USA, Canada, Australia, New Zealand and indeed even in China despite it being the most populous country. By contrast, in India, the second most populous, close to half the children are underweight with a constantly increasing level of hungry or starving people of all ages.

It is ironic that while over 800 million people will die young from malnutrition, even more will die younger than necessary to obesity if current trends continue.

The importance of feed quality

[A]

Fig. 78 The chief difference between [A] the Kage sheep (left) on very sparse feed in Katmandu Valley plus [B] the sheep and goats (centre left) feeding on rough native mountain grasses in Western Nepal—and [C] the much heavier and more productive sheep grazing on improved grassland in New Zealand (bottom left)—is very much due to *the quality and quantity of feed available to each.*

[B]

[C]

In particular, the green grassland [left] has a clover content fixing valuable (free) nitrogen to the use of plants and animals. This nitrogen green carries on up into the distant hills.

We have a very serious world imbalance.

While China has positively addressed the problem of a burgeoning population, significantly, India has not. To its credit however, India has steadily increased its grain and rice harvest to the point where it has recently achieved a small surplus to internal requirements but, might it be that if all of its population could afford to buy adequate food, it would promptly fall back into a grain deficit again?

Greater use of soil and water—and more equitable distribution of food: these hold the answer for the forseeable future—how difficult can that be to apply? Some wealthy countries give huge aid, but the underfed Nations must themselves make it work and overcome that basic inertia so easily blamed on climatic, educational, cultural, religious and cost factors which have been allowed to block the path of human progress. Even in food rich New Zealand there are undernourished children, but that is a problem of parental mismanagement, of drugs, alcohol and the unskilled trapped in poverty, a social problem, not related to underdevelopment and overpopulation—they are the major problem.

Recent Developments and New Techniques

While it has become important to lift the production of food in developed countries and to push their frontiers of agricultural and horticultural science creating new opportunity in reasonably well developed agricultural nations, such progress also has potential for transmission to some of the undeveloped. There is the advantage of lower cost for them due to already proven methodology and hopefully stress-free implementation [e.g., the research has already been done]. This book however, is chiefly concerned with seed, a huge subject in itself, so we will not be tempted to stray into the vastness of agricultural and horticultural science in general.

Recent developments with [or associated with] seed which could lead to those ideals are for instance:

- A major breakthrough in an important part of technology may soon be achieved. Plant breeding scientists are pursuing the exciting prospect of identifying the DNA segment of *Rhizobium* nitrogen fixing bacteria, and/or leguminous plants—which allows rhizobia to access atmospheric nitrogen. Then when identified and isolated, inserting that quality into the DNA of [non-leguminous] grain, cereal and other cash crops. Eventually, it may be possible to include non-leguminous grasses. If equipped with nitrogen fixing capabilities, the new transgenic crops should be able to generate their own nitrogen and not only survive—but indeed thrive in those marginal conditions formerly devoid of nitrogen where application of expensive artificial nitrogen fertiliser remains simply unaffordable.
  Such development would expand cropable land exponentially and change the world.

While it might seem "straightforward" to the layperson, it would require deeply complex and successful transfer of DNA signalling from microbes to plants—which if achievable, would make a major impact on world food supplies—too late however, for the 6 million children under the age of 5 years who have already died this last year from malnutrition—or food related disease around the world.

- The discovery that at least one species of *Bradyrhizobium* not only fixes nitrogen on the roots of tropical legumes, but can also colonize the root systems of rice becoming an endophyte producing a beneficial increase [>20% lift in crop yield] on plots inoculated with this species.[2] If this phenomenon can be successfully developed for commercial rice production around the tropical world, increases in rice yields would be huge.

- Scientists are testing new vaccines against malaria with promising results already recorded. Because approximately 1 million people die of malaria annually, mostly in Africa, this may help revive enthusiasm for irrigation, in turn providing better crops with higher production of grain and seed. MVI Path (Malaria Vaccine Initiative) advise that the leading treatment may produce results by 2015 of many vaccines under development

- Engineers are developing machinery which not only drills seed with precision but also applies whatever coating, pelleting, inoculation or additives are required just prior to placement in the soil. This would appear to remove the need for any drying of product but would have to create a reasonably dry pellet—or at least free of any stickiness, to permit single selection and placement. Such a highly specified and versatile machine is likely to be expensive and more suitable for contractors than for farmers or growers themselves. The contractor could supply everything—seed, inoculation, additives, coatings, fertilizer and precision drilling itself, creating a significant bulk purchaser able to pass on economies of size.

- Biological control of seed and soil borne pathogens by inoculation of seed with cultures of fungal spores and/or microorganisms which have shown ability to protect the seed have been investigated in the past with variable degrees of success. Fungal products were sold in the early 1900's to control seedling disease but were eventually replaced by more stringent effective chemical fungicides. More recently, with resistance to chemical control increasing, biological control is again becoming important. New research indicates that this method has important potential once the sorting out of natural predation specificity plus simplified culturing techniques have been achieved. Instead of insect resistance to chemicals, this aims at plant resistance to insects.

- Plant Breeders Rights Legislation provides security for private agronomists or companies employing them to ensure a period of protection of the investment they have made in any successful cultivar they have developed. The developer still has to sell enough of its product successfully to retrieve its costs, some do not. But around the world, due to more intensive plant breeding and modification—by selection, by genetic adaption and by DNA engineering, many more valuable plant varieties have been produced; armed now with patent "Rights" protection [also called Intellectual Property], the intensity and volume of new plant products will increase—probably dramatically.

Independent Evaluation

Because of the proliferation of new plant species, it is becoming more important that an independent organization be appointed to catalogue all new cultivars in each participating country, provide details about their merits in comparison with parent or basic stock, apply an unbiased [independent] professional rating as to their effectiveness in respect of the qualities they are claimed to have—and apply a score reflecting the fair market price of each product related to its claimed quality.

Farmers need this information to allow purchase of the best products from an already wide range which is only going to get larger and more complex. Each agronomic firm, of course, promotes their own plant bred products—they can "blind" the farmer with "ploidy science", if he isn't already confused by the claims of competing seed company sales/technicians. Unbiased reporting is clearly a Department of Agriculture function, which organization has been stripped of "referee" and advisory roles in several countries by successive politicians of both left and right persuasions. This has certainly been true of successive New Zealand governments who's policy of cutting every perceived surplus cost in the productive sector with dilution of effectiveness by name changing to agglomerations such as Ministry of Primary Industries to embrace wider responsibilities with capped resources has been matched only by the magnitude of its misunderstanding. Due to population demographics there is always a predominance of City politicians who blissfully "pass" critically important rural affairs legislation with no knowledge of Agricultural Science whatsoever—they "toe the party line" which often involves superficial cost cutting—but ignores benefits [some of which are crucial]. With an increasingly complex agriculture, farmers and growers need accurate advisory information. When they grow the best, they earn more and the government earns more tax.

Reporting sincerely and accurately has reached somewhat sinister proportions in the USA where scientists are required to obtain permission of plant breeding firms before they can publish a technical paper on any of the breeder's cultivars. The plant breeder can say "no", and the research/information is stifled. This suggests that "GMO grain" going into the food chain may not have any technical information published about it at all except what the plant breeder wants the public to know. In the *"Scientific American"* (Aug. 2009), scientists and the public alike are calling for removal of this legal impediment to independent investigation and publication. Any cultivar which has the right to be sold to the public must be able to be submitted to independent professional examination protecting both farmers and consumers concerning products which have the capacity to reduce or damage production of food and/or health of the international community.

On the other side of this argument of course is the importance of accurate analysis and fair reporting by the independent authority or investigating scientists whom US plant breeders claim may have a personal bias against any particular firm. Some, like Monsanto are giant manufacturers of agricultural and horticultural chemicals as well. A few carelessly selected words could "kill" the commercial value of a promising new cultivar which has been years in the making and at considerable cost.

This is easily remedied by allowing the researcher/breeder to view the evaluation of a specific cultivar and if in disagreement with it, to request an independent professional opinion—a referee's opinion who's appraisal should be final.

World Expansion

Precision Seed Coaters of Yuma, Arizona, USA published a well presented brochure which was distributed to its clients some several years ago. It included a perceptive observation that modern day coated seed was pioneered in New Zealand during the 1950's where rugged hill country [similar to stress conditions found in many US land reclamation areas] created the often dry and infertile conditions in which coated seed was seen to best advantage. It went on to explain that many thousands of acres of South Island (New Zealand) High Country had been successfully established in clovers and grasses mainly due to aerial application of coated seed.

The article went on to explain that this coating technology gradually developed to include a wide range of forage, then turf grasses and spread from New Zealand to North America and Europe where today almost any type of seed which the market requires is now available as coated seed.

That was an accurate summary. Provided we exclude the pelleting of sugar beet seed commenced in USA many years ago by iconic "Germains"–the above analysis appears basically correct and the New Zealand author of this book is pleased to have participated in that pioneering work on inoculated coated clovers and aerial grassland establishment in New Zealand as did contributor John Lloyd. The original pioneer of seed inoculation in New Zealand was however, Athol Hastings of DSIR.

Author [Bennett] and contributor [Lloyd] both visited USA when the Celanese Corporation purchased the CSL technology [which did not include NIT—new inoculation technique], then developed it further as Celpril (with it's former HQ in Manteca, California).

Some 30 years later, the situation has changed.

New Zealand is no longer a leader in this field. A large number of new companies have become established worldwide—in Canada, USA, Sweden,

The Netherlands, France and UK—but particularly throughout the USA. These are specifically involved in inoculation and coating, pelleting, film processing, coloring, encrusting and chemical protection of seed—some in processing alone and others in manufacturing sophisticated new machinery for processors including the Rotary coater which currently dominates the field plus superbly designed spraying, drying, screening and handling machinery.

Question:

What message can we take from this proliferation of processors, specially in the USA, now arguably leader in the field of seed processing, being also at the forefront of agricultural and horticultural excellence being the world's largest producer of several key agricultural products like soybean, a legume.

Answer:

This must surely be the way the world will proceed. It is now part of established agriculture with much potential for further development.

Thus it is highly likely that other nations will follow. It is all part of *"precision agriculture"* including seed processing, indeed many US partnerships and franchises with firms in Europe—The Netherlands, Germany and France have already been established as well as in China, Brazil, India, Malaysia and Indonesia. There will probably be many more by the time this book is published.

To list but a few American firms that offer seed coating and related treatments:

[In no special order]
Seed Support, Florida
Precision Seed Coaters, Arizona
American Seed Coating, Georgia
Landec Ag., Indiana
US Rare Minerals Inc., Oregon
Germains Seed Technology, California and North Dakota
Incotech Inc., California
Pennington, Madison
Agricoat LLC, California
Seed Coaters, Arizona
Chromatec Inc., Michigan
Croda Crop Care, Delaware
Agtech Global Seed Technology
Universal Coating Systems, Oregon and Kansas
Smith Seed Services, Oregon
Summit Seed Coating, Idaho
California Seed Coaters—[Holtville]
Celpril, California (closed in 2009 after 30 years pioneering).

This impressive growth in technology has happened in the last 30 years (50 years if we include the early inoculation pioneering in New Zealand).

While many nations have a different type of agriculture to the USA, and have to deal with different geographical features, it is more likely most will have more of the *"rugged hill country"* referred to [above] where inoculation and coating of seed will show greater value in evolving forage grasslands than it has done in the US where processing of crop seed is more solidly in focus. Seed establishment under hostile conditions is where the greatest potential lies.

Expansion can only happen where key basic requirements can be met.

In developing societies, before the technology of seed treatment can be employed practically, there must be security of investment and effort—when initial research has identified local requirements.

Fundamentals must be accessible, land tenure—owned, leased [with right of credit for improvements] or assigned with the right to fence it from interference, the provision of loan finance, of water, the availability of fertilizer, seed, inoculants, coatings, and of livestock plus a place to live—all free from the corruption, crime and oppression which have sadly and for many years, defined undeveloped societies in many locations around the world, to the detriment of everyone involved.

The Future

If we knew what the future held, it would of course no longer be "the future"! That we do not know what the future holds for us is the fuel of human enterprise—universal determination to succeed and an unquenchable curiosity which, as science becomes more widely understood, will reveal new dimensions at an increasing pace as we have witnessed over the latter century. IT [internet technology] to-day—probably personal flight soon, which would possibly solve the traffic problem.

We can however make some reasonably educated guesses.

- Microbiologists may develop *rhizobia* "Superbug strains" by selection and/or genetic engineering which will not only "fix" as much atmospheric nitrogen as their host plants can use, but also generate a surplus by substantial nodule discharge creating a significant subsoil supply of nitrogen fertilizer on which other plants may feed.

- Included in this Superbug DNA may be a superior capacity to infect root systems predominantly in spite of huge competition from indigenous strains [with high infective ability but of low nitrogen fixing capability]—the latter which of course exist now.

- Scientists may develop nitrogen fixing capability into crop and forage plants via modified *Rhizobium* and/or plant mechanisms.

- Simplified "on farm" methods of *Rhizobium* culturing will be developed and supplied as Kit sets direct to farmers and growers. These kit sets will contain some "Mother culture" from which growers can produce their own inoculum cultures for application to already coated or pelleted seed right at time of sowing. This for some [those with much at stake] will move the responsibility of freshness of *Rhizobium* away from distant processors and suppliers, to the site of sowing. Milled modified peat (as a medium for both *Rhizobium* growth and application to seed) will be replaced by a *rhizobia* enhancing broth—probably a polymer [supplied in drums] in which *rhizobia* would be multiplied in semi sterile conditions over several weeks then applied to seed [as a predominantly *Rhizobium* culture] in the barn, in a drum mixer or in the drill box at sowing. At application [e.g., inoculation] the pre-coated seed will be lightly mist sprayed with this rapidly drying {as in fast drying acrylic paint} rhizobia-rich and enhancing polymer which, once applied and exposed to air will be dry within 3 or 4 minutes. After that time it may be safely loaded into farm drills, aircraft hoppers or if pelleted—into precision drills.

Only the kit set with instructions plus some Mother culture will be required by the farmer, grower or contractor, plus of course access to the "magic polymer" in drums—or maybe just polymer solids from which the polymer solution can be formulated—such lightweight [solids only] being important where transportation cost is significant.

- Seed coated or pelleted or filmed and chemically protected will be bought and sold internationally, coloured for identification, not only tested for purity and germination but also for identification, seed/coat ratios, additive effectiveness and compliance with importers international standards. This will require an expansion of services by Seed Testing Stations in many locations around the world part financed by a levy paid by sellers related to the extent and number of tests [several tests for large volumes]. Eventually, an international standard for all qualities will be agreed.

- Developing countries will make much more use of solar powered electric fencing to keep domestic livestock in and unwanted animals and humans out.

A cautioning orange coloured hemp, nylon or polypropylene light net (up to 2 m high) with selected levels of electrified tinsel interwoven in the rope strands (as is well established now in agriculture), fitted onto lightweight bamboo, wood or insulated aluminium or steel pins driven into the ground at suitable intervals which netting fence can be moved to new grazing—or to extended boundaries, to allow better utilization of pastures and crop.

Electricity will be provided by batteries locked or secured inside the electrified fence, charged by one of several methods, i.e., By battery charger from a standard power source; By a petrol, diesel, waterfall, river or manually driven generator; By fixed solar panels or by transformer moderated connection to a mains supply. With access to the basics, nomadic sheep and goat herders could soon be taught how to make such netting and electrify it—in the same way fishermen make fishing nets. Sheep and goat herders certainly have the time, a need to deal with the boredom and would not have to round up and enclose animals each night against predators if they had electrically charged netting.

Fig. 79 Author Bennett visited this nomadic sheep and goat herd in Himalayan Nepal in 1959. The animals are shifted to new winter grazing periodically by these young Nepalese men who come from villages at lower altitude. In the middle foreground are basketweave mats which are encircled around groups of animals at night for protection from predators—leopards, jackals and bears.

Because electric fencing is re-locatable where battery power is used—electrified netting with a pulse energizer would likely deter most predators which could be a major step forward for these shepherds provided a manually operated battery charger is included. This photo is repeated from illustration on page 126 Chapter 3 because the importance of transportable fencing is the significant issue here clearly demonstrated by this scene. Fencing alone represents a mammoth task just in Nepal. It is much more so worldwide and must be preceded by land tenure establishment.

In developing regions, community groups will combine their resources of labour, finance, security and protection when it has been shown that good productive grasslands can be established by following a proven formula. This will allow access to seed, inoculants, fertilizer, irrigation and fencing—hopefully also security of land ownership or occupation. To make this shared operation work, a fair system of contribution and equivalent reward must be established to which all shareholder participants must agree to abide as in a private Company to be paid at various levels of contribution of assets, skills and effort. There would be no payment outside the shareholders and all payment would be independently approved. Later, as individuals succeed they will farm individually. Such development group to start as a commune, but a private one.

The key to this "door of opportunity" is Government provision of small professional research teams to identify in any promising location the needs and the barriers to farm production and methods for dealing with both.

This combining of resources for better outcomes has already started. It is in its infancy in Nepal where villagers contribute one or more sheep or goats to make up flocks such as shown in the illustration (Fig. 79) with fit young men contributing their labour as shepherds day and night in shifts moving the flock to new grazing being rewarded from increase in both numbers and maturity of the combined asset. With formalized land tenure, the next step is to set up a company to manage all the assets including the land.

Generation of energy from any light source is already commercialized in wristwatches where the battery is recharged by light—particularly sunlight.

Currently in New Zealand one can purchase a 20 watt solar panel unit weighing 2 kilograms for less than $250. In the near future solar panels are very likely to be easily transportable, less expensive and more efficient, powering batteries for night lights, radio reception and electric fences in remote locations, and for livestock security.

- "Pasture pellets" or "Supercote" crop pellets (both the name and the products yet to be fully refined) will have a big future. Forage seed [pasture pellets] are likely to be comprised of slow release fertilizer, compressed to achieve optimum weight/size ratios, containing inoculant where required, other chemicals for specific conditions, aerodynamically shaped, surface hardened with pharmaceutical type binders, colour coded for identification, containing (2 or more) of the preferred seed, discharged at high velocity into wet or damp soil to achieve a depth of 5 or 10 millimetres. The applicator used will be a multi-barrel compressor driven air gun with (say) 4 barrels wide by 4 barrels deep configured to space the 16 seed pellets (for a hill country oversowing) approximately 10 cm apart thus covering 40 cm x 40 cm = 1600 cm^2 per single discharge applied at a height of about 20 metres flying at 50 mph via unmanned helicopter drones with GPS, height and horizontal sensors able to lift about the same weight as fixed wing topdressing aircraft—but the drone weighing much less and without continuous pilot weight. It would have to carry compressed air—however, may also be part elevated with lighter than air gasses).

These pellets will have all the nutrient requirements for pasture establishment plus the first two years of growth, without any other top dressing.

This concept may also eventually replace conventional farm drills on arable land as it will seed a crop, paddock or field much faster than drills and without touching and compressing the cultivated soil. Paddocks can be seeded this way when wet, which is not desirable for heavy wheeled vehicles. These drones will be controlled from the ground by powerful wireless regulators—predominantly operated by contractors but also by large scale farming operations.

In developing nations, the same principle (all-in-one pellets) will be used, but manually applied by taking a pellet [from a backpack with shute to the front] placing it in a pointed steel or brass trocar (hollow behind the point to hold a pellet) and while tramping over the hills is jabbed into the ground at desired intervals, a handle squeeze opening the spearhead and releasing the pellet at (say) 2 cm depth. Ideally, pellets will feed by gravity to the injector head and a width of 10 or 20 people in line traversing a hillside will cover the ground reasonably quickly. An important job for physically fit but marginally skilled people maximizing on human resources, probably contractor hired.

- Plant breeding is currently in its infancy. Hundreds of new cultivars providing a wide range of enhancements will be produced around the world to cope with both universal and local conditions for which an international [and "online"] catalogue providing a fully descriptive directory will be established. It will carry several language options stating prices, quantities available and enhancement claims—with independent test results if any, seed testing and quarantine details, and be constantly updated for which, of course, the Internet is ideally suited. It is likely to be managed by a competent International Authority funded by the *Seed Associations* of participating nations.

These are just some of the exciting prospects for the future indicated by mankind's achievements to date. The speed of technical development has now also become a phenomenon in itself.

References

(1) During, C., N.A. Cullen and G.M. Bennett. 1962. The establishment of Pasture on Yellow-Brown Loams near Te Anau New Zealand. New Zealand Journal of Agricultural Research 5, No. 3 and 4, June and August 1962.
(2) Giraud, E. et al. 2007. Science 316 (5829)1307–1312.

Index

Enumeration of viable rhizobia, 63
Expired drum confectionery patents provide
   pelleting skills, 253

F

Factory Production, 275, 278, 285
FAO, vii
Farm Advisory Division, Department of
   Agriculture, NZ, 298
Farm Advisory Officer, viii, ix, xiv, xxiii
Food Growth Considerations, 294
Forage Grass Coatings, 225
Ford's Nursery, Oamaru, 202
Forestry precision pelleting, 187(Fig. 44), 199
Formula for success, 98
Formulae, 235, 236, 245, 249–252, 266
Freeze dried capsule, 62(Fig. 14)
Fruitgrowers Chemical Co., 15, 86, 128, 151

G

Geoff Taylor, 155
Germination, 166–173, 175, 176, 178, 181, 183
Germination of Coated Seed, 92
Granulation of crop seed, 108
Grass Seed Additives, 227
Grass Seed Oversowing, 173, 176
Grass species, 184, 185
Grasses in Hill and High Country
   Oversowing, 164
Grasslands Association Conferences, 291(Fig.
   73)
Grasslands Memorial Trust, xxvi

H

Hale C.N. (Dr.), 16
Handling, 223, 226, 230, 244, 250, 256, 257,
   267, 273, 275, 277, 282–284, 289
Handling Care, 273
Harmful rhizobia, 17
Herridge David (Dr.), 37
Himalayas, viii, 125
History (of NZ Dairy Industry), viii

I

I.R. McDonald, 141
ICMP, 51, 54, 55, 61
ICST, 34, 35, 38, 39, 63, 71
Identifying Nodulation success and failure,
   160
Importance of quality and quantity of feed
   for sheep, 304(Fig. 78)
Improving dispersion, 68
Independent Evaluation, 307

Industry names, 219
Ineffective strains, 7, 8
Inoculant carriers, 40, 45
Inoculant manufacture, 24, 34, 44, 46, 48, 56,
   58, 76
Inoculant Manufacturers, 105, 112
Inoculant quality, 34, 36–38, 77
Inoculants, 20, 22, 27, 29, 31, 32, 34–39, 42, 45,
   46, 48–53, 56–63, 66–68, 71
International opinion, 26
International seed sales, v
International standards, 34
Internet, 221, 228, 240, 242, 267, 291, 292
Invermay, 96, 130, 133, 136, 141, 146, 151
Irrigation and a Warning, 301

J

J.S. Dunn, 202
John Paterson, 87

K

K.R. Middleton, 155, 158

L

Laboratory Quality Control, 137
Lands and Survey Dept. NZ, 296
Lawn seed, 170, 171, 172(Fig. 41)
Legume root nodules, 7(Fig. 3)
Lignite, 28, 29, 31, 40, 44, 47, 48
Limitations of Commercial Exclusivity, 245
Lincoln University, 132, 134
Lincoln University (NZ), 3
Lloyd J.M., iii, 15
Lowther W.L. (Dr.), 23, 46
Lucerne (alfalfa), 4(Fig. 1), 8
Lucerne and Fungicide research, 156
Lucerne success and related cost, 158(Table
   12)
Lucerne Yield, 155

M

Machinery, 227, 236, 242, 248, 249, 251, 261,
   264–266, 268, 274, 279, 280, 284, 289, 292
Miscellaneous additives, 231
Miscellaneous Equipment, 272
Mix Accurately, 121
Mixing inoculated seed with superphosphate,
   274(Table 18)
Mixing Just Prior to Sowing, 115
Mixing with fertilizer, 120
Modern Pellet Contents, 200
Mother culture, 32, 48, 52–54
MPN method, 63, 64, 66, 71, 76, 77